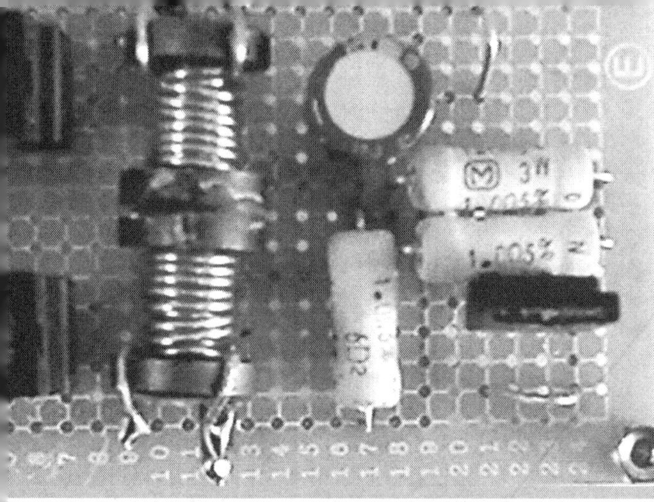

Handbook of Power Management Circuits

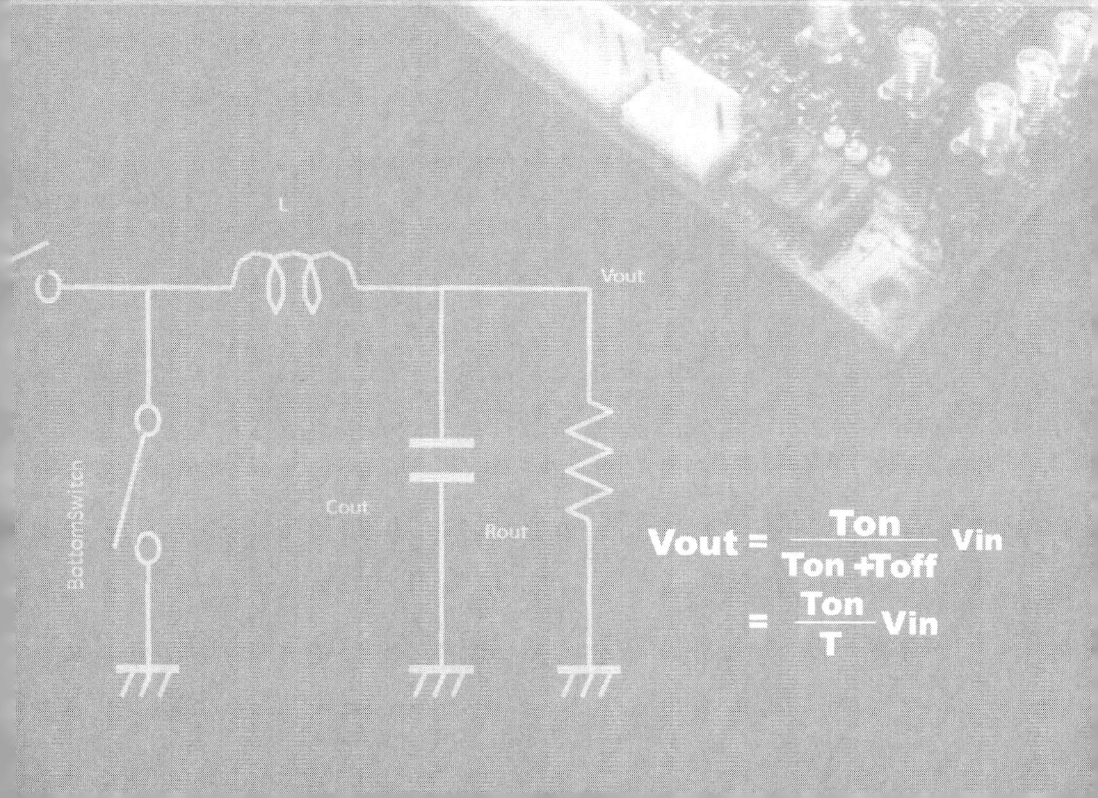

Handbook of Power Management Circuits

edited by

Haruo Kobayashi
Takashi Nabeshima

Pan Stanford Publishing

Published by

Pan Stanford Publishing Pte. Ltd.
Penthouse Level, Suntec Tower 3
8 Temasek Boulevard
Singapore 038988

Email: editorial@panstanford.com
Web: www.panstanford.com

British Library Cataloguing-in-Publication Data
A catalogue record for this book is available from the British Library.

Handbook of Power Management Circuits

ISBN 978-981-4613-15-6 (Hardcover)
ISBN 978-981-4613-16-3 (eBook)

Printed in the USA

Contents

Preface

Electronics and electrical engineering may be only one part of physics. However, during the last 100 years, they have advanced rapidly and changed our lives drastically. Their roles can be classified into the following categories: (i) information and signal processing, (ii) information storage, (iii) communication, and (iv) energy and power. In this book, we focus on the fourth category—energy and power, or power electronics—which is becoming more and more important to make the earth green.

The book is intended for tutorials on power supply circuits for engineers and graduate students in circuit design fields as well as power electronics, and it covers a wide range of power supply circuits. The authors of all chapters have been engaged in research and development of their contents, and hence each chapter has its own originality, reflecting the authors' experiences. It is noteworthy that the power supply circuits as well as power amplifier circuits are different from analog, mixed-signal, and RF-integrated circuit design, and even circuit designers who have good background of analog, mixed-signal, and RF circuit design often get puzzled when they start to get involved in power supply circuits. In the 1997 IEEE International Solid-State Circuits Conference, there was a panel discussion session entitled "RF Designers are from Mars, Analog Designers are from Venus." Here I would like to add the following statement "Power Supply Designers are from Mercury and Power Amplifier Designers are from Jupiter."

This handbook is organized in two parts. In Part I, basics of power supply circuit have been reviewed systematically. In Chapter 1, basics of power supply circuit are introduced. The first hurdle to understand the DC–DC converter is the circuit behavior of an inductor. For example, current can be made to flow from lower- to higher-voltage nodes through the inductor, and thanks to the inductor, the DC–DC converter efficiency can theoretically be 100% in ideal conditions.

In Chapter 2, a buck converter—the most important DC–DC converter—for low-voltage applications is described elaborately.

First, buck, boost, and buck–boost DC–DC converters are introduced. Then two operation modes, that are, continuous current mode (CCM) and discontinuous current mode (DCM) are explained. Then their operating principle, circuit analysis with transfer function, closed-loop operation, design consideration such as error amplifier design, are discussed. An example of power supplies in a computer system is also discussed.

In Chapter 3, isolated DC–DC converters with a transformer (isolation of large voltage and current conversion, minimizing voltage and current stresses, multiple outputs, flyback converter, forward converter, push–pull converter, half-bridge converter, full-bridge converter of various types) are explained. These are used for handling relatively large power, and even beginners can understand them by a careful read, although they may find them difficult to understand at first.

Chapter 4 covers modeling and analysis of switching converters, such as state–space average model, averaged device model, and CCM and DCM models as well as transfer function. In Chapter 5, control schemes of switching converters are described, such as a self-oscillating hysteretic PWM control and a current mode control as well as a voltage mode PWM control, including some content based on the authors' research.

Chapter 6 describes passive components (inductor, transformers, and capacitors) and explains the fundamental physics behind inductors and transformers. It then introduces capacitors for switching converters, such as aluminum electrolytic capacitor, tantalum electrolytic capacitor, film capacitor, and ceramic capacitor as well as characteristics and applications of various capacitors.

In Part II, several selected topics are introduced individually. In Chapter 7, on-chip voltage converters are explained for large-scale integration (LSI) designer, such as voltage-reference circuit (bandgap reference circuit, or BGR), voltage-down converters, and voltage-up converters. On-chip voltage converters are very important for low-power operation in large-scale integrations (VLSIs).

Chapter 8 describes applications of DC–DC AC–DC switching converters and some of them are recent research results of the author: non-inverted buck–boost DC–DC converter with dual delta-sigma modulators and non-isolated AC–DC direct converter.

In Chapter 9, single-inductor multi-output DC–DC converters are introduced. The single-inductor multi-output DC–DC converters are attractive for small size but their control is difficult. Their several configurations and control methods are also described.

Chapter 10 shows a small, low-power boost regulator optimized for energy-harvesting applications. Recently, interests of energy-harvesting applications are booming up and an example of boost-converter design for this purpose is introduced. Chapter 11 introduces wireless power delivery for 3D system integration and for non-contact wafer-level test focusing on the author's experience and interest. Chapter 12 shows high-power GaN–HEMT amplifier for cellular base stations. A lot of attention is now being paid to GaN HEMT, and several power amplifier architecture, design, implementation, and measurement examples with this technology are introduced. Chapter 13 describes power supply circuits with capacitors and switches.

We hope that this book will be helpful for electronics engineers from various fields in understanding these interesting and important areas and the readers will enjoy reading all the chapters.

Finally, we would like to thank Dr. Masashi Ochiai for reviewing the manuscript and providing valuable comments.

Haruo Kobayashi

Takashi Nabeshima

Winter 2015

Chapter 1

Power Supply Circuit Fundamentals

Jun-ichi Matsuda and Haruo Kobayashi
Division of Electronics and Informatics, Gunma University,
1-5-1 Tenjin-cho, Kiryu, Gunma 376-8515, Japan
jun-ichi.matsuda@nifty.com, k_haruo@el.gunma-u.ac.jp

This chapter introduces the fundamentals of power supply circuits.

1.1 Introduction

1.1.1 Why Do We Study Power Electronics?

The field of electronics has advanced very rapidly and has changed our lives significantly in less than 100 years. The roles of electronics in our lives include:

- information and signal processing;
- information storage;
- communication; and
- energy and power.

We here focus on energy and power, or power electronics.

Handbook of Power Management Circuits
Edited by Haruo Kobayashi and Takashi Nabeshima

ISBN 978-981-4613-15-6 (Hardcover), 978-981-4613-16-3 (eBook)
www.panstanford.com

Soichiro Honda, the founder of Honda Motor Co., Ltd. said, "The purpose of our technology is to make people's lives happier." We believe that this should also apply to power electronics technology, and the objectives should be as follows:

- Better daily human life
- Prosperity of related industries
- Long-term preservation of the ecology of the earth

We need global thinking and long-term thinking in power electronics:

- **Global thinking:** Suppose we plan a solar cell power generation project for the Sahara Desert. Solar energy generated in the Sahara Desert could supply the energy needs of the whole world. However, such a project, involving power generation, power transmission, power transformation, and storage of electricity, cannot be realized by a single company or even a single country. We can easily see that international collaboration and global thinking is required.
- **Long-term thinking:** Let us review the history of the Industrial Revolution. The first Industrial Revolution (1760–1830s) started in the U.K., and the technologies developed included dynamics-mechanics and motive power, which triggered a boom in railway and sea transport. The second Industrial Revolution (1865–1890s) was based on material science developed in Germany and the United States, which impacted heavy chemical, steel, and automotive industries. The third Industrial Revolution (1990–2000s), which we are enjoying, involved information and communication technology. It started in the United States and is centered on mathematical engineering, digital technology, communications, computers, semiconductors, the Internet, and digital home appliances. The fourth Industrial Revolution (2010s–) that has just started involves the environmental and energy revolution. New challenges include electric vehicles (EVs) (which are not just from Toyota, GM, Volkswagen, Renault–Nissan, but involve hundreds of small companies), the smart grid, and power semiconductors (Si metal–oxide–semiconductor field-effect transistors [MOSFETs], insulated-gate bipolar transistors [IGBTs], SiC, GaN). Governments in many countries support

these areas. Semiconductor technology should be directed to become an important player in solving environmental and energy technology/industry problems. Audio, games, cellular phones, automotives, and digital electronic home appliances have made our daily lives convenient and happy, but now we have to pay attention to environmental problems to realize sustainable progress. This is not a temporal or transient fashion or trend; it is very important, or even essential.

1.1.2 Positioning of Power Supplies

1.1.2.1 Switching-mode power supplies

Switching-mode power supplies produce stabilized DC power, controlled by a switching mechanism, from an input such as the commercial AC power supply or a DC power supply. This field includes high-speed switching using semiconductor devices at frequencies between several tens of kilohertz to several megahertz and featuring a small size, light weight, and high efficiency. Efficient switching-mode power supplies are essential for the efficient usage of electrical energy.

1.1.2.2 History of switching-mode power supplies

The switching-mode power supply was invented in the 1960s by NASA for rocket applications to realize power supply circuit implementations with a small size and high efficiency. In the 1970s, the switching-mode power supply prevailed in the consumer and industry markets, especially in communication equipment. In Japan, many power supply and component manufacturers started to produce switching-mode power supplies. Domestic standardization activities as well as joint international standardization activities started at that time. One important aspect of switching-mode power supply technology is small-size implementations. Advances in related technologies and industries are being seen.

1.1.2.3 Applications and products using switching-mode power supplies

They are widely used in most electronic machinery, such as home appliances, audiovisual equipment, entertainment equipment, power transfer infrastructure, lighting and display applications, EVs, medical equipment, measuring instruments, control equipment, factory automation equipment, office automation equipment, communication equipment, and computer systems. Power supply technologies concern high-voltage applications, electricity storage, control, energy conversion, high-frequency circuits, harmonics, energy distribution, communication, low-power analog and digital circuits, power load leveling, point of load, compatibility, and components. Markets for switching-mode power supplies demand special minerals and chemical substances used in electronic circuits, compliance with regulations, and contribution to a low-carbon-footprint society, clean energy, environment-friendliness, high efficiency, small size, high performance, low noise, long life, safety, high reliability, and low price.

1.1.2.4 Power supply technological classification

Power supplies are classified based on input voltage, regulations restricting harmonic current, formation (implementation), noise level, switching frequency, cooling methods, safety regulation, and isolation between the first and second stages.

Components of switching-mode power supplies include switching devices, rectifying devices, control ICs, aluminum electrolytic capacitors, ceramic capacitors, film capacitors, power modules, printed boards, transformers, coils, choke coils and photocouplers.

1.1.2.5 Electric power flow from generation to consumption and related technologies

- Electricity generation converts energy from coal, oil, nuclear energy, geothermal power, solar thermal power, solar photovoltaic power, hydroelectric power, wind power, ocean wave power, tidal current power, and tidal power to electrical energy. The basic laws of physics involved include electromagnetic induction, electrochemical reactions, the photovoltaic effect, the Seebeck (thermoelectric) effect, etc.
- Electricity transmission technologies include DC power transmission (which Thomas A. Edison strongly argued for), AC power transmission (due to ease of converting between

different voltages using transformers), microwave power transmission, smart grids, and wireless power transmission.

- Electricity storage is realized with capacitors and batteries.
- Electricity distribution is the final stage in delivery of electricity to end users. A distribution system network carries electricity from the transmission system and delivers it to consumers.
- AC–DC converters are widely used to convert from a commercial AC supply voltage to DC supply voltages used for operating electric equipment.
- Power factor (PF) correction circuits improve the PF, which is defined as the ratio of the real power, flowing to the load, to the apparent power in the circuit. Real power is the capacity of the circuit for performing work in a particular time. Apparent power is the product of the current and voltage of the circuit.
- DC–DC converters may be classified as series regulators, switching regulators (with three basic possible topologies), and switched-capacitor (or charge pump) circuits.

1.1.3 Power Supply Circuit Basics

1.1.3.1 Why are power supply circuits required?

Let us consider the design of complementary metal-oxide-semiconductor (CMOS) digital large-scale integration (LSI). The dynamic power consumption of CMOS digital circuits (such as inverters, NAND gates, NOR gates) is given by fCV_{dd}^2, where f is the output-node toggle frequency, C is the capacitive load, and V_{dd} is the supply voltage; we see that as V_{dd} becomes smaller, power consumption becomes less. For a power amplifier, on the other hand, its output power is the product of output voltage and output current. Power loss due to parasitic resistance becomes large if the output current is large, and hence we set the output voltage to such a value as to keep the output current as small as possible. Also, many analog circuits require some minimum level of supply voltage to obtain high performance. Inside one LSI, it is desirable to have multiple supply voltages; however, it is usually desirable to have only one input supply voltage (such as a battery), so power supply circuits need to efficiently convert the input supply voltage to multiple supply voltages.

1.1.3.2 Importance of power supply technology progress

Technology progress in power supplies has a significant impact on society as well as on industry. The number of power supplies in the world is huge, and improving their efficiency by even 1% could save a lot of energy. Also they will be used for a long term; using supplies that are not energy efficient will be more costly in terms of power usage in the long term. Recently energy-harvesting techniques, which "generate finite energy virtually for free," are attracting much attention, and they are expected to significantly contribute to human life and the earth environment.

Power supply circuit devices are classified as active and passive devices, and both are important. The technology trend is that progress in active devices as well as circuit and control technology helps relax the requirements for passive devices.

- Switching transistor: Its figure of merit (FOM) can be the product of the drain–source resistance R_{ds} and gate capacitance Q_g. Here R_{ds} is the drain–source resistance when the drain–source voltage V_{ds} is close to zero (the MOSFET is deeply in the triode region).
- Diode (or rectifier device)
- Semiconductor devices for controller
- Inductor
- Transformer
- Capacitor

1.1.3.3 Transistor roles

Roles of MOS transistors in electronic circuits can be classified as follows:

- Switch (MOS transistor operates in the cutoff or triode region @ $V_{ds} \approx 0$).
- Signal amplification (MOS transistor operates in the saturation region).
- Current source (MOS transistor operates in the saturation region).
- Variable resistor (MOS transistor operates in the triode region). Recall that "transistor" means "trans" + "resistor."

1.1.3.4 Basic physics of power circuits

Power supply circuit design and analysis are based on the physics laws of Ohm's law, Kirchhoff's current and voltage laws, the first law of thermodynamics (the law of conservation of energy, adopted for thermodynamic systems), and the second law of thermodynamics (which means that a perpetual-motion machine cannot be realized). They require knowledge of electronics, electricity, and magnetism. The balance of current and voltage is important.

Technologies for power supply design are circuit, control, modeling, device, and components (such as semiconductor, inductor, and capacitor).

1.1.3.5 Control technology

Control technology is very important in power supply circuits. Several US and European semiconductor companies have obtained a lot of profit from their power supply control ICs. Markets (such as microprocessors) demand for fast response and low ripple control.

Pulse width modulation (PWM) control is its basic, and there are several control methods; their other keywords are pulse frequency modulation (PFM), pulse density modulation (PDM) or delta–sigma modulation, feedback control, feed-forward control, voltage-mode control, current-mode control, hysteresis control, and digital control. Combinations of these are also possible.

Output load is often modeled as a resistor R; note that a small R means a heavy load because a large current flows through a small R.

1.1.3.6 Modeling

Modern control theory by R. E. Kalman introduced the state-space approach in 1960. The switching-mode power supply includes switching or time-discontinuous operation, and it was difficult to apply the state-space approach to it. However, the idea of a state-space averaging approach was introduced to switching-mode power supply circuit design and analysis, and it has been well investigated and used widely for practical applications [1].

Design trade-offs for switching-mode power supplies include output voltage ripple, response, efficiency, and size. If a large inductor L is used with a fixed switching frequency, a low output voltage ripple can be obtained, but the response to a load change is slow. If high-frequency switching is used, a fast response and low output voltage ripple can be obtained with a small inductor L and capacitor C (hence a small size), but switching loss becomes large; then switch devices with low switching loss are desirable to realize a small size, high efficiency, low output voltage ripple, and fast response. Sometimes soft switching can be also used to obtain low switching loss. It is important to understand these design trade-offs as they are the essence of power supply circuit design. In other words, the overall system-level design viewpoint is important.

1.1.3.7 Inductor L

The switching-mode power supply circuit uses an inductor L, which is an essential and superior passive device, but it is difficult for beginners to be familiar with its basic operation.

(1) In radio frequency (RF) circuits, its frequency domain characteristics are considered, and we understand that its impedance $Z = j\omega L$; we see that as the frequency f increases, $|Z|$ increases. Also it has a 90° phase lead.

(2) In power supply circuits, time domain characteristics of L (for current flow through L) are considered, and it is used as an energy storage element $\left(E = \frac{1}{2}LI^2\right)$ in case of linear magnetic characteristics.

(3) In baseband analog circuits, an inductor L is rarely used, because it occupies a large chip area and hence it is costly.

- Fundamental principle of inductor operation (1)

Q. Can current flow through an inductor from lower-voltage to higher-voltage nodes?

A. Yes, current can flow through an inductor from lower-voltage to higher-voltage nodes. In this case the current decreases as time goes on, and then, after some time, its direction is reversed (Fig.1.1). Many analog circuit designers who start designing power supply circuits don't understand this fact.

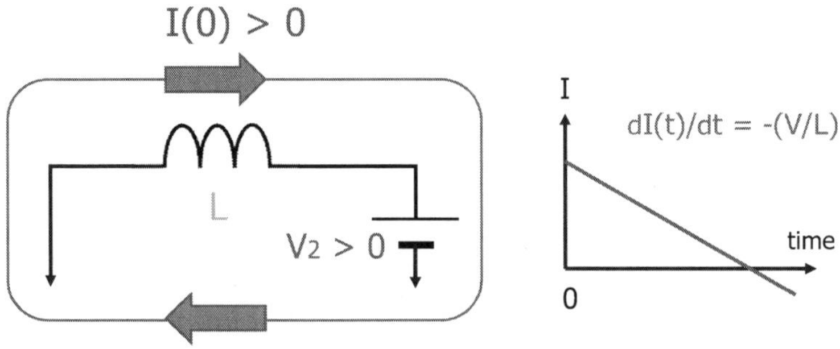

Figure 1.1 Current can flow through an inductor from lower-voltage to higher-voltage nodes.

- Fundamental principle of inductor operation (2)

Q. Can current flow through an inductor from higher-voltage to lower-voltage nodes?

A. Of course current can flow from higher-voltage to lower-voltage nodes. In this case the current increases (and hence the energy stored in L increases) as time goes on (Fig. 1.2).

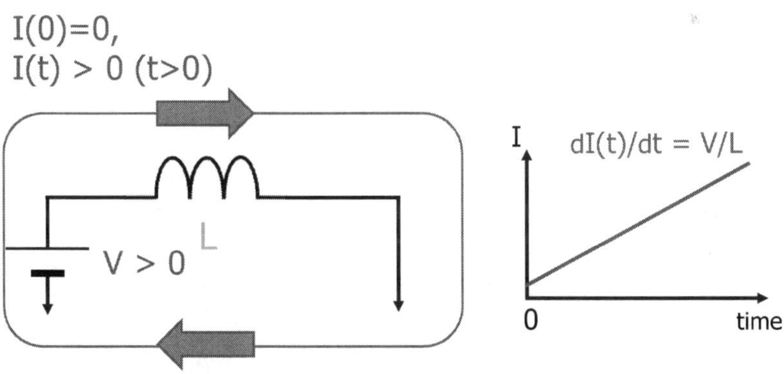

Figure 1.2 Current can flow through an inductor from higher-voltage to lower-voltage nodes.

- Fundamental principle of inductor operation (3)

Q. How does current flow through an inductor loop without a resistive component?

A. The inductor loop can be used as a current memory cell. In a superconductor circuit, the resistance is zero (Fig. 1.3). A long time ago, during research on designing computers using superconductors, inductor loop current memory was one of the memory cell circuit candidates.

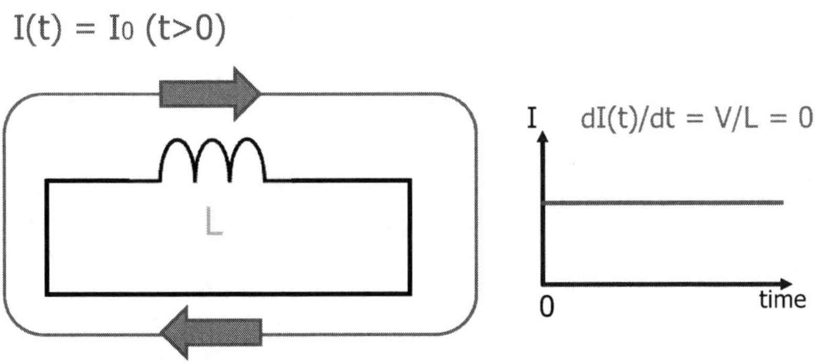

Figure 1.3 Constant current flows in an inductor loop, which is a current memory.

1.1.3.7.1 *Current flow through an inductor*

Energy can be transferred from a voltage source to an ideal inductor without loss but not to a capacitor without loss. Also an inductor can transfer its energy to a voltage source without loss. This is why a switching-mode power supply uses inductors. The following propositions will explain it in detail.

Proposition 1: The voltage source and the inductor have a good relationship.

Energy can be stored from a voltage source V into an inductor L without loss.

Suppose the inductor current is zero at $t = 0$, its stored energy in L is $\frac{1}{2}(V^2/L)t^2$ at time t, and all energy is transferred from the voltage source to the inductor without loss (Fig. 1.4).

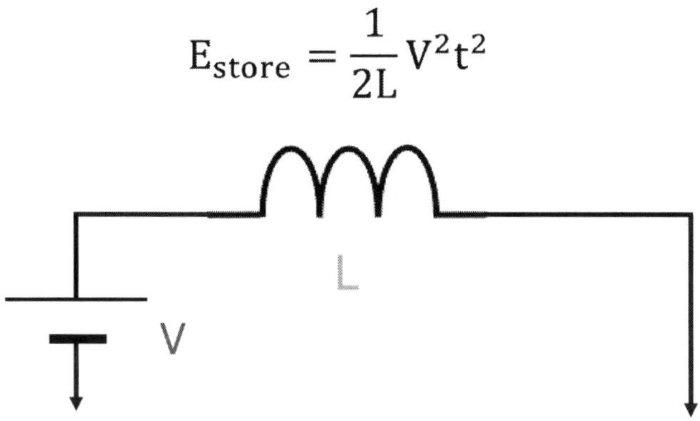

Figure 1.4 Ideally a voltage source can provide energy to an inductor indefinitely without loss.

Proposition 2: The voltage source and the capacitor do **not** have a good relationship.

Energy **cannot** be stored from a voltage source into a capacitor without loss (Fig. 1.5). Let us consider a circuit that consists of a voltage source V, a resistor R, and a capacitor C, and at $t = 0$, the voltage across the capacitor is zero. Then at the steady state the stored energy in C is given by $\frac{1}{2}CV^2$ and the dissipated energy through R is $\frac{1}{2}CV^2$, which does not depend on the value of *R*.

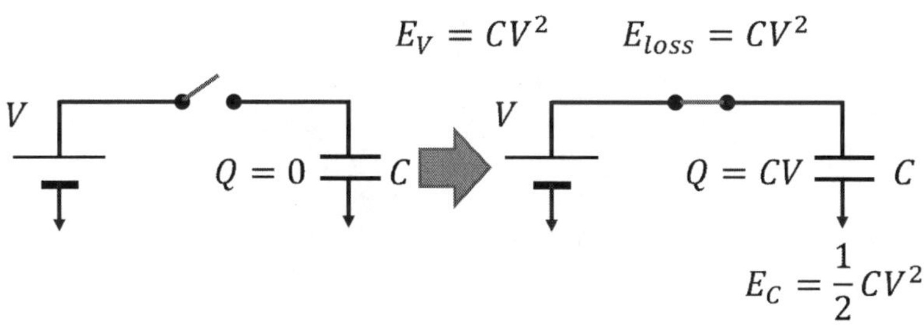

Figure 1.5 A voltage source V can provide energy up to $(1/2)CV^2$ into a capacitor C, and there power loss is essential in principle.

Proposition 3: The current source and the capacitor have a good relationship.

Energy can be stored from a current source into a capacitor without loss (Fig. 1.6).

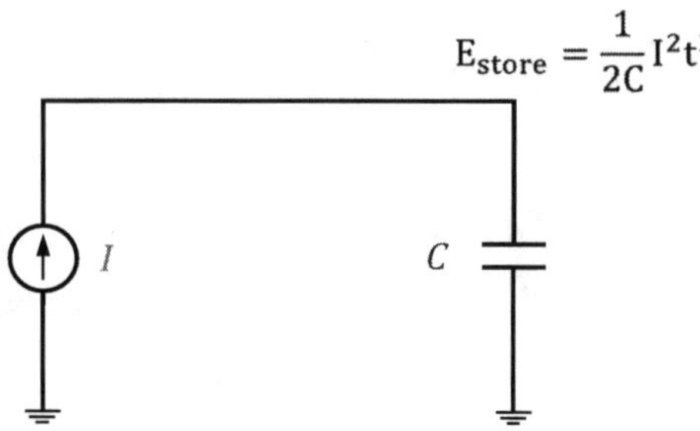

Figure 1.6 Ideally a current source can provide energy to a capacitor indefinitely without loss.

Proposition 4: The current source and the inductor do not have a good relationship.

Energy cannot be stored from a current source into an inductor without loss (Fig. 1.7).

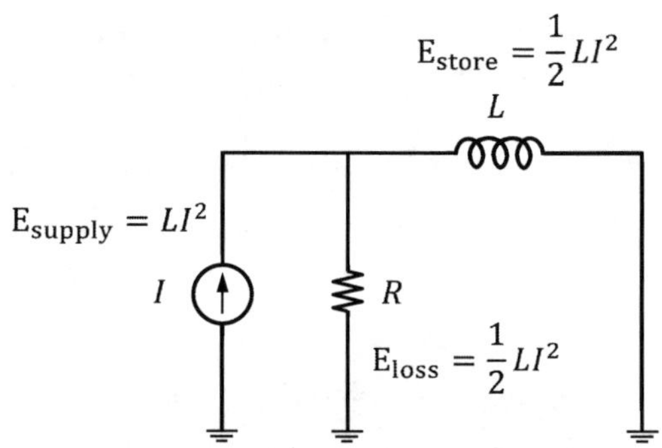

Figure 1.7 A current source I can provide energy up to $(1/2)LI^2$ into an inductor L, and there power loss is essential in principle.

Let us consider a circuit that consists of a current source I, a resistor R, and an inductor L, and at $t = 0$, the inductor current is zero. Then at the steady state the stored energy in L is given by $\frac{1}{2}LI^2$ and the dissipated energy through R is $\frac{1}{2}LI^2$, which does not depend on the value of R.

The above propositions can be interpreted intuitively as follows:

Fact 1: The inductor current L cannot change suddenly, but it changes smoothly so that an inductor with current I can be approximated by a current source I.

Fact 2: Similarly a capacitor with voltage V can be approximated by a voltage source V.

- The voltage source and the inductor can be approximated by a voltage source and a current source; this combination agrees with Kirchhoff's voltage law and Kirchhoff's current law, which explains proposition 1 (Fig. 1.8).

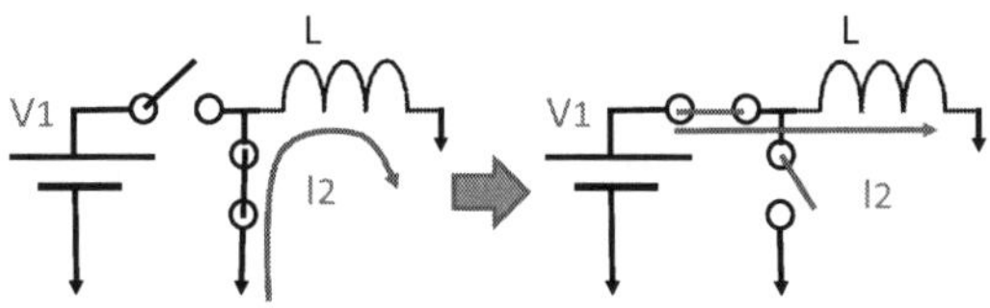

Figure 1.8 A voltage source and an inductor have a good relationship.

- The voltage source V_1 and a capacitor with voltage V_2 can be approximated by $V_1 + V_2$, which is against Kirchhoff's voltage law. This explains proposition 2 (Fig. 1.9).

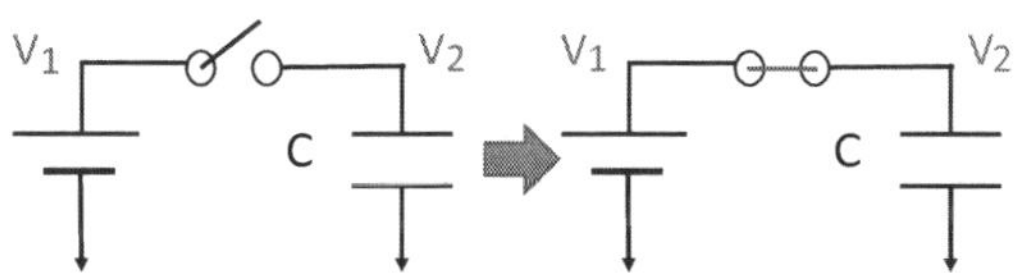

Figure 1.9 A voltage source and a capacitor do not have a good relationship.

- The current source and the capacitor can be approximated by a current source and a voltage source; this combination agrees with Kirchhoff's voltage law and Kirchhoff's current law, which explains proposition 3 (Fig. 1.10).

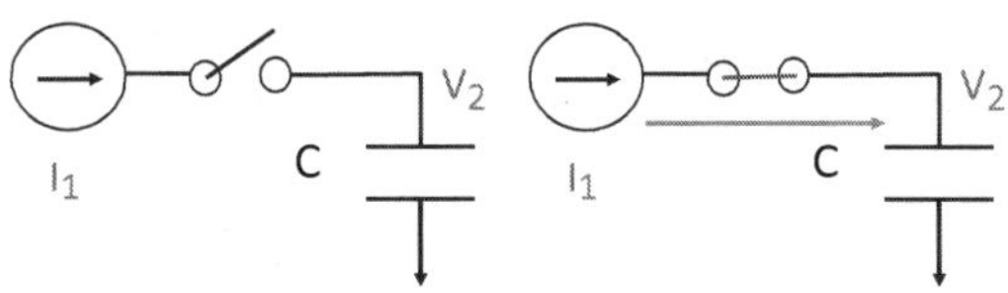

Figure 1.10 A current source and a capacitor have a good relationship.

- A current source I_1 and an inductor with a current I_2 can be approximated by a current source I_2, which is against Kirchhoff's current law. This explains proposition 4 (Fig. 1.11).

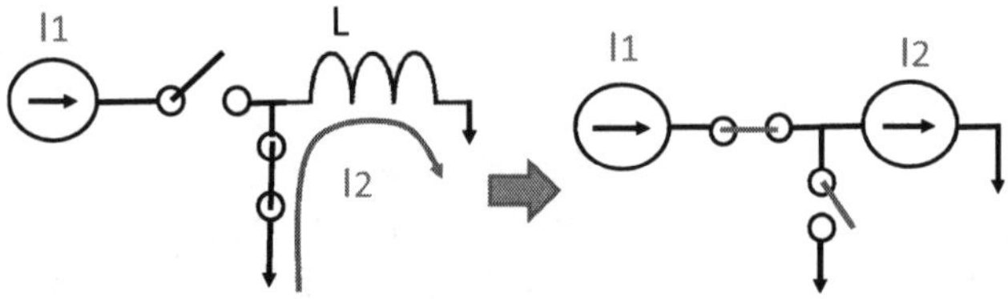

Figure 1.11 A current source and an inductor do not have a good relationship.

1.1.3.8 Duality of *C* and *L*, Voltage and Current

Look at the following equations: $I = C\dfrac{dV}{dt}$; $V = L\dfrac{dI}{dt}$

We see there are dualities between C and L as well as between voltage and current. In baseband analog integrated circuit design (such as operational amplifier design), usually an inductor L is not used; only C is used. However, in switching-mode power supplies, both L and C are used, and we can understand that this allows a wide range of circuit techniques to be exploited.

1.1.3.8.1 *RCL circuit*

The inductor L causes overshoot and ringing. Note that an RC circuit is a first-order system (which does not have ringing characteristics), while an RCL circuit is a second-order system. When the inductor L is relatively large, the response can have ringing characteristics (Fig. 1.12).

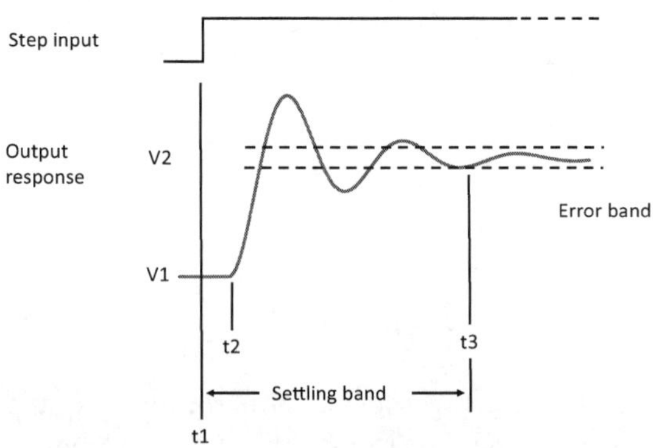

Figure 1.12 An inductor can cause overshoot or ringing for step response.

1.1.3.9 Why are switching-mode power supplies highly efficient?

The reasons are:

- They use switches.
- They use inductors as well as capacitors
- The input energy source is a voltage source instead of a current source.

Current flows from the input voltage source V to the inductor L, and the inductor current (and hence the stored energy in L) increases without loss as time goes on. Then we change the switches on/off, and the inductor current flows into an output capacitor C (in other words the stored energy is provided to the output capacitor without loss). If there are no parasitic elements (such as switching loss or parasitic resistance), the input energy can be provided to the output capacitor or output load without loss (with 100% efficiency). On the other hand, the switched-capacitor or charge pump power supply circuit, which consists of capacitors and switches without inductors, cannot achieve 100% efficiency in principle even if the circuits (such as switches) are ideal due to proposition 2.

Recently efficiency of power supply circuits has become very high. For example, it may seem that there is no significant difference between efficiencies of 96% and 98%. However, actually there is a significant difference; an efficiency of 96% means a power loss of 4%, while an efficiency of 98% means a power loss of 2%. In other words the power loss of a circuit with 98% efficiency is half that of a circuit with 96% efficiency, which may ease cooling requirements (and hence permit low-cost, small-size circuitry).

1.1.3.9.1 *Ideal switch power loss is zero*

Let us consider a switch. When the switch is off, the current I through the switch is zero. When the switch is on, the voltage V across the switch is zero. Note that the dissipated power $P = VI$, and thus in both cases, the power P is equal to zero (Fig. 1.13).

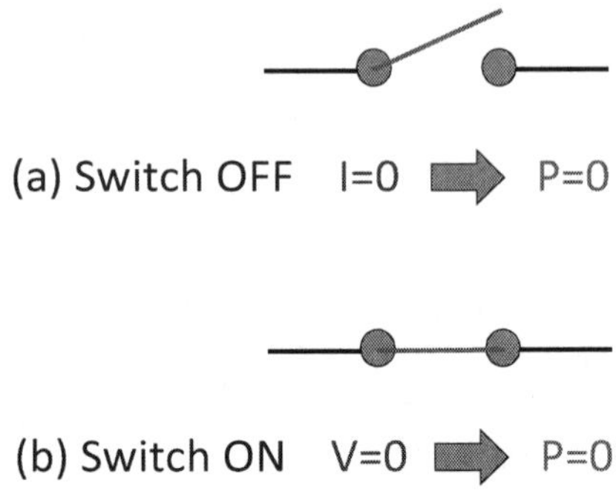

Figure 1.13 Power loss of an ideal switch is zero.

1.1.3.9.2 *Real switches have finite power loss*

Real switches have two types of power loss, conduction loss and switching loss.

(i) Conduction loss: When a switch is on, its on resistance is not zero but finite. Hence the voltage V across it is finite, and a current I flows through it. Then a power ($P = IV$) is dissipated in the switch, which is called conduction loss. As the on resistance of the switch becomes smaller, the conduction loss becomes smaller (Fig. 1.14).

(ii) Switching loss: When the switch transitions from on to off or off to on, the transition time is not zero but finite in an actual switch. During these transitions, there are finite amounts of voltage and current and hence finite power dissipation, which is called switching loss. As the switching transition time approaches zero, the switching loss also approaches zero (Fig. 1.15).

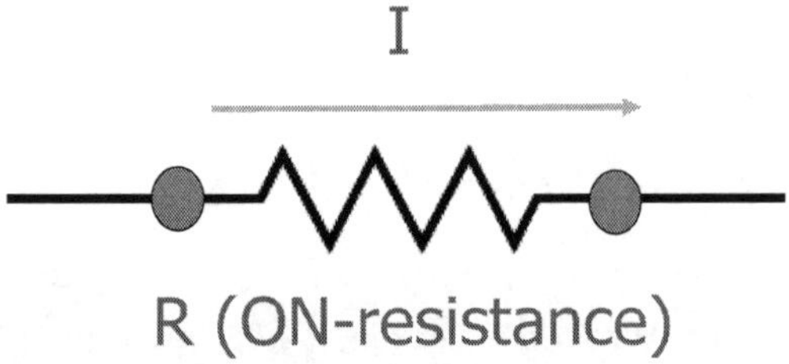

Figure 1.14 Conduction loss $P = RI^2$ when the switch is on.

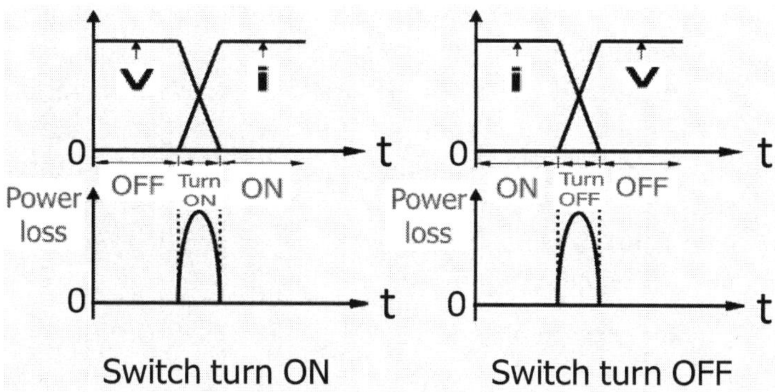

Figure 1.15 Switching loss.

1.1.3.10 Intrinsic power loss due to switch on/off transitions and soft switching

We note that even if the switch had zero on resistance and zero transition time, the switch on/off transition can cause power dissipation in the circuits that are connected through the switch.

Let us consider the circuits in Fig. 1.16, where circuit 1 and circuit 2 are connected through the switch.

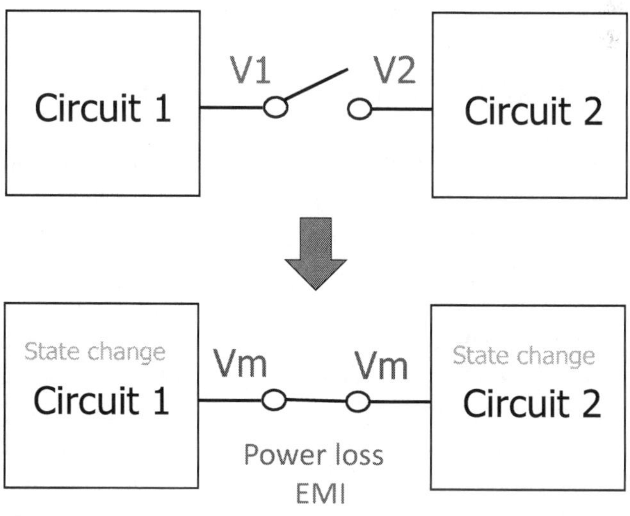

Figure 1.16 Switch turns on when $V_1 \neq V_2$.

(1) Suppose the switch is off, and V_1 and V_2 are not equal when the switch is turned on. Then V_1 and V_2 are changed and they are equal to V_m. The internal states of circuit 1 and circuit 2 are changed, causing power loss as well as switching noise. However, if $V_1 = V_2$ when the switch is turned on, the internal states of circuit 1 and circuit 2 are not changed; there is no energy loss and no switching noise; this is called zero-voltage switching (ZVS) (Fig. 1.17).

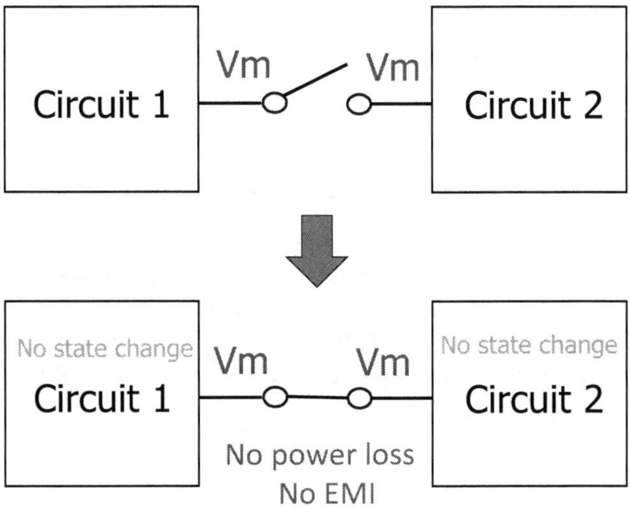

Figure 1.17 Switch turns on when $V_1 = V_2$ (ZVS).

Zero-voltage switching: Turn on the switch when the voltage across the switch is zero. This is similar to opening a door when nobody is waiting in front of the door: it has no noticeable effect.

(2) Suppose the switch is on, and a finite current I is flowing through the switch when the switch is turned off. The current I becomes zero, the internal states of circuit 1 and circuit 2 are changed, and there is power loss as well as switching noise. If $I = 0$ when the switch is turned off; on the other hand, the internal states of circuits 1 and 2 are not changed: there is no energy loss and no switching noise; this is called zero-current switching (ZCS) (Fig. 1.18).

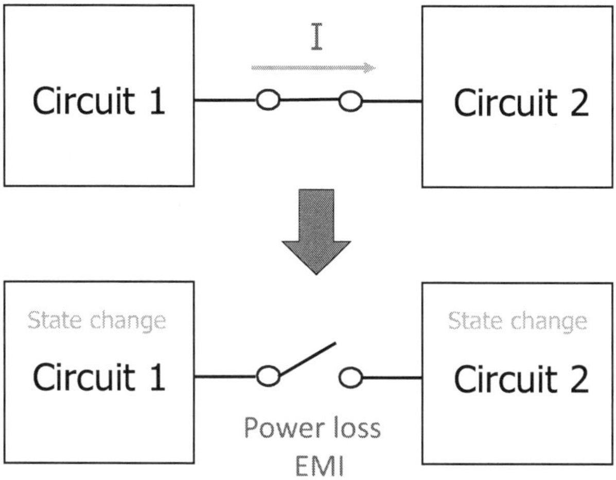

Figure 1.18 Switch turns off when $I \neq 0$.

Zero-current switching: Turn off the switch when no current is flowing through it. This is similar to closing a door when nobody is going through it: in this case, closing the door has no noticeable effect (Fig. 1.19).

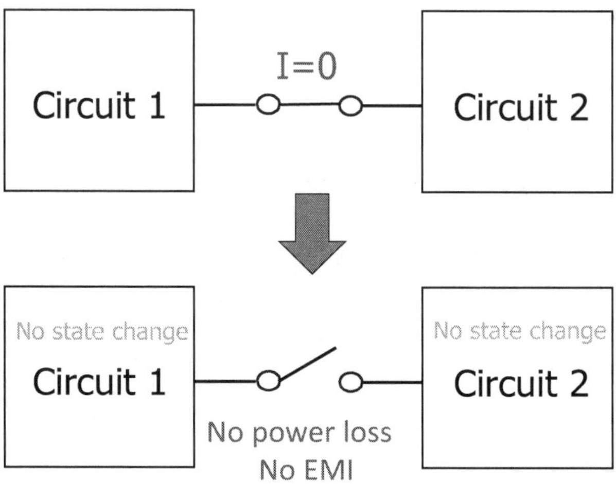

Figure 1.19 Switch turns off when $I = 0$ (ZCS).

ZVS and ZVC are called soft switching, which is the switch turning off/on such that the circuit internal states are not changed; in such a case there is no power loss and no electromagnetic interference noise generated.

1.1.3.11 Switching frequency and circuit technology

We reviewed the history of circuit technology progress and found that "passive elements have been replaced with active devices." In switching-mode power supply circuits, the switching frequency is gradually increasing thanks to advances in switching (active) devices, and this allows the sizes of L and C to be smaller. In the future, power supply circuits will be composed of rich active devices and light passive devices due to device, control, and circuit improvements.

1.1.3.12 Difference between analog and power supply circuits

There are several differences between analog integrated circuit and power supply circuit designs.

- Power supply circuits use inductors and transformers, whereas analog circuits do not. In power supply circuits, capacitors, inductors, and transformers (passive elements), as well as active devices, are very important. On the other hand, in analog integrated circuit design, passive elements occupy a large chip area and hence are costly, their usage is not welcome, and their values as well as qualities are limited.
- In analog circuits, "symmetry is beauty" and symmetric circuits (such as differential circuits) and symmetric layout are important, while this is not necessarily true for power supply circuits.
- In power supply circuits, the drain–source on resistance R_{ds} has to be as low as possible when the drain–source voltage V_{ds} is close to zero (the MOSFET is deeply in the linear region). The modeling of this operating region is important (accuracy is needed). On the other hand, the MOSFET characteristics in the saturation region are important in analog circuit design.
- The size of the switch or power MOS is very large, which is implemented by using many MOSFETs in parallel, and drive circuitry design for power supply circuits is very challenging.

- The word "circuit" means that current circulates or it returns; this concept is very important in power supply circuits. Of course this is also true for analog integrated circuit design.
- In power supply circuits, designers have to pay close attention to thermal effects, temperature, aging, reliability, and component life. For example, currently the life of capacitors is shorter than light-emitting diodes (LEDs), which is a problem for LED driver circuits.
- International conference paper presentations on power supply circuits come from many countries and many organizations. However, those for analog/RF integrated circuits come from only a limited number of groups.
- In universities there are basic courses related to power supply circuit design. *Electric circuit courses* are well suited to power electronics circuit design and analysis, while *electronic circuit courses* are well suited to analog integrated circuit design and analysis. *Control engineering* is well suited to design and analysis of both power electronics and analog integrated circuits; interestingly enough, most circuit design professors realize this fact, but many control professors do not. *Semiconductor device courses* are of course important to both.

1.1.4 Future Direction

Hot topics related to power supply circuits are as follows. LED drivers are important, as often constant current rather than constant voltage is supplied to LEDs in series. Technologies related to wireless energy transfer, EVs, and renewable energy, as well as energy harvesting are also recent topics.

There are many techniques developed for low-power system-on-chip (SoC) design; the same function works with less than 1% of the usual power (significant power reduction techniques have been developed). It is our belief that from this analogy, there is a lot of room for energy reduction in our daily life and in manufacturing activities.

1.2 Basics

In the following sections power supplies focused on a small (laptop) computer application are described. The power supplies consist of an AC-to-DC converter having power factor correction (PFC), a bus converter having a transformer to isolate from the AC line input and to supply appropriate voltages to multiple outputs, and buck converters. Understanding the basics of these converters in advance is beneficial. Therefore in this section let us consider (1) the principles of inductor volt-second (or magnetic flux linkage) balance and capacitor charge balance in a buck converter, (2) a transformer equivalent circuit, and (3) the general expression of the PF.

The above converters use semiconductor switching devices. The devices have power losses during turn-on and turn-off transitions of the switching. Since the power losses can be significant at a high switching frequency, let us consider (4) the power loss (or switching loss) in a buck converter as well.

1.2.1 Inductor Volt-Second (or Magnetic Flux Linkage) Balance and Capacitor Charge Balance in a Buck Converter

Let us consider buck converter operation in a steady state. Figure 1.20 shows a buck converter circuit. Here the switch S and the diode D are ideal: nondissipative in the on state, no leak in the off state, and no switching losses. Figure 1.21 shows steady-state inductor voltage $v_L(t)$ and current $i_L(t)$ waveforms in the buck converter. When S is in the on state, the value of $v_L(t)$ becomes the input voltage V_i minus the output voltage V_o. Then the time differential (or slope) of the $i_L(t)$ waveform is obtained by dividing $V_i - V_o$ by the inductance L:

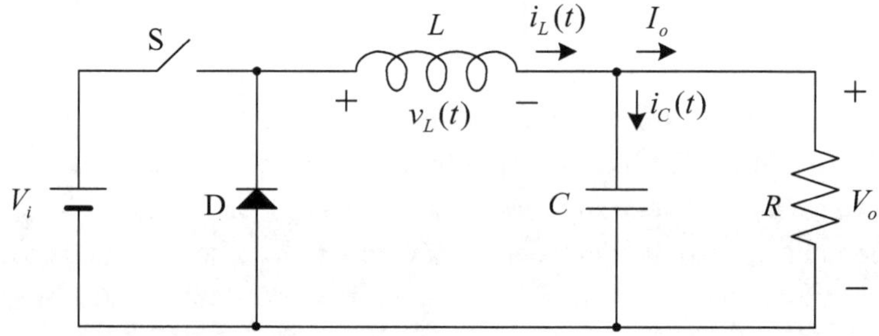

Figure 1.20 Buck converter circuit.

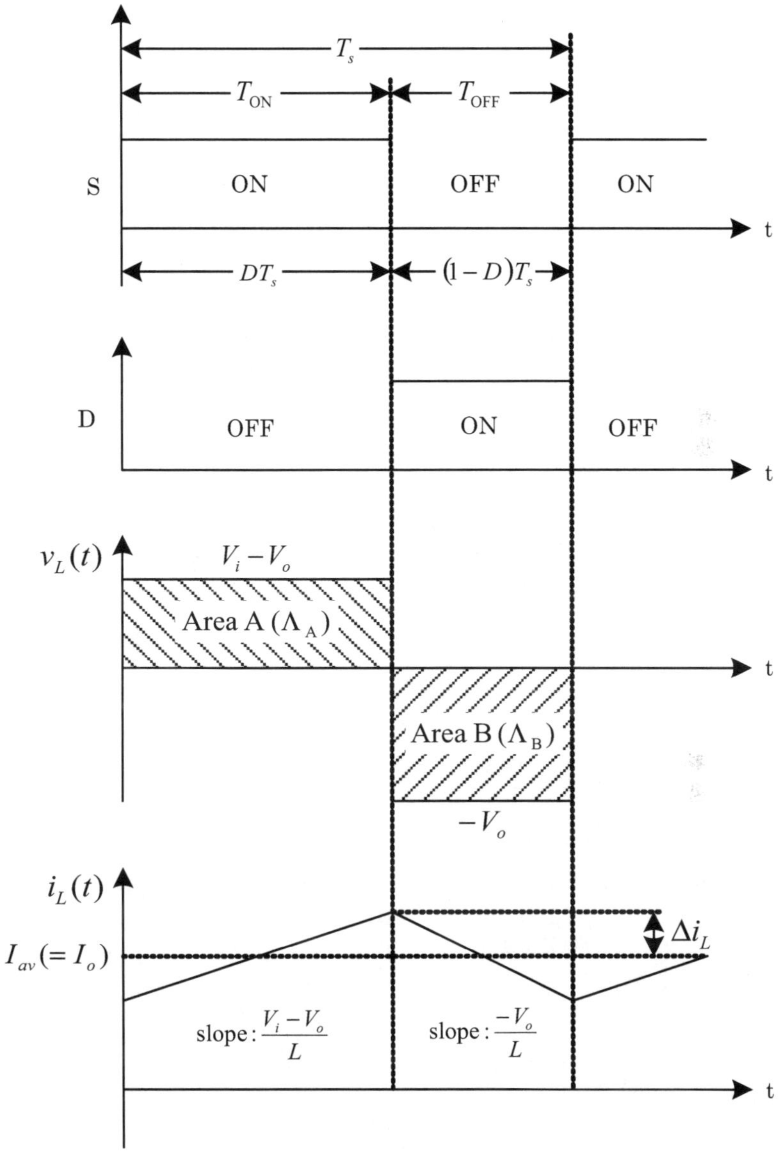

Figure 1.21 Steady-state inductor voltage and current waveforms in a buck converter.

$$\frac{di_{\mathrm{L}}(t)}{dt} = \frac{V_{\mathrm{i}} - V_{\mathrm{o}}}{L} \tag{1.1}$$

Since the right-hand side of the above equation is constant in the steady state, the $i_L(t)$ linearly increases from a value of $i_L(0)$ with time t, and it attains to the peak value of $i_L(DT_s)$ at time $t = DT_s$ where D is the duty cycle defined as

$$D = \frac{T_{ON}}{T_{ON} + T_{OFF}} \tag{1.2}$$

Here T_{ON} is the switching-on time and T_{OFF} the switching-off time. The denominator of Eq. 1.2 is equal to the switching period T_s.

When S is in the off state, the value of $v_L(t)$ becomes the negative output voltage $-V_o$, and then the slope of the $i_L(t)$ waveform is expressed as

$$\frac{di_L(t)}{dt} = -\frac{V_o}{L} \tag{1.3}$$

From Eq. 1.3 the $i_L(t)$ linearly decreases, and during the off state it changes from $i_L(DT_s)$ to $i_L(T_s)$, which is equal to $i_L(0)$ in the steady state.

The peak-to-peak ripple $2\Delta i_L$ of the $i_L(t)$ is obtained by using Eq. 1.1 as

$$2\Delta i_L = \left(\frac{V_i - V_o}{L}\right)(DT_s) \tag{1.4}$$

where Δi_L is the peak-to-average ripple. The average value I_{av} of the $i_L(t)$ is its DC component, and its waveform, symmetrical to I_{av}.

In the steady state, since $i_L(0) = i_L(T_s)$, as indicated above, integration of $v_L(t)$ over one switching period T_s becomes zero:

$$\int_0^{T_s} v_L(t)dt = \int_0^{T_s} L\frac{di_L(t)}{dt}dt = L\left[i_L(T_s) - i_L(0)\right] = 0 \tag{1.5}$$

Using Eqs. 1.1, 1.3, and 1.5 yields

$$(V_i - V_o)(DT_s) - (V_o)(1 - D)\, T_s = 0 \tag{1.6}$$

This states that in Fig. 1.21, area A, the magnetic flux linkage Λ_A increased in the inductor, is equal to area B, the magnetic flux linkage Λ_B decreased in the inductor, that is, the total area is zero. This is called the principle of inductor volt-second (or magnetic flux linkage) balance, because the integration has the units of volt-seconds. From this equation we obtain the following relationship for the input and output voltages:

$$V_o = DV_i \tag{1.7}$$

Since $0 \le D \le 1$, V_o is less than V_i and V_o is linearly proportional to D at a constant value of V_i. This inductor volt-second balance in the steady state can also be used in other converter analyses (boost, buck–boost, etc.).

Figure 1.22 shows steady-state capacitor voltage $v_C(t)$ and current $i_C(t)$ waveforms in the buck converter. For a well-designed buck converter, almost all the AC components of $i_L(t)$ flow to the capacitor and its DC component to the load resister, since the value of the capacitance C is set so large that its impedance at the switching frequency is much smaller than the load impedance R. In this case, the inductor average DC current component I_{av} is equal to the output DC current I_o. The positive capacitor current increases $v_C(t)$ owing to charging of the capacitor. Integration of the positive capacitor current over a half switching period $T_s/2$, area A in Fig. 1.22, yields the charge Q_A increased in the capacitor:

$$Q_A = \frac{1}{2}(\Delta i_L)\left(\frac{T_s}{2}\right) \tag{1.8}$$

Inversely the negative capacitor current decreases $v_C(t)$ owing to discharging of the capacitor. Similarly integration of the negative current over a half switching period $T_s/2$, area B in Fig. 1.22, yields the charge Q_B decreased in the capacitor:

$$Q_B = -\frac{1}{2}(\Delta i_L)\left(\frac{T_s}{2}\right) \tag{1.9}$$

In the steady state, since the values of Q_A and $-Q_B$ are the same from Eqs. 1.8 and 1.9, integration of $i_C(t)$ over one switching period is zero:

$$\int_0^{T_s} i_C(t)dt = Q_A + Q_B = \int_0^{T_s} C\frac{dv_C(t)}{dt}dt = C\left[v_C(T_s) - v_C(0)\right] = 0 \tag{1.10}$$

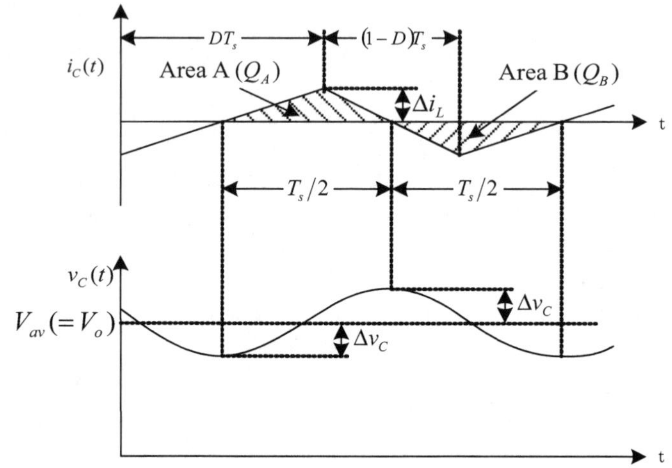

Figure 1.22 Steady-state capacitor voltage and current waveforms in the buck converter.

This equation also includes $v_C(T_s) = v_C(0)$ in the steady state. This is called the principle of capacitor charge balance, which can also be used in other converter analyses (boost, buck–boost, etc.).

The capacitor voltage has the average DC component V_{av} leading to the output voltage V_o and the AC component with the peak-to-average ripple Δv_C (or output voltage ripple). Since $Q_A = C(2\Delta v_C)$, we can obtain the Δv_C by using Eq. 1.8 as

$$\Delta v_C = \frac{\Delta i_L T_s}{8C} \tag{1.11}$$

Elimination of Δi_L in Eq. 1.11 using Eqs. 1.4 and 1.7 yields

$$\Delta v_C = \frac{V_i(1-D)DT_s^2}{16LC} \tag{1.12}$$

Equation 1.12 does not include parasitic components, an equivalent series resistance (ESR), and an equivalent series inductance (ESL) in the capacitor. More detailed analyses including those parasitic components are described in Chapter 2.

1.2.2 Transformer-Equivalent Circuit

In switching-converter applications a transformer is incorporated when following items are required: (1) DC isolation between the converter input and output for safety, (2) a large step-up or step-down conversion ratio, and (3) multiple outputs as well. To analyze switching-converter circuits including a transformer, we have to understand the basic operation of a transformer.

Let us consider a multiple-winding transformer having a turns ratio of $n_i : n_{o,1}: n_{o,2}: \bullet \bullet\ n_{o,n}$ and an equivalent circuit of the transformer. Figure 1.23a shows a circuit of a transformer and Fig. 1.23b its equivalent circuit that will be used for circuit analyses in Chapter 3. The equivalent circuit consists of an ideal transformer having perfect coupling between windings and no losses and a magnetizing inductance of L_m connected in parallel with the ideal transformer. The ideal transformer has the relationships for an input (or primary) voltage $v_i(t)$ and output voltages $v_{o,1}(t)$, $v_{o,2}(t)$, ••, $v_{o,n}(t)$:

$$\frac{v_i(t)}{n_i} = \frac{v_{o,1}(t)}{n_{o,1}} = \frac{v_{o,2}(t)}{n_{o,2}} = \bullet\ \bullet = \frac{v_{o,n}(t)}{n_{o,n}}, \tag{1.13}$$

and for an input (or primary) current $i_{i,i}(t)$ flowing into the ideal transformer part and output currents $i_{o,1}(t)$, $i_{o,2}(t)$, ••, $i_{o,n}(t)$ as well:

$$n_i i_{i,i}(t) + n_{o,1} i_{o,1}(t) + n_{o,2} i_{o,2}(t) + \bullet\bullet + n_{o,n} i_{o,n}(t) = 0. \tag{1.14}$$

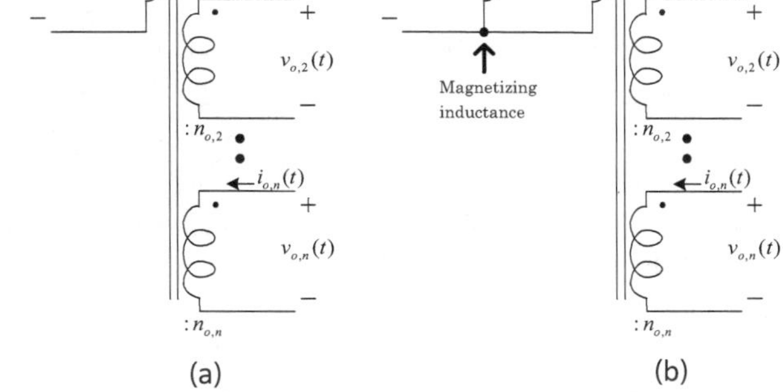

Figure 1.23 (a) Multiple-winding transformer circuit and (b) its equivalent circuit containing a magnetizing inductance and an ideal transformer.

Eliminating n_i, $n_{o,1}$, $n_{o,2}$, ••, $n_{o,n}$ from Eqs. 1.13 and 1.14 yields

$$v_i(t)i_{i,i}(t) = -\left(v_{o,1}i_{o,1}(t) + v_{o,2}i_{o,2}(t) + \bullet\bullet + v_{o,n}i_{o,n}(t)\right) \tag{1.15}$$

This equation states that the input power is distributed to each of the output ports, depending on output voltages needed, and is equal to the total output power. Here the negative sign in Eq. 1.15 comes from the output current directions defined in Fig. 1.23b.

The magnetizing inductance obeys the following relationship:

$$v_i(t) = L_m \frac{di_{i,m}(t)}{dt} \tag{1.16}$$

where $i_{i,m}(t)$ is the magnetizing current. Therefore the magnetizing inductance has zero impedance under the DC condition, while it has a very large impedance under high-frequency AC conditions (or high-switching-frequency conditions). Under the AC conditions, since the $i_{i,m}(t)$ is much smaller in magnitude than the input current $i_i(t)$, we have $i_i(t) \cong i_{i,i}(t)$, and behavior of the transformer is nearly ideal. Here note that $i_{i.m}(t)$ is determined by integration of the input voltage (or the voltage applied to the magnetizing inductance):

$$i_{i,m}(t) - i_{i,m}(0) = \frac{1}{L_m}\int_0^t v_i(\tau)d\tau \tag{1.17}$$

and it is independent of $i_i(t)$.

In the steady state the magnetizing inductance has no DC current component and holds the principle of inductance volt-second balance. Namely, integration of $v_i(t)$ over one switching period is zero. If there is a net increase of the $i_{i,m}(t)$ during each switching period, excessive large currents flow to the magnetizing inductance and lead to transformer saturation. Since this saturation causes the transformer to short out, we have to avoid it in real operation.

Practical transformers have some leakage inductance. This leakage inductance comes from imperfect flux linkage between windings, and it may be modeled with small inductors in series with the windings in the case of the two-winding transformer. Therefore the leakage inductance expresses nonidealities leading to switching losses, increased peak transistor voltage, and reduction of cross regulation. However, it does not affect basic converter operation.

1.2.3 General Expression for the Power Factor

PF expresses efficiency of the real energy transmission from a source to a load at a given location in a system shown in Fig. 1.24 and is defined as the ratio of real (or average) power P_{re} to apparent power P_{ap} measured at the location:

$$\text{PF} = \frac{P_{\text{re}}}{P_{\text{ap}}} \tag{1.18}$$

PF has a value between zero and one. At the location the voltage $v(t)$ is supplied by the source and the current $i(t)$ is determined by the load. If $v(t)$ and $i(t)$ waveforms are periodic with a period T, P_{re} becomes

$$P_{\text{re}} = \frac{1}{T}\int_0^T v(t)i(t)dt \tag{1.19}$$

Here $v(t)$ and $i(t)$ can be expressed as Fourier series:

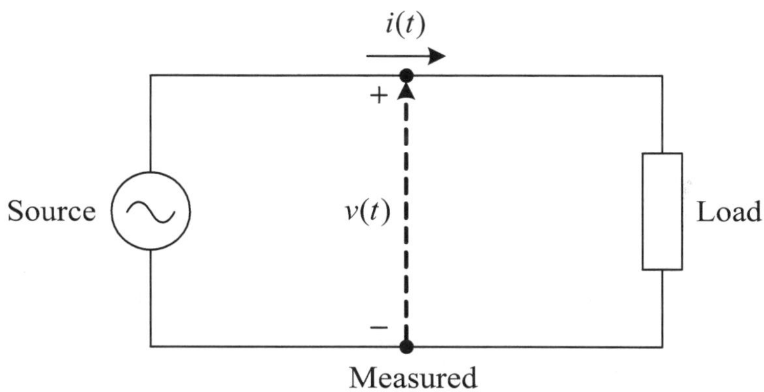

Figure 1.24 Measurement of energy transmission in a power supply system.

$$v(t) = V_0 + \sum_{n=1}^{\infty} V_n \cos(n\omega t - \phi_n) \tag{1.20}$$

$$i(t) = I_0 + \sum_{n=1}^{\infty} I_n \cos(n\omega t - \theta_n) \tag{1.21}$$

where $\omega (= 2\pi/T)$ is the angular frequency. Substituting Eqs. 1.20 and 1.21 into Eq. 1.19 yields

$$P_{\rm re} = \frac{1}{T}\int_0^T \left(V_0 + \sum_{n=1}^{\infty} V_n \cos(n\omega t - \phi_n) \right) \left(I_0 + \sum_{n=1}^{\infty} I_n \cos(n\omega t - \theta_n) \right) dt \tag{1.22}$$

In this integration the cross-product terms having different $n\omega t$ are zero, so that $P_{\rm re}$ becomes

$$P_{\rm re} = V_0 I_0 + \sum_{n=1}^{\infty} \frac{V_n I_n}{2} \cos(\phi_n - \theta_n) \tag{1.23}$$

This equation states that only the same-frequency components of voltage and current contribute to energy transmission to the load.

$P_{\rm ap}$ is defined as the product of the root-mean-square (rms) voltage $V_{\rm rms}$ and current $I_{\rm rms}$:

$$P_{\rm ap} = V_{\rm rms} I_{\rm rms} \tag{1.24}$$

Here $V_{\rm rms}$ and $I_{\rm rms}$ are defined as

$$V_{\rm rms} = \sqrt{\frac{1}{T}\int_0^T v^2(t)dt}\ , \tag{1.25}$$

$$I_{\rm rms} = \sqrt{\frac{1}{T}\int_0^T i^2(t)dt} \tag{1.26}$$

Substituting Eqs. 1.20 and 1.21 into Eqs. 1.25 and 1.26, respectively, yields

$$V_{\rm rms} = \sqrt{V_0^2 + \sum_{n=1}^{\infty} \frac{V_n^2}{2}} \tag{1.27}$$

$$I_{\rm rms} = \sqrt{I_0^2 + \sum_{n=1}^{\infty} \frac{I_n^2}{2}} \tag{1.28}$$

Therefore $P_{\rm ap}$ is

$$P_{\rm ap} = \sqrt{\left(V_0^2 + \sum_{n=1}^{\infty} \frac{V_n^2}{2} \right) \left(I_0^2 + \sum_{n=1}^{\infty} \frac{I_n^2}{2} \right)} \tag{1.29}$$

The presence of harmonics in voltage and current waveforms increases the values of $V_{\rm rms}$, $I_{\rm rms}$, and $P_{\rm ap}$ as well.

If the load has only resistance R, the phase difference between voltage and current is zero: $\varphi_{\rm n} = \theta_{\rm n}$, and the magnitude of the current is proportional to that of the voltage in spite of any voltage waveform supplied: $I_n = V_n/R$. In this case since both P_{re} and P_{ap} become V^2_{rms}/R from Eqs. 1.23 and 1.29, respectively, PE is equal to one from Eq. 1.18. This states that all harmonics carry the energy to the resistive load.

More generally let us consider that a fundamental sinusoidal voltage is supplied to a nonlinear dynamical and reactive load. The load causes the phase difference between the voltage and current, and P_{re} is determined by only the fundamental voltage and current components:

$$P_{re} = \frac{V_1 I_1}{2}\cos(\phi_1 - \theta_1) \tag{1.30}$$

The rms current I_{rms} is the same as Eq. 1.28, while the rms voltage V_{rms} is

$$V_{rms} = \frac{V_1}{\sqrt{2}} \tag{1.31}$$

Therefore P_{ap} becomes

$$P_{ap} = \frac{V_1}{\sqrt{2}}\sqrt{I_0^2 + \sum_{n=1}^{\infty}\frac{I_n^2}{2}} \tag{1.32}$$

Substituting Eqs. 1.30 and 1.32 into Eq. 1.18 yields

$$\text{PF} = \text{DF}\cos(\phi_1 - \theta_1) \tag{1.33}$$

where DF is called the distortion factor:

$$\text{DF} = \left(\frac{\frac{I_1}{\sqrt{2}}}{\sqrt{I_0^2 + \sum_{n=1}^{\infty}\frac{I_n^2}{2}}}\right). \tag{1.34}$$

This PF is composed of two terms. The first term, DF, is the ratio of the fundamental rms current component to the total rms current component, including harmonics. The second term is called the displacement factor. This factor is cosine of the phase difference between the fundamental voltage and current components. Therefore PF is affected by the phase difference and harmonics as well. Here the harmonics derive energy from the source, yet they do not deliver the energy to the load. They cause power losses of $I^2_{rms}R$

in the system. Here R is the total series resistance in the system: load, source, and transmission wire resistances as well.

DF is related to the total harmonic distortion (THD), defined as the ratio of the rms value of the waveform except for the fundamental component to the rms fundamental component:

$$\mathrm{THD} = \frac{\sqrt{\sum_{n=2}^{\infty} I_n^2}}{I_1} \tag{1.35}$$

From Eq. 1.34 with $I_0 = 0$ and Eq. 1.35 the relationship between DF and THD is

$$\mathrm{DF} = \frac{1}{\sqrt{1+(\mathrm{THD})^2}} \tag{1.36}$$

As an example of this case, let us look at a conventional diode bridge rectifier. Figure 1.25 shows a circuit and waveforms of the rectifier. The AC input current flows in a short duration, and its peak positions are near those of the voltage. Therefore the fundamental component of the current is in phase with the voltage and the displacement factor, almost one. However, the low-order current harmonics are relatively large compared to the fundamental component. This leads to a reduction of the DF. In this case the DF usually ranges from 55% to 65% [2]. The PF is accordingly similar to the DF in value. To improve the PF, PFC circuits are discussed in [3] and [4].

1.2.4 Switching Loss

Semiconductor switching devices are used in switching regulator power supplies. These devices cause energy loss during the turn-on and turn-off switching transitions, because charges required to drive the switching devices are inserted to or removed from them, so they affect the switching transition time and are dissipated in the devices. The energy loss can be significant in a high switching frequency. In this section, using a buck converter having a MOSFET and a diode as switching devices we consider switching losses caused by (1) MOSFET switching, (2) diode reverse recovery, and (3) MOSFET output and diode junction capacitances. Further we briefly discuss influence of parasitic series inductances on switching loss during the turn-off switching transition as well. Here conduction losses of the devices are neglected for simplicity.

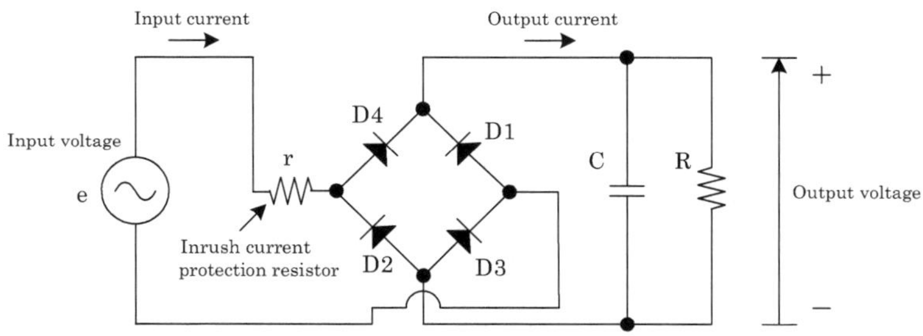

(a)

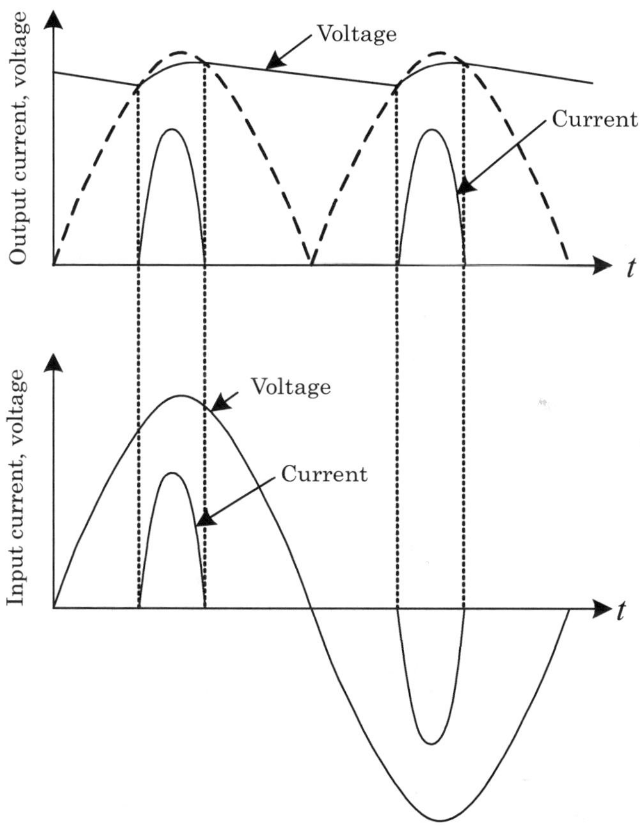

(b)

Figure 1.25 (a) Circuit and (b) waveforms of a conventional diode bridge rectifier.

1.2.4.1 MOSFET switching

Let us consider switching loss caused by MOSFET switching in a buck converter circuit shown in Fig. 1.26a. Here we use an ideal diode that has no current flow, no junction capacitance, and no reverse recovery caused by excess minority carriers when reverse-biased and no forward voltage drop when forward-biased. The diode and inductor form a clamped inductive load to the MOSFET. Figure 1.26b shows a waveform driving the gate of the MOSFET and capacitances in the MOSFET as well, where the drain–source (or drain–substrate junction) capacitance of the MOSFET is neglected. Figure 1.27a shows simple turn-on transition piecewise-linear waveforms of the MOSFET and diode for the buck converter circuit in Fig. 1.26. The gate voltage $v_{GS}(t)$ begins to rise from time $t = t_0$, and when it attains the threshold voltage V_T at time $t = t_1$, the drain current $i_{DS}(t)$ begins to flow. For $t_1 < t < t_2$, the gate current $i_G(t)$ charges the gate-to-source capacitance C_{GS}, and then $i_{DS}(t)$ rises, while the diode current $i_{Di}(t)$ decreases, because these currents have the following relationship:

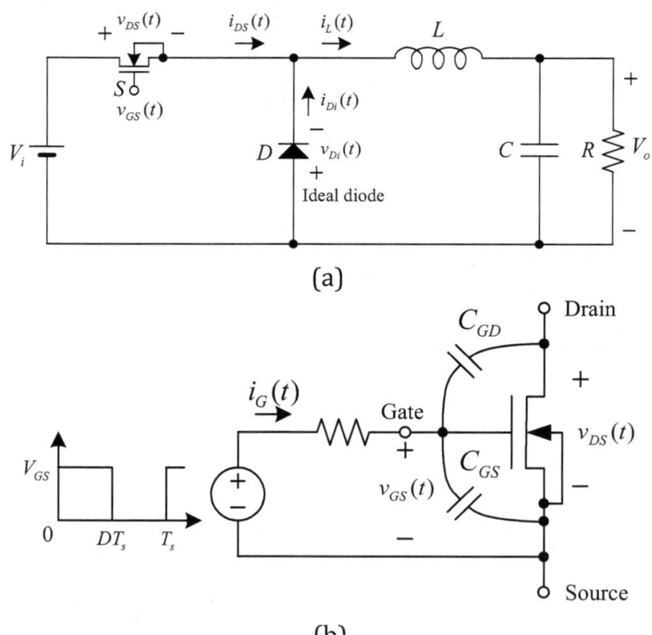

Figure 1.26 Buck converter circuit having a MOSFET and an ideal diode as switching devices. (a) Buck converter circuit and (b) capacitances in the MOSFET and a waveform driving its gate.

$$i_{DS}(t) + i_{Di}(t) = i_L(t) \tag{1.37}$$

where the inductor current $i_L(t)$ is constant during the turn-on transition. Then the inductor current path is changed from the diode to the MOSFET. During this time interval, the drain voltage $v_{DS}(t)$ maintains the input voltage V_i, since the diode is forward-biased. At time $t = t_2$, $i_{DS}(t)$ attains to $i_L(t)$, and $i_{Di}(t)$ becomes zero, and then the diode gets into the reverse-biased state. After that, for $t_2 < t < t_3$, $v_{DS}(t)$ decreases with the diode voltage $v_{Di}(t)$, and at time $t = t_3$, $v_{DS}(t)$ becomes nearly zero, and $v_{Di}(t)$, nearly $-V_i$, because these voltages have the following relationship:

$$v_{DS}(t) - v_{Di}(t) = V_i \tag{1.38}$$

During this time interval, the gate current $i_G(t)$ discharges the gate–drain or Miller capacitance C_{GD} with the constant gate voltage (or the gate plateau voltage [5]) determined by the constant inductor current i_L. For $t > t_3$, $v_{GS}(t)$ increases until V_{GS} indicated in Fig. 1.26b, and then $v_{DS}(t)$ becomes zero, and $v_{Di}(t)$, $-V_i$.

The instantaneous power $p_{ON}(t)$ dissipated in the MOSFET during the turn-on transition is also shown in Fig. 1.27a. The energy loss W_{ON} during this transition is expressed as the integration of $p_{ON}(t)$:

$$W_{ON} = \frac{1}{2} i_L V_i (t_3 - t_1) \tag{1.39}$$

Figure 1.27b shows simple turn-off transition piecewise-linear waveforms of the MOSFET and diode for the buck converter circuit in Fig. 1.26. Waveforms of the turn-off transition are qualitatively similar to those of the turn-on transition with the time axis reversed. From time $t = t_4$, $v_{GS}(t)$ begins to fall, and at $t = t_5$ it reaches the gate voltage determined by the inductor current i_L that is also constant during the turn-off. For $t_5 < t < t_6$, $v_{DS}(t)$ rises with $v_{Di}(t)$ under the constant $v_{GS}(t)$, and then C_{GD} is charged. For $t_6 < t < t_7$, $i_{DS}(t)$ falls, while $i_{Di}(t)$ raises with the $v_{GS}(t)$ reduction. Then inductor current path changes from the MOSFET to the diode, and C_{GS} is discharged. During this time interval, $v_{DS}(t) = V_i$, since the diode is forward-biased.

The instantaneous power $p_{OFF}(t)$ dissipated in the MOSFET during the turn-off transition is also shown in Fig. 1.27b. The energy loss W_{OFF} during this transition is expressed as the integration of $p_{OFF}(t)$:

$$W_{\mathrm{OFF}} = \frac{1}{2} i_{\mathrm{L}} V_{\mathrm{i}} (t_7 - t_5) \tag{1.40}$$

Therefore the average power loss P_{SW} induced by the turn-on and turn-off is expressed as

$$P_{\mathrm{SW}} = (W_{\mathrm{ON}} + W_{\mathrm{OFF}}) f_{\mathrm{s}} \tag{1.41}$$

where f_s is the switching frequency. The switching-mode power loss is proportional to the switching frequency.

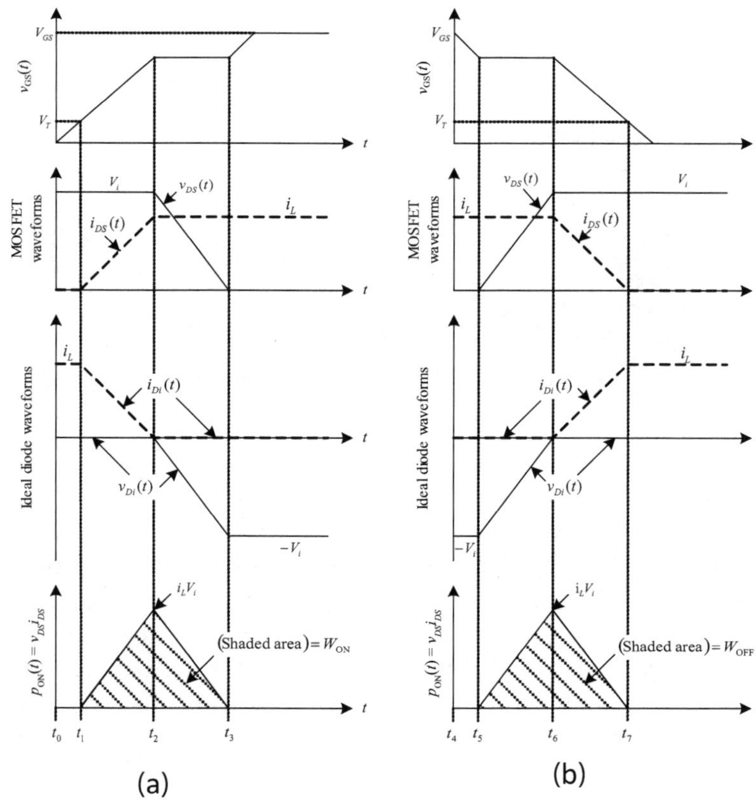

Figure 1.27 (a) Turn-on and (b) turn-off transition piecewise-linear waveforms of the MOSFET and diode for the buck converter circuit in Fig. 1.26.

1.2.4.2 Diode reverse recovery

In the previous section we considered MOSFET switching losses in a buck converter having a MOSFET and an ideal diode. In this section, let us consider MOSFET switching losses caused when using a p-n$^-$-n$^+$ (or pin) diode instead of the ideal diode. Figure 1.28 shows a buck converter circuit having a MOSFET and a pin diode. Here the switching transition time of the MOSFET is much shorter than that of the diode, so the main switching loss is caused by the diode. The diode obtains a high breakdown voltage thanks to a wide depletion region formed in the n$^-$ low-doping region in the off (or reverse-biased) state, while in the on (or forward-biased) state it yields a very low on resistance under conductivity modulation caused by excess minority carriers stored in the n$^-$ region. However, when the diode is turned off, the excess minority carriers are removed. Then the current induced by the excess minority carriers causes a MOSFET switching loss. This transition process from the on state to the off state is called diode reverse recovery.

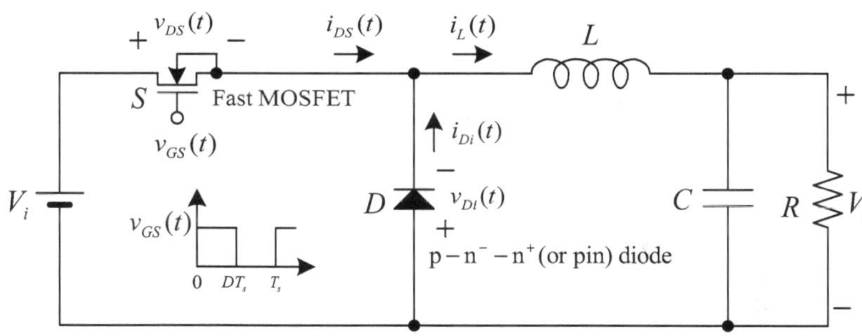

Figure 1.28 Buck converter circuit having a MOSFET and a p-n$^-$-n$^+$ (or pin) diode.

Figure 1.29 shows MOSFET turn-on transition piecewise-linear waveforms for the buck converter circuit in Fig. 1.28. Initially the diode conducts the inductor current in the off state of the MOSFET. At time $t = t_0$ the drain current $i_{DS}(t)$ of the MOSFET begins to rise, and then the diode current $i_{Di}(t)$ begins to fall with a slope limited by stray inductances present in the external circuit. At time $t = t_1$, $i_{Di}(t)$ becomes zero. For $t_1 < t < t_2$, $i_{Di}(t)$ continues to fall with the slope, and then it becomes negative. Since $i_{Di}(t)$ and $i_{DS}(t)$ have the same relationship as Eq. 1.37 under the constant $i_L(t)$ during this transition, $i_{DS}(t)$ continues to rise over i_L. Here the drain voltage $v_{DS}(t)$ of the MOSFET is constant at the input voltage V_i because the diode is still in the on state. During this time interval the stored carriers in the n$^-$ region flow out and recombine there, and at time

$t = t_2$, they are exhausted in the vicinity of the p-n$^-$ junction. For $t_2 < t < t_3$, the diode becomes reverse-biased, and then $i_{Di}(t)$ charges the depletion capacitance of the p-n$^-$ junction and rises, while $i_{DS}(t)$ falls. At time $t = t_3$, $i_{Di}(t)$ becomes zero, and $i_{DS}(t)$, i_L, so that the diode reverse recovery process finishes. In the meantime $v_{DS}(t)$ and the diode voltage $v_{Di}(t)$ fall, and at time $t = t_3$, they become zero and $-V_i$, respectively. Here the area within the negative portion of $i_{Di}(t)$ is the diode-recovered charge $-Q_r$, and the time interval length $t_3 - t_1$ is the reverse recovery time t_r. Since $i_{DS}(t)$ supplies the charge $-Q_r$, the MOSFET dissipates the power during the turn-on transition.

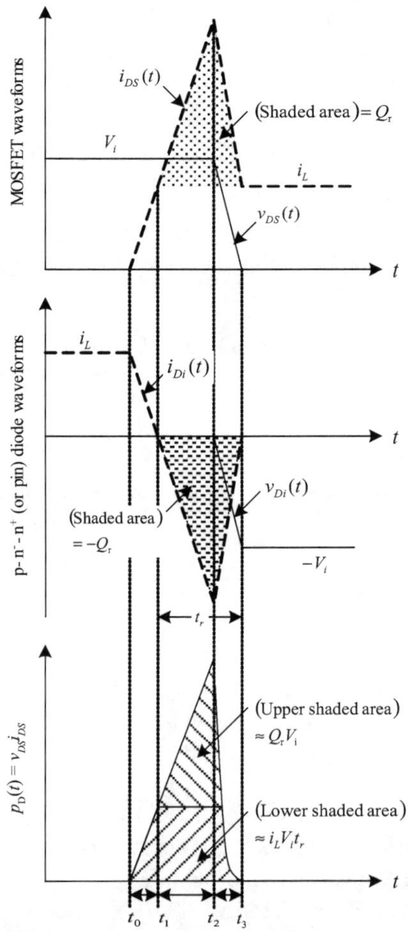

Figure 1.29 MOSFET turn-on transition piecewise-linear waveforms for the buck converter circuit in Fig. 1.28.

The instantaneous power loss $p_D(t)$ in the MOSFET induced by the diode-recovered charge is also shown in Fig. 1.29. The energy W_D lost during the turn-on transition is expressed as the integration of $p_D(t) = i_{DS}(t)\, v_{DS}(t)$ over the turn-on transition:

$$W_D = \int_{\substack{\text{turn-on}\\ \text{transition}}} i_{DS}(t) v_{DS}(t) dt \tag{1.42}$$

If $(t_2 - t_1) >> (t_3 - t_2)$, $v_{DS}(t)$ can be equal to the input voltage V_i during the turn-on transition, and then the integration is approximated as

$$W_D \approx \int_{\substack{\text{turn-on}\\ \text{transition}}} (i_L(t) - i_{Di}(t)) V_i dt = Q_r V_i + i_L V_i t_r \tag{1.43}$$

The average power loss P_D induced by the diode reverse recovery becomes $W_D f_S$. This power loss can be reduced by using faster diodes designed for low recovered charge.

1.2.4.3 MOSFET output and diode junction capacitances

Let us consider switching loss caused by the drain–source (or output) capacitance C_{DS} of the MOSFET and the p-n$^-$ junction capacitance C_J of the diode in a buck converter circuit. Figure 1.30 shows these capacitances in a buck converter circuit. When the MOSFET turns on, the energy of (1/2) $C_{DS}V_i^2$ stored in C_{DS} in the off state of the MOSFET is dissipated by the MOSFET. In the meantime the energy of (1/2) $C_J V_i^2$ is stored in C_J by the drain current $i_{DS}(t)$, and then the same amount of energy is dissipated by the MOSFET. When the MOSFET turns off, the energy of (1/2) $C_{DS}V_i^2$ is stored in C_{DS} by the inductor current $i_L(t)$, while the energy of $(1/2)C_J V_i^2$ stored in C_J in the on state of the MOSFET is transported to the inductor. During the turn-off transition, since both the energy storage in C_{DS} and the energy transport from C_J are induced by the inductor current $i_L(t)$, there is no energy loss. Therefore the energy lost during one switching period is

$$W_C = \frac{1}{2}(C_{DS} + C_J)V_i^2 \tag{1.44}$$

and the average power loss induced by those capacitances becomes $W_C f_S$.

The incremental drain–source capacitance C_{ds} of the MOSFET strongly depends on the drain–source voltage v_{DS}, and except for a small v_{DS} region it can be approximated as

$$C_{ds}(v_{DS}) \approx \frac{C_0}{\sqrt{v_{DS}}} \tag{1.45}$$

where C_0 is a constant. Since the charging current i_c in C_{ds} is expressed as

$$i_c = C_{ds}(v_{DS})\frac{dv_{DS}}{dt} \tag{1.46}$$

the energy stored in C_{ds} (v_{DS}) during the turn-off transition is

$$W_{C_{ds}} = \int_{\substack{\text{turn-off}\\ \text{transition}}} v_{ds} i_c dt \approx \int_0^{V_i} v_{ds} \frac{C_0}{\sqrt{v_{ds}}} dv_{ds} = \frac{2}{3} C_0 V_i^{3/2} = \frac{1}{2}\left(\frac{4}{3} C_{ds}(V_i)\right) V_i^2 \tag{1.47}$$

From the Eq. 1.47, C_{DS} in Eq. 1.44 can be replaced with a capacitance having the value of $(4/3)C_{ds}(V_i)$ from the viewpoint of energy loss.

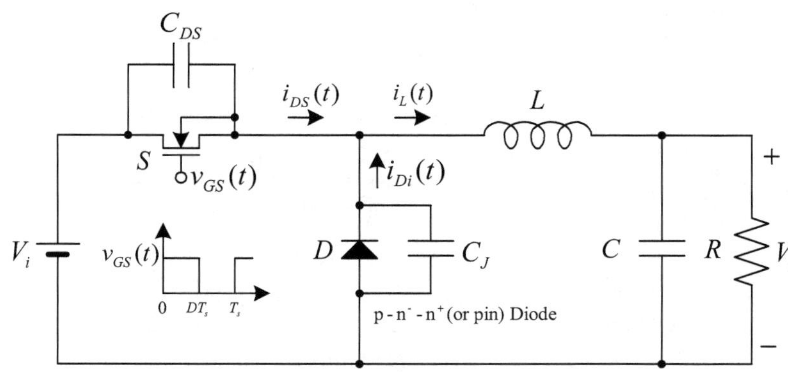

Figure 1.30 Drain–source capacitance of the MOSFET and p-n⁻ junction capacitance of the pin diode in a buck converter circuit: the energy stored in the capacitances is dissipated by the MOSFET during the turn-on transition.

1.2.4.4 Parasitic series inductances

In real circuits packaged, parasitic series inductances are interconnection inductances and transformer leakage inductances in isolated converters as well. The energy stored in those inductances is dissipated by the transistors in series with the inductances during the transistor turn-off transition, leading to switching loss. Interconnection inductances can cause significant switching loss in high-current applications and give excessive high-voltage stress to the transistors as well. Leakage inductances lead to important switching loss in many transformer-isolated converter applications.

References

1. R. W. Erickson and D. Maksimonic, Chapter 7, in *Fundamentals of Power Electronics, Second Edition,* Springer Science + Business Media, 2001.
2. R. W. Erickson and D. Maksimovic, Chapter 16, in *Fundamentals of Power Electronics, Second Edition*, Springer Science + Business Media, 2001.
3. R. W. Erickson and D. Maksimovic, Chapter 18, in *Fundamentals of Power Electronics, Second Edition*, Springer Science + Business Media, 2001.
4. A. I. Pressman, K. Billings, and T. Morey, Chapter 15, in *Switching Power Supply Design, Third Edition*, McGraw-Hill Professional, 2009.
5. B. J. Baliga, Chapter 6, in *Fundamentals of Power Semiconductor Devices,* Springer Science + Business Media, 2008.

Chapter 2

Buck Converter for Low-Voltage Application

Takashi Nabeshima
Department of Electrical and Electronics Engineering, Oita University, 700 Dannoharu, Oita 870-1192, Japan
nabesima@oita-u.ac.jp

2.1 Introduction

Switching power converters are widely used as DC power supplies for various kinds of electronic equipment because of their higher efficiency and smaller size compared to linear regulators. The most prominent feature of a switching converter is its higher conversion efficiency, since it uses a semiconductor as a switching device, while a conventional linear regulator uses a semiconductor as a dissipative element. Figure 2.1 shows basic circuit configurations of three types of DC-to-DC converters. These circuits consist only of a switch, a diode, an inductor, and a capacitor. The state of the switch S and the node voltage *V*a in the steady-state operation are sketched in Fig. 2.2. For the constant-switching-frequency operation, the ratio $T_{\text{on}}/T_{\text{s}}$ is an important parameter. We here define the duty cycle D as

Handbook of Power Management Circuits
Edited by Haruo Kobayashi and Takashi Nabeshima

ISBN 978-981-4613-15-6 (Hardcover), 978-981-4613-16-3 (eBook)
www.panstanford.com

$$D = \frac{T_{on}}{T_{on} + T_{off}} = \frac{T_{on}}{T_s} \quad (0 \le D \le 1) \tag{2.1}$$

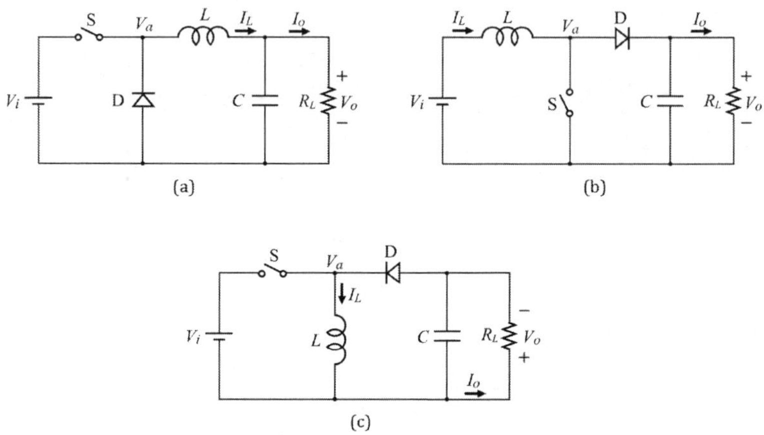

Figure 2.1 Three typical DC-to-DC converters. (a) A buck converter, (b) a boost converter, and (c) a buck–boost converter.

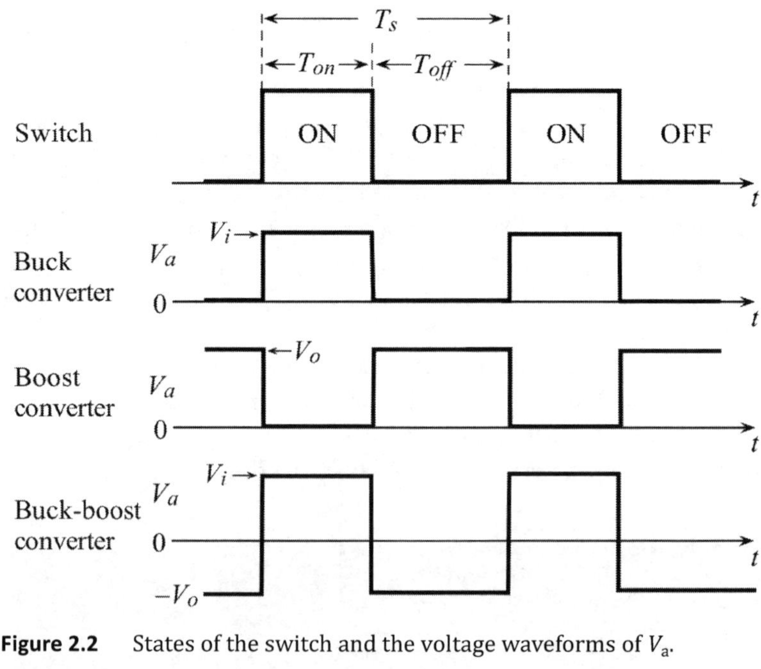

Figure 2.2 States of the switch and the voltage waveforms of V_a.

To obtain an ideal voltage conversion ratio, V_o/V_i, for three converters, all circuit elements and the output ripple voltage are assumed to be ideal and zero, respectively. During the on state of the switch S in the buck converter, the voltage $V_i - V_o$ is applied to the inductor winding, and $-V_o$ is applied during the off state. Since the change in the inductor current during the on state is equal to that during the off state in the steady-state operation, the following equations are obtained:

$$\frac{V_i - V_o}{L} T_{on} = \frac{V_o}{L} T_{off} \tag{2.2}$$

$$\frac{V_o}{V_i} = \frac{T_{on}}{T_{on} + T_{off}} = D \tag{2.3}$$

For the boost converter, the voltage V_i is applied to the inductor winding, and $V_o - V_i$ is applied in the off state. Then

$$\frac{V_i}{L} T_{on} = \frac{V_o - V_i}{L} T_{off} \tag{2.4}$$

$$\frac{V_o}{V_i} = \frac{T_{on} + T_{off}}{T_{off}} = \frac{1}{1-D} \tag{2.5}$$

Similarly, the following equations are obtained for the buck-boost converter:

$$\frac{V_i}{L} T_{on} = \frac{V_o}{L} T_{off} \tag{2.6}$$

$$\frac{V_o}{V_i} = \frac{T_{on}}{T_{off}} = \frac{D}{1-D} \tag{2.7}$$

A comparison of the voltage conversion ratios of the three types of the converters is illustrated in Fig. 2.3. Note that the operation of the converter can be divided into two modes, continuous inductor current mode (CCM) and discontinuous inductor current mode (DCM). The above analyses and results are in the case of CCM. In DCM, the inductor current in the off state reaches zero before the switch is turned on, and stays at zero until the next switching cycle. This mode may occur under the conditions of a small load current, a small inductance L, or a low switching frequency. Details of the voltage conversion ratio and dynamic characteristics in DCM are described in Chapter 4, Section 4.4.

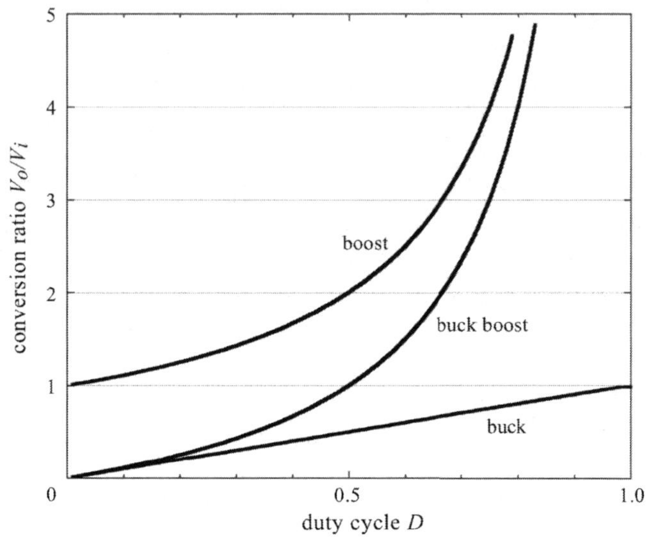

Figure 2.3 Voltage conversion ratio V_o / V_i.

We now focus on a small (laptop) computer application. The power supply used in a computer system is generally divided into three sections from the AC line input to the loads, as illustrated in Fig. 2.4. At the AC line input, an AC-to-DC converter having power factor correction (PFC) is employed so that the input current tracks the applied voltage as closely as possible. For this purpose a boost-type converter is commonly used. The bus converter in the middle isolates the bus line from the AC line input and converts the PFC converter output to an appropriate DC bus voltage, such as 12 V. For this purpose, a transformer is quite suitable, not only to electrically isolate, but also to realize a large voltage conversion ratio. Several types of bus converters are described in Chapter 3. Nonisolated converters close to the load, called point-of-load (POL) converters, and voltage regulator modules (VRMs), especially for microprocessors. Typical output voltages of these converters are in the range of 0.8–3.3 V, and load currents are from less than 1 A to 100 A.

In this chapter, basics of a nonisolated buck converter, which directly provides a low DC voltage to the load, such as a microprocessor, digital signal processor (DSP), chip set, etc., are reviewed.

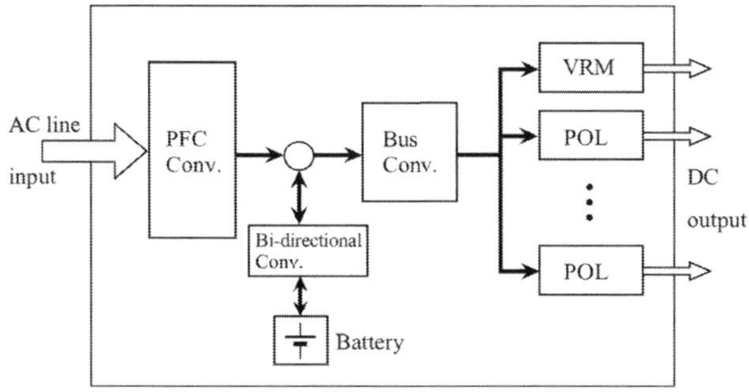

Figure 2.4 Schematic of the power supply in a computer system.

2.2 Operation and Circuit Analysis

Figure 2.5 shows the circuit of a buck converter used for a typical POL converter. Switches S_1 and S_2 are complementary, driven by the pulse voltage signal, as illustrated in the bottom of Fig. 2.5. Since the forward voltage drop of the diode, such as 0.6 V, seriously affects the conversion efficiency of the converter, especially for low-voltage output applications, the synchronous rectifier S_2 is practically used instead of the diode in the buck converter of Fig. 2.1a. Furthermore this converter operates in CCM under any operating condition and DCM no longer exists. The voltage v_1 at the switching node and currents i_{S1}, i_{S2}, and i_L passing through both switches and the inductor L are depicted in Fig. 2.6.

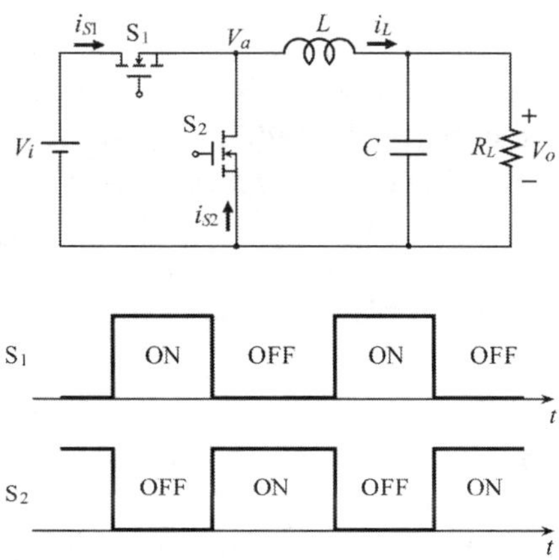

Figure 2.5 Circuit configuration of a buck converter and states of switches S_1 and S_2.

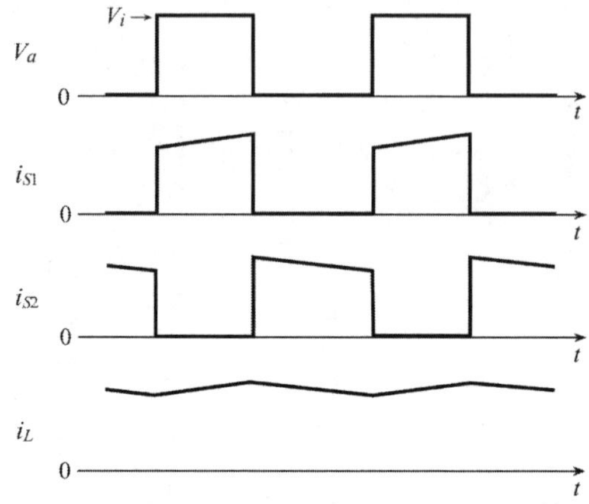

Figure 2.6 Waveforms of voltage and currents.

2.2.1 Operation of a Buck Converter

2.2.1.1 S_1 in the on state

When the switch S_1 is turned on and the switch S_2 is turned off, the input current passing through S_1 and the inductor flows into the output CR circuit, rising linearly. The magnetic energy is stored in the inductor until S_1 is turned off. The slope of the inductor current di/dt is $(V_i - V_o)/L$.

2.2.1.2 S_1 in the off state

After S_1 is turned off, S_2 turns on and the inductor is disconnected from the DC source. Since the inductor current i_L cannot change instantaneously, the voltage polarity across the inductor immediately reverses and the current starts to flow through S_2. The inductor current in this period falls linearly as the stored energy in it is transferred to the output CR circuit. The slope of the inductor current di/dt is $-V_o/L$.

2.2.2 Circuit Analysis of a Buck Converter

It is important to study the basic characteristics of a converter to predict its steady-state and dynamic performance under various operating conditions. For example, when a converter is used as a power stage of the voltage regulator, design engineers should initially examine the open-loop transfer function of the converter to realize the stable regulator system. The open-loop transfer function is also key to predicting the dynamic performance of the input voltage susceptibility and the load current transient response.

Before starting the analysis of the converter, let's introduce the following assumptions on circuit devices:

(1) Semiconductor device: The metal–oxide–semiconductor field-effect transistors (MOSFETs) S_1 and S_2 used as switching devices are represented as on resistances r_{S1} and r_{S2} in the on state and as open circuits in the off state. All other parameters, such as parasitic capacitance between terminals of the MOSFET and the turn-on and turn-off times, are neglected.

(2) Passive devices: The inductor L is a linear device and is represented by the series circuit of the inductor and resistance r_L of the winding. The capacitor C is also a linear device and is represented by the series circuit of the capacitor and an equivalent series resistance (ESR) r_C.

It is obvious that switching converters are nonlinear circuits, and it is not easy to obtain mathematical expressions of voltages or currents in converter circuits in general. The state-space averaging method is a well-known technique that approximates the switching converter as a continuous linear system [1]. This technique is effective as long as the switching frequency is much higher than the corner frequency of the LC network. In Chapter 5 a more intuitive approach introducing an averaged device model is proposed to analyze converter circuits.

As can be seen from the switching operation in Fig. 2.5, two equivalent circuits, shown in Fig. 2.7, are obtained on the basis of the state of the two switches. Defining the duty cycle of the switch, as shown by Fig. 2.2, the averaged equivalent circuit of the buck converter is obtained by the circuit shown in Fig. 2.8. The average output voltage V_o and the inductor current I_L in the steady state are expressed as follows:

$$V_o = \frac{R_L}{R_L + r} D V_i = \frac{1}{1 + r/R_L} D V_i \tag{2.8}$$

$$I_L = \frac{V_o}{R_L} = \frac{D V_i}{R_L + r} \tag{2.9}$$

where $r = r_S + r_L = r_{S1}D + r_{S2}(1 - D) + r_L$.

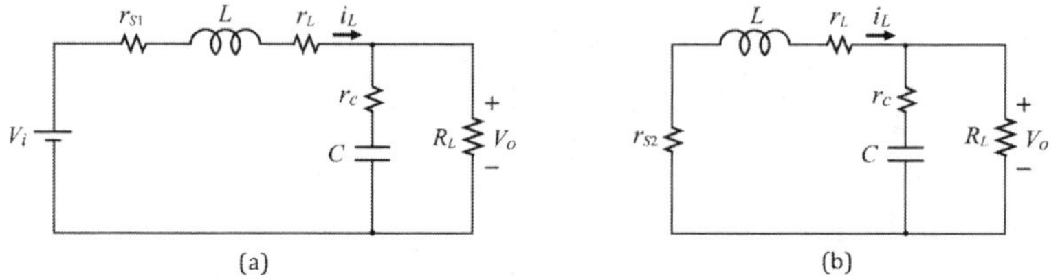

Figure 2.7 Equivalent circuit of a buck converter. (a) On state and (b) Off state.

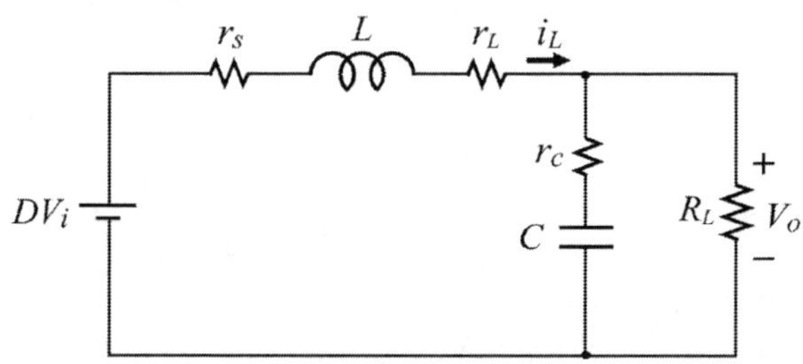

Figure 2.8 Averaged equivalent circuit.

Note that the inductor winding resistance, r_L, and the ESR of the capacitor are taken into account, while these are neglected in Chapter 4, Section 4.3. In addition, DCM is not considered here since a synchronous switch is used instead of a diode.

2.2.2.1 Output ripple voltage

The inductor current and output voltage obtained from the averaged equivalent circuit are averaged values in one switching cycle, and their behaviors can be predicted in a frequency region much lower than the switching frequency. On the other hand, the instantaneous inductor current, i_L, is a sum of the DC component expressed by Eq. 2.9 and the ripple current caused by the switching operation, as shown in Fig. 2.5. Since the impedance of the output capacitor C at the switching frequency is much lower than that of the load resistance R_L, the ripple current mostly flows through the capacitor. As a result, a small ripple voltage appears on the output. When we examine the output voltage ripple, we need to analyze the ripple current using equivalent circuits shown in Fig. 2.7.

The average current flowing through the capacitor is zero in the steady state, and therefore the capacitor current i_{cr} during the on state is approximated by a linear equation as follows:

$$i_{cr} \simeq \frac{(1-D)V_i}{L}\left(t-\frac{DT_s}{2}\right). \tag{2.10}$$

Similarly, i_{cr} during the off state is approximated as

$$i_{cr} \simeq -\frac{DV_i}{L}\left(t-DT_s-\frac{(1-D)T_s}{2}\right) \tag{2.11}$$

Though the capacitor used in circuit analysis is modeled as a series circuit of the ESR and the ideal capacitor shown in Fig. 2.7, we here consider the capacitor as a three-component model depicted in Fig. 2.9 for a discussion of output ripple voltage. The inductance L_c is an equivalent series inductance (ESL). The ripple voltage v_{orip} appearing on the output is obtained from the circuit in Fig. 2.10 and the above equations as

$$\begin{aligned} v_{orip} &= v_{cr} + v_{cL} + v_{cc} \\ &= r_c i_{cr} + L_c \frac{di_{cr}}{dt} + \frac{1}{C}\int i_{cr} dt \end{aligned} \tag{2.12}$$

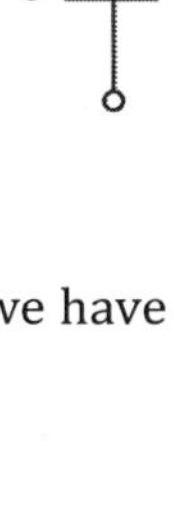

Figure 2.9 Capacitor model.

From the above equation, we have following expressions for the ripple voltage.

$[0 \le t \le DT_s]$

$$v_{orip} \simeq \frac{(1-D)V_i}{L}\left[\frac{t}{2C}(t-DT_s)+r_c\left(t-\frac{DT_s}{2}\right)+L_c\right] \tag{2.13}$$

$[DT_s \le t \le T_s]$

$$v_{\text{orip}} \simeq -\frac{DV_i}{L}\left[\frac{1}{2C}\left(t-DT_{\text{s}}\right)\left(t-T_{\text{s}}\right)+r_{\text{c}}\left(t-DT_{\text{s}}-\frac{(1-D)T_{\text{s}}}{2}\right)+L_{\text{c}}\right] \tag{2.14}$$

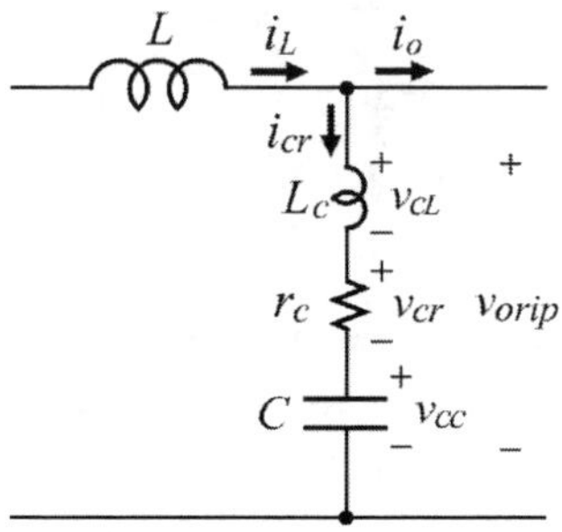

Figure 2.10 Output circuit for ripple voltage analysis.

Figure 2.11 shows calculated waveforms of the ripple current and voltages of three components of the capacitor. The step voltage v_{cL} caused by the ESL is easily obtained as $L_{\text{c}}\, V_{\text{i}}/L$ by subtracting Eq. 2.14 from Eq. 2.13. However, the existence of the ESL would produce spike voltages on the capacitor terminals for the boost and buck–boost converters due to the large di_{cr}/dt at the switching action. Details of the capacitor used in power supplies are described in Chapter 6, Section 6.2.

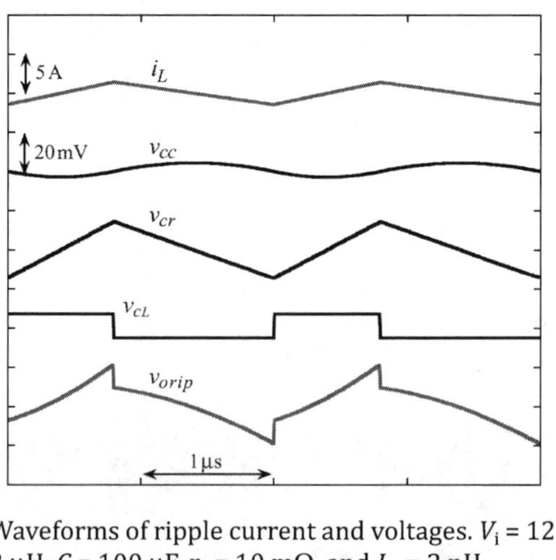

Figure 2.11 Waveforms of ripple current and voltages. V_{i} = 12 V, D = 0.4, L = 2 μH, C = 100 μF, r_{c} = 10 mΩ, and L_{c} = 2 nH.

2.2.2.2 Transfer function

Figure 2.8 reveals that an averaged equivalent circuit is a linear one and low-frequency AC analysis is readily applied to the circuit where high-frequency components such as switching ripples are neglected. Notice that the averaged model for the switching converter is valid up to half the switching frequency. When employing a voltage feedback from the output to the duty cycle to regulate the output voltage, it is necessary to survey the nature of the transfer function in order to design a proper feedback controller. As described in detail in Chapter 4, Section 4.3, the following open-loop transfer function for small perturbations of the output voltage ΔV_o and the duty cycle ΔD is obtained:

$$\frac{\Delta V_o(s)}{\Delta D(s)} = \frac{V_o}{D}\left(1+\frac{r_{S2}+r_L}{R_L}\right)\frac{(1+r_c Cs)}{P(s)} \tag{2.15}$$

where

$$P(s) = \left(1+\frac{r_c}{R_L}\right)LCs^2 + \left\{\left[\left(1+\frac{r_c}{R_L}\right)r + r_c\right]C + \frac{L}{R_L}\right\}s + 1 + \frac{r}{R_L} \tag{2.16}$$

As can be seen from Eq. 2.15, the open-loop transfer function has a zero in the left half of the s plane due to the ESR. This zero contributes to improve the stability of the regulator when the voltage feedback is applied, since this term leads the phase. The polynomial $P(s)$ also appears in the denominator of another transfer function, such as the input voltage susceptibility or the output impedance under operation without voltage feedback.

2.3 Closed-Loop Operation

2.3.1 Voltage Regulator

When the converter is controlled to regulate the output voltage, the duty cycle of the switch becomes a control variable. Figure 2.12 displays a schematic diagram of the regulator system. The element β in the figure is a voltage divider, usually composed of two resistors to adjust the output voltage to the reference voltage V_{ref}. The output voltage of the error amplifier, V_{oe}, is fed to a pulse width modulator (PWM) to generate a pulse with a proper duty cycle.

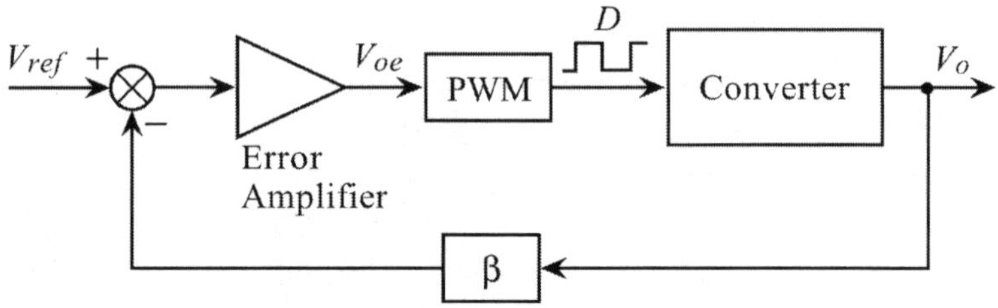

Figure 2.12 Block diagram of the regulator system.

The open-loop transfer function of a buck converter is a typical second-order system, as shown by Eq. 2.15, and therefore classical control techniques can be applied to design the feedback controller. Figure 2.13 shows a calculated example of the Bode plot of the transfer function between the duty cycle to the output voltage. Parameters used in the example are listed in Table 2.1. Solid lines in the figure display the case of an ESR of 10 mΩ, and dashed lines correspond to the case without an ESR. It is observed in this figure that the phase lag with the case of an ESR of 10 mΩ at high frequency is decreased compared to that without an ESR.

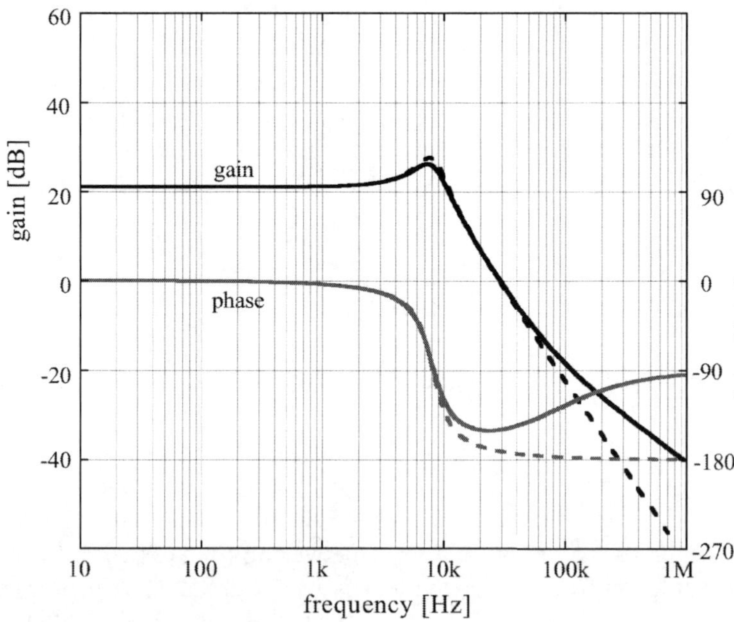

Figure 2.13 Frequency response of an open-loop transfer function.

Table 2.1 Parameters used in Fig. 2.13

Input voltage, V_i	12 V
Inductance, L	2.0 μH
Capacitance, C	200 μF
ESR of capacitor, r_c	10 mΩ
Resistance of winding, r_L	20 mΩ
Resistance of switch Q_1, r_{d1}	10 mΩ
Resistance of switch Q_2, r_{d2}	10 mΩ
Load resistance, R_o	0.5 Ω

2.3.2 Design Consideration of Feedback Circuit

In recent applications of the low-voltage power supply for high-performance large-scale integrations (LSI)s, excellent transient responses are often required so that transient output voltages are within given tolerances. For these requirements, the regulator should be designed to provide not only better steady-state accuracy but also smaller transient voltages for rapid change of the input voltage or load current. The buck converter is a second-order system, as expressed by Eq. 2.15, and its phase around the corner frequency of the LC network rapidly shifts with frequency, as depicted in Fig. 2.13. This will cause a large ringing on the output transient voltage when a high-rate proportional feedback is applied. For these reasons, a phase compensation method is commonly used in the feedback controller.

It is known that the phase-lag compensation improves the steady-state response or the stability margin but often results in a longer rise time because of reduced bandwidth of the feedback control system. The phase-lead compensation usually improves the rise time and overshoot but increases bandwidth. A more practical and versatile arrangement is, therefore, to use a combination of the lag and lead compensations. As a matter of fact, the combination of the lag and lead compensations, so-called lag–lead compensation, contains all the advantages of both schemes. This phase compensation method is a practical technique and is very similar to a proportional integral derivative (PID) control.

Figure 2.14a shows a schematic of the phase lag–lead compensation circuit using an operational amplifier, with input arm impedance Z_i for the phase lead and feedback arm impedance Z_f for the phase lag. In practice, each impedance block is realized by a simple RC network, as shown in Fig. 2.13b, in which the capacitor C_f is used to reduce high-frequency switching noise. Since the capacitance C_2 is generally much larger than that of C_f, transfer functions of these networks are expressed by

$$\left.\begin{aligned} Z_i(s) &= \frac{(1+sT_{ip})R_i}{1+sT_{iz}} \\ Z_f(s) &= \frac{1+sT_{fz}}{sC_2(1+sT_{fp})} \end{aligned}\right\} \tag{2.17}$$

where $T_{ip} = R_1C_1$, $T_{iz} = (R_1 + R_i)C_1$, $T_{fz} = R_2C_2$, and $T_{fp} = R_2C_f$.

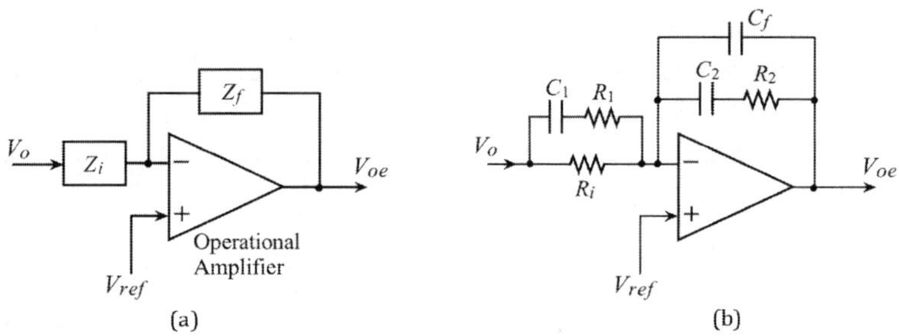

Figure 2.14 Error amplifier with phase compensation.

When the operational amplifier is assumed to be ideal, the transfer function $G_c(s)$ of the error amplifier in Fig. 2.14 is simply expressed by

$$\left.\begin{aligned} G_c(s) &= -\frac{(1+sT_{iz})(1+sT_{fz})}{sT_p(1+sT_{ip})(1+sT_{fp})} \\ &= -\frac{(1+s/\omega_{iz})(1+s/\omega_{fz})}{(s/\omega_p)(1+s/\omega_{ip})(1+s/\omega_{fp})} \end{aligned}\right\} \tag{2.18}$$

where $T_p = R_iC_2$, and $T_{fp} = R_2C_f$.

As can be seen from the above equation, the error amplifier in Fig. 2.14 has three poles at 0, ω_{ip}, and ω_{fp} and two zeros at ω_{fz} and

ω_{iz}, as sketched by the linear asymptotic approximation of the gain curve in Fig. 2.15. This circuit is often called type 3 error amplifier [2]. The steady-state error on the output voltage is minimized by the infinite gain of $G_c(s)$ at $\omega = 0$. For a relatively slow variation of the output voltage due to either line input or load changes, the mid-frequency gain plays a role to reduce it. On the other hand, the dynamic performance for rapid variation of the line input or load condition strongly depends on the transfer characteristics of the error amplifier in the high-frequency region. Generally speaking, the wider bandwidth of the loop gain provides a faster transient response. Therefore, it is very important to design a proper error amplifier for satisfying the given specifications on the dynamic performance of the regulator.

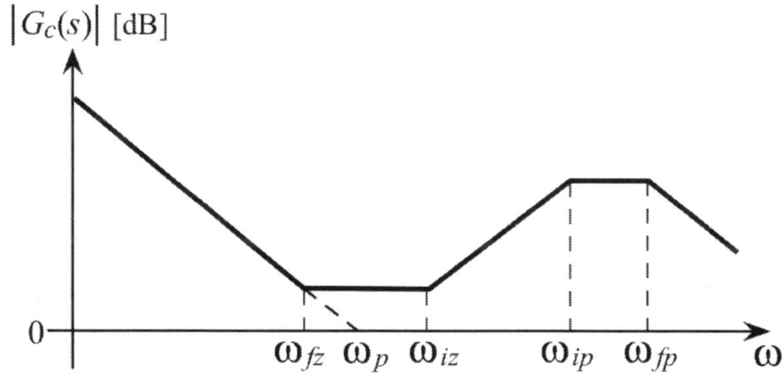

Figure 2.15 Gain versus frequency of an error amplifier with phase compensation networks.

The operational amplifier used in the error amplifier with phase compensation networks discussed above is assumed to be ideal. This means the gain is infinite and independent of frequency. All real operational amplifiers, however, have a finite gain with a certain bandwidth, and it will cause some instability problem in poorly designed circuit parameters. Defining $G_{op}(s)$ as an open-loop transfer function of the operational amplifier itself, the $G_c(s)$ is rewritten as follows:

$$G_c(s) = -\frac{Z_f(s)}{Z_i(s)} \cdot \frac{1}{1 + \frac{1}{G_{op}(s)}\left(1 + \frac{Z_f(s)}{Z_i(s)}\right)} \tag{2.19}$$

For example, the operational amplifier is approximated as a first-order lag element with a gain bandwidth (GBW) product G_b, as shown in Fig. 2.16.

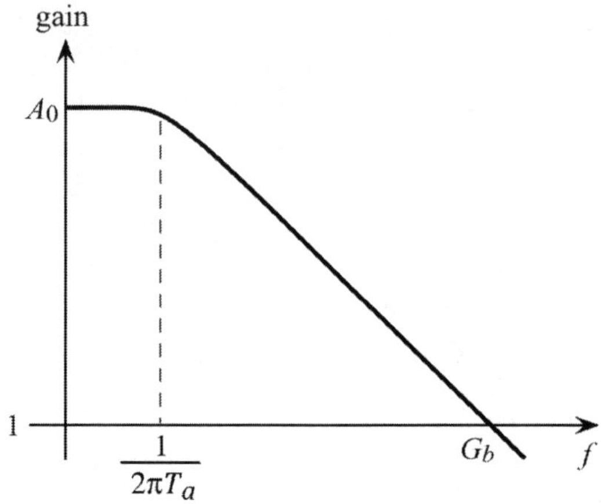

Figure 2.16 Frequency response of open-loop gain for a first-order lag circuit.

The open-loop transfer function of the operational amplifier is then given by

$$G_{\mathrm{op}}(s) = \frac{A_0}{1 + T_{\mathrm{a}} s} \quad \left(T_{\mathrm{a}} = \frac{A_0}{2\pi G_{\mathrm{b}}} \right) \tag{2.20}$$

From Eqs. 2.19 and 2.20, the transfer function of the error amplifier $G_c(s)$ with compensation networks is obtained as follows:

$$G_{\mathrm{c}}(s) = \frac{Z_{\mathrm{f}}(s)}{Z_{\mathrm{i}}(s)} \cdot \frac{1}{1 + \dfrac{1 + T_{\mathrm{a}} s}{A_0}\left(1 + \dfrac{Z_{\mathrm{f}}(s)}{Z_{\mathrm{i}}(s)}\right)} \tag{2.21}$$

Figure 2.17 shows an example of frequency responses of a transfer function for cases of ideal and actual operational amplifiers. Pole and zero frequencies in the figure are chosen at 50 kHz, 1 MHz and at 500 Hz, 5 kHz, respectively. It is observed that both responses in the low-frequency region, that is, lower than 10 kHz in the figure, are almost identical. However, responses for the amplifier with 5 MHz of a GBW plotted by solid lines are quite different from those of the ideal one in the high-frequency region. For example, the phase lead for the ideal amplifier is about 25 degrees, while that of the actual one is –4 degrees at 100 kHz. This means that the phase compensator no longer leads the phase in a frequency region higher than 100 kHz, even if the gain is still larger than 20 dB.

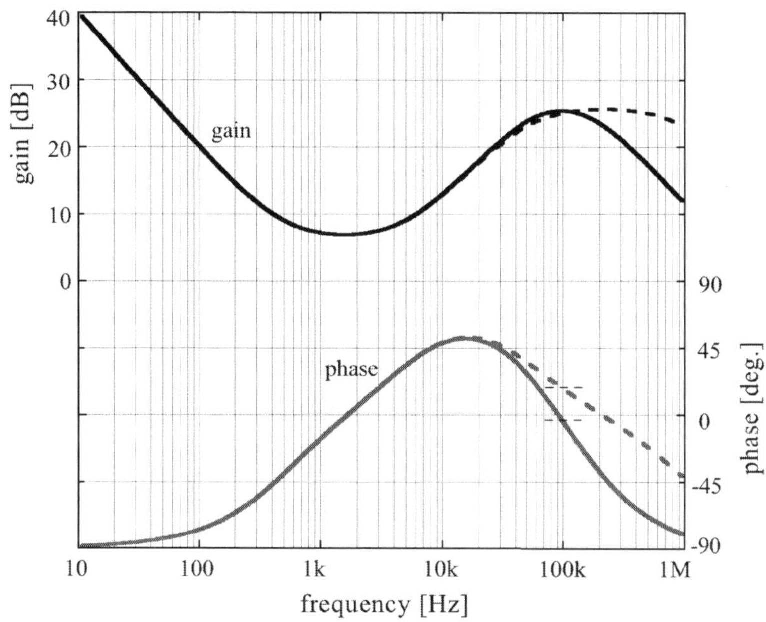

Figure 2.17 Comparison of frequency responses of the error amplifier for ideal and actual operational amplifiers. Dashed lines are the ideal case, and solid lines are the actual case with a 5 MHz GBW.

Calculated frequency responses of the total open-loop transfer function for a buck converter employing the above error amplifier are plotted in Fig. 2.18. The use of two different operational amplifiers and ESRs is compared in the figure. Frequency responses in (a) and (b) using an ideal operational amplifier show the phase margins of 71 degrees and 24 degrees for 10 mΩ and 1 mΩ of an ESR, respectively. When using a typical operational amplifier with a 5 MHz GBW product, on the other hand, phase margins become smaller as 36 degrees and –2 degrees, respectively. As can be seen from these results, improper design of the phase lead network with such a conventional operational amplifier easily causes poor dynamic performance or instability of the feedback system, especially when the resonant frequency of the power stage shifts to high.

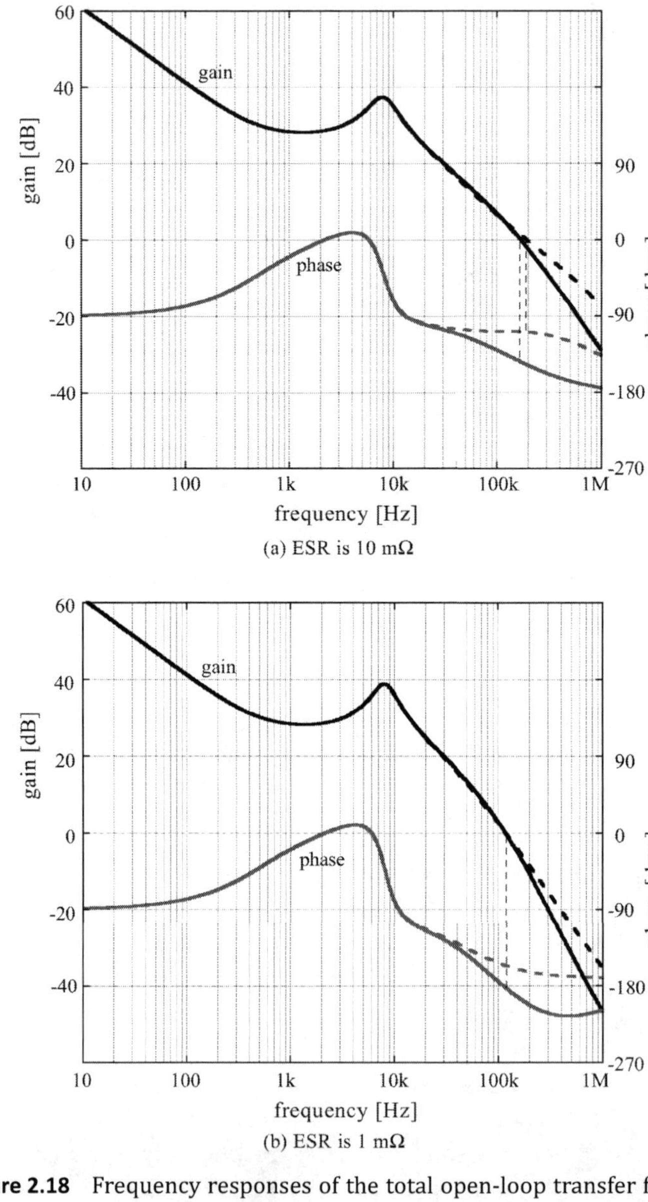

Figure 2.18 Frequency responses of the total open-loop transfer function. Dashed lines are the ideal case, and solid lines are the actual case with a 5 MHz GBW.

References

1. R. D. Middlebrook and S. Cuk, A general unified approach to modeling switching converter power stage, *IEEE PESC'76 Rec.*, 18–34, 1976.
2. A. I. Pressman, K. Billings, and T. Morey, *Switching Power Supply Design, Third Edition*, 588–593, McGraw-Hill, 2009.

Chapter 3

Isolated DC–DC Converters

Kimihiro Nishijima
Department of Electrical and Electronics Engineering,
Oita University, 700 Dannoharu, Oita 870-1192, Japan
nishijima@oita-u.ac.jp

3.1 Introduction

Power supplies often incorporate transformers into their designs in order to realize some or all of the following functions:

- Isolation: An isolation transformer ensures that the converter does not have a direct electrical path from the power input side to the power output side, which helps meet safety requirements.
- Large voltage and current conversion: A large step-up or step-down conversion can be achieved by the transformer's turns ratio.
- Minimization of voltage and current stresses: Voltage and current stresses on devices are reduced by the transformer's turns ratio when performing a large step-up or step-down conversion.

Handbook of Power Management Circuits
Edited by Haruo Kobayashi and Takashi Nabeshima

ISBN 978-981-4613-15-6 (Hardcover), 978-981-4613-16-3 (eBook)
www.panstanford.com

- Multiple outputs: Transformers with multiple windings can be used to supply multiple, different output voltages.

The basic operation of the following converters is explained here, assuming ideal circuit elements (metal–oxide–semiconductor field-effect transistors [MOSFETs], diodes, capacitors, and inductors) and ignoring parasitic power losses: flyback converter, forward converter, push–pull converter, half-bridge converter, and full-bridge converter. In equivalent circuits as shown in following figures, average voltages and currents during each state are descripted.

3.2 Flyback Converter

Figure 3.1 shows an ideal flyback converter circuit, and Fig. 3.2 shows a flyback converter circuit with the transformer-equivalent circuit. This is an isolation-type buck–boost converter (Fig. 2.1). When the switch is closed, the energy is stored in the magnetizing inductor of the transformer; when the switch is open, the energy stored in the transformer is released to the output.

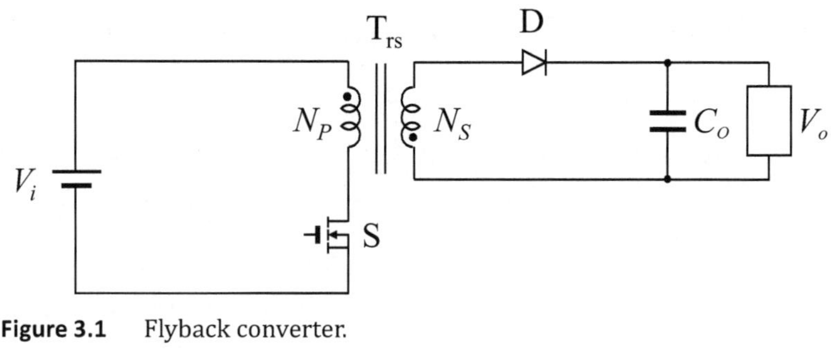

Figure 3.1 Flyback converter.

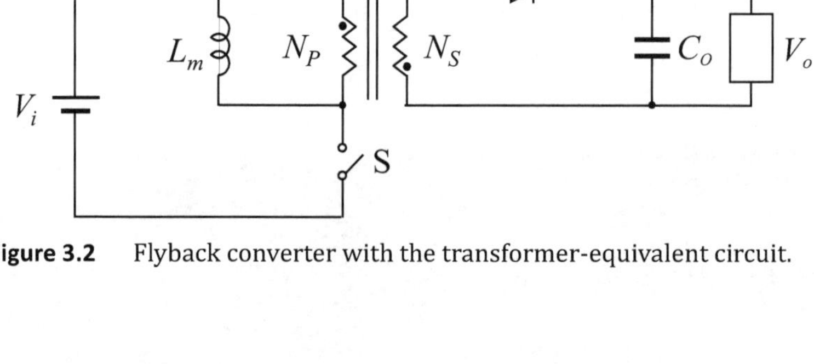

Figure 3.2 Flyback converter with the transformer-equivalent circuit.

Figure 3.3 shows two equivalent circuits for each operation state, and Fig. 3.4 shows typical switching waveforms of a flyback converter.

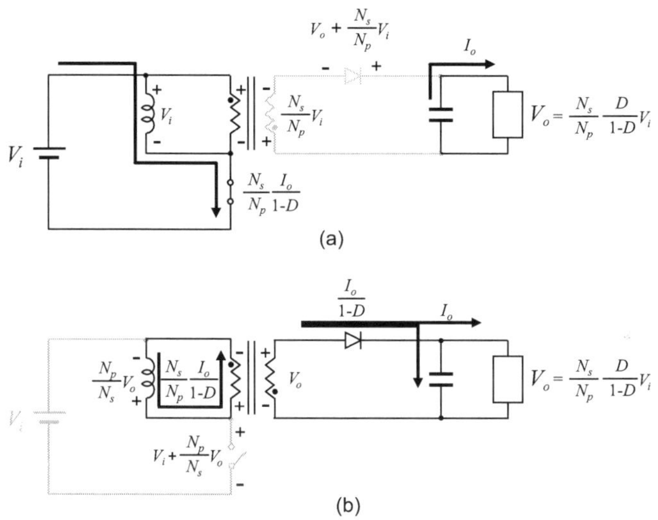

Figure 3.3 Equivalent circuits for each operation state. (a) $t_0 - t_1$ and (b) $t_1 - t_2$.

When the switch S is closed, the input voltage V_i is directly applied in the positive direction across the transformer primary winding, N_p, and the transformer secondary winding, N_s, generates a voltage of $V_{Ns} = N_s/N_p)V_i$ in the opposite direction. In this state, the rectifier diode D of the secondary side is off with a reverse bias of $V_D = (N_s/N_p)V_i + V_o$. Additionally, energy is stored in the magnetizing inductor, L_m, and the magnetizing inductor current, i_{Lm}, is increased linearly with a slope of $di_{Lm}/dt = V_i/L_m$.

When S is open, the winding voltage polarities are reversed, the secondary winding, N_s, connects to the output-smoothing capacitor, C_o, through the rectifier diode D, and the energy stored in the magnetizing inductor is released to C_o and the load. In this state, since the secondary winding voltage is equal to the output voltage, V_o, the primary winding voltage becomes $V_{Np} = (N_p/N_s)V_o$. Therefore the magnetizing inductor current is decreased linearly with a slope of $di_{Lm}/dt = -(N_p/N_s)V_o/L_m$. Also, the voltage stress of switch S is $V_{sm} = (N_p/N_s)V_o + V_i$.

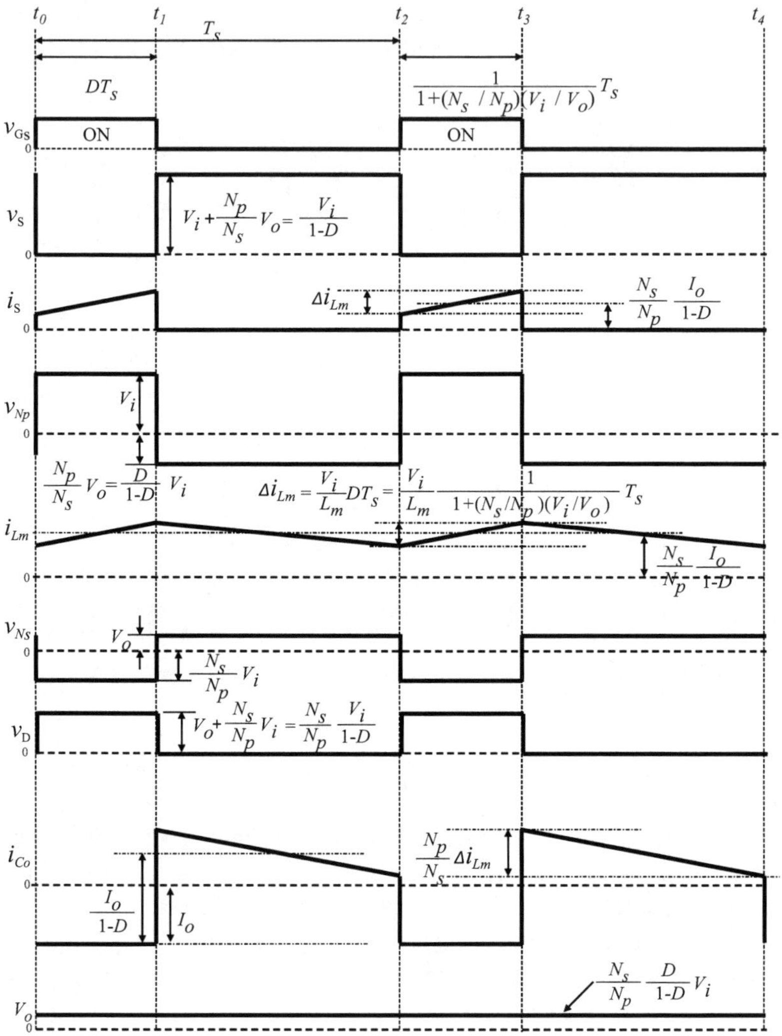

Figure 3.4 Typical waveforms of a flyback converter.

By utilizing steady-state analysis, the voltage conversion ratio of the flyback converter is given by

$$\frac{V_o}{V_i} = \frac{N_s}{N_p}\frac{D}{1-D} \tag{3.1}$$

where D is the duty cycle ($D = T_{on}/T_s$, T_{on}) is the on time of switch S, and T_s is the switching period.

An advantage of the flyback converter is that it is the simplest and cheapest isolated converter topology. However, a disadvantage of the converter is that the output ripple current is very high, as shown in Fig. 3.4. The output-smoothing capacitor, C_o, is only charged by the magnetizing inductor current during the switch's on period, although it is constantly discharged by the output load current, I_o. In the steady state, since the discharged energy of C_o is equal to the charged energy,

$$I_o T_s = \left(\frac{N_p}{N_s} I_{Lm} \right)(1-D)T_s \tag{3.2}$$

where I_{Lm} is the average magnetizing inductor current. From Eq. 3.2, the average magnetizing inductor current of the flyback converter is given as

$$I_{Lm} = \frac{N_s}{N_p} \frac{I_o}{1-D} \tag{3.3}$$

Therefore the peak-to-peak current ripple of the output-smoothing capacitor, Δi_{Co}, is as follows:

$$\Delta i_{Co} = \frac{N_p}{N_s} \left(I_{Lm} + \frac{1}{2} \Delta i_{Lm} T_{on} \right) = \frac{I_o}{1-D} + \frac{1}{2} \frac{N_p}{N_s} \frac{V_i}{L_m} DT_s \tag{3.4}$$

As a result, the flyback converter is suitable for low-power applications such as AC adapters and chargers of various kinds of portable equipment.

Another disadvantage of the flyback topology is that a large voltage spike and noise from the switching elements are generated by the presence of the transformer leakage inductor. In the equivalent circuit, the leakage inductor is in series with each winding of the transformer. Therefore the magnetic energy is stored in the leakage inductor of the primary side during the switch's on time, but the stored energy cannot release to the secondary side during the switch's off time due to the open circuit of primary side. As a result, when the switching element changes from the on state to the off state, a large switching surge is generated by the ringing between the transformer leakage inductance and the output parasitic capacitance of the switching element.

To prevent the switching surge and noise, an auxiliary circuit, called a snubber circuit, is connected across the switching elements. Figure 3.5 shows an example of using a resistor/capacitor/diode

(RCD) snubber circuit. When the switching element changes from the on state to the off state, the energy stored in the leakage inductor can be absorbed by the snubber capacitor, C_S, through the snubber diode, D_S. Because the capacitance of the snubber capacitor is sufficiently larger than the output parasitic capacitance of the switching element, the drain-to-source voltage spike is suppressed. However, the absorbed energy is dissipated by the snubber resistor, R_S, which contributes to one of the main power losses of the flyback converter.

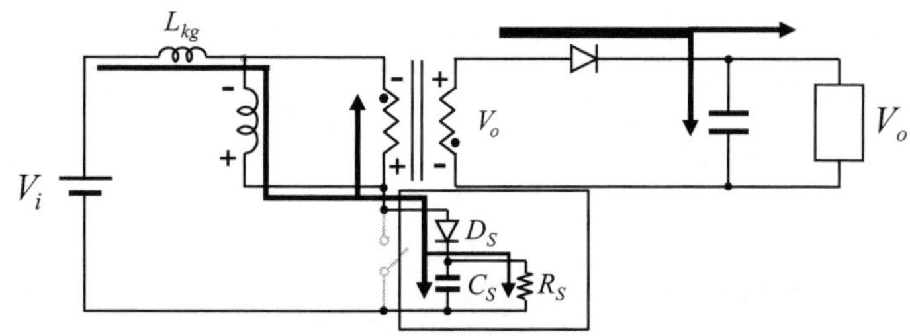

Figure 3.5 Flyback converter with an RCD snubber circuit.

3.3 Forward Converter

3.3.1 Single-Switch Forward Converter

Figure 3.6 shows an ideal single-switch forward converter circuit, and Fig. 3.7 shows a single-switch converter circuit with the transformer-equivalent circuit. A single-switch forward converter is an isolated-type buck converter. When the switch is closed, energy is stored in the output inductor; when the switch is open, the stored energy is released to the output through the freewheel diode.

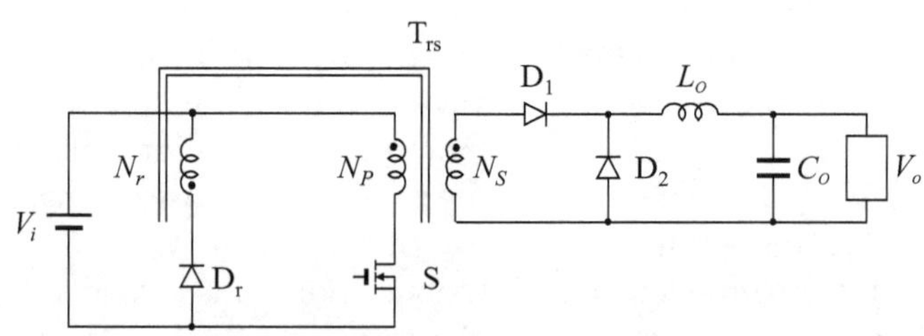

Figure 3.6 Ideal single-switch forward converter.

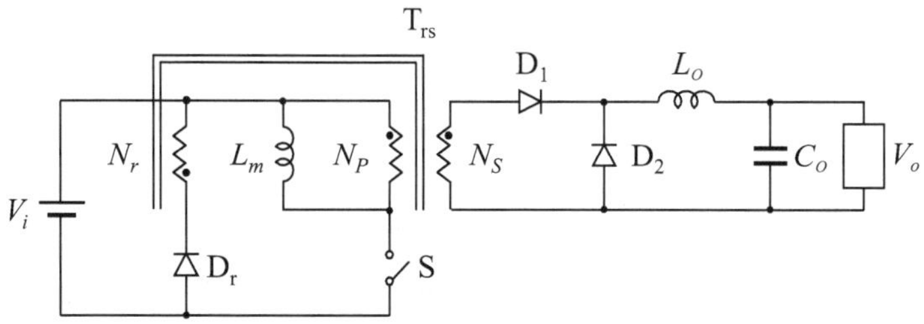

Figure 3.7 Single-switch forward converter with the transformer-equivalent circuit.

Figure 3.8 shows three equivalent circuits for each operation state, and Fig. 3.9 shows typical waveforms of the forward converter.

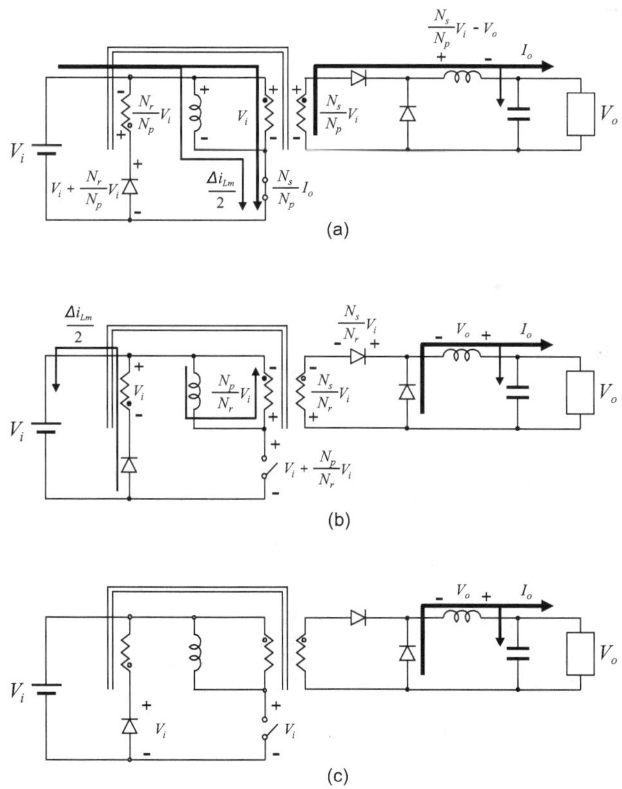

Figure 3.8 Equivalent circuit for each operation state. (a) $t_0 - t_1$, (b) $t_1 - t_2$, and (c) $t_1 - t_3$.

When the switch S is closed, the input voltage, V_i, is directly applied to the positive side of the primary winding, N_p, so that the secondary winding, N_s, generates a voltage of $V_{Ns} = (N_s/N_p)V_i$ in the positive direction. Because the diode D_1 is forward biased and the diode D_2 is reverse biased, the current flows to the output-smoothing capacitor, C_o, and the load through the secondary winding N_s, rectifier diode D_1, and output inductor, L_o. Next, the energy flows to L_o, whose current is increased linearly with a slope of di_{Lo}/dt

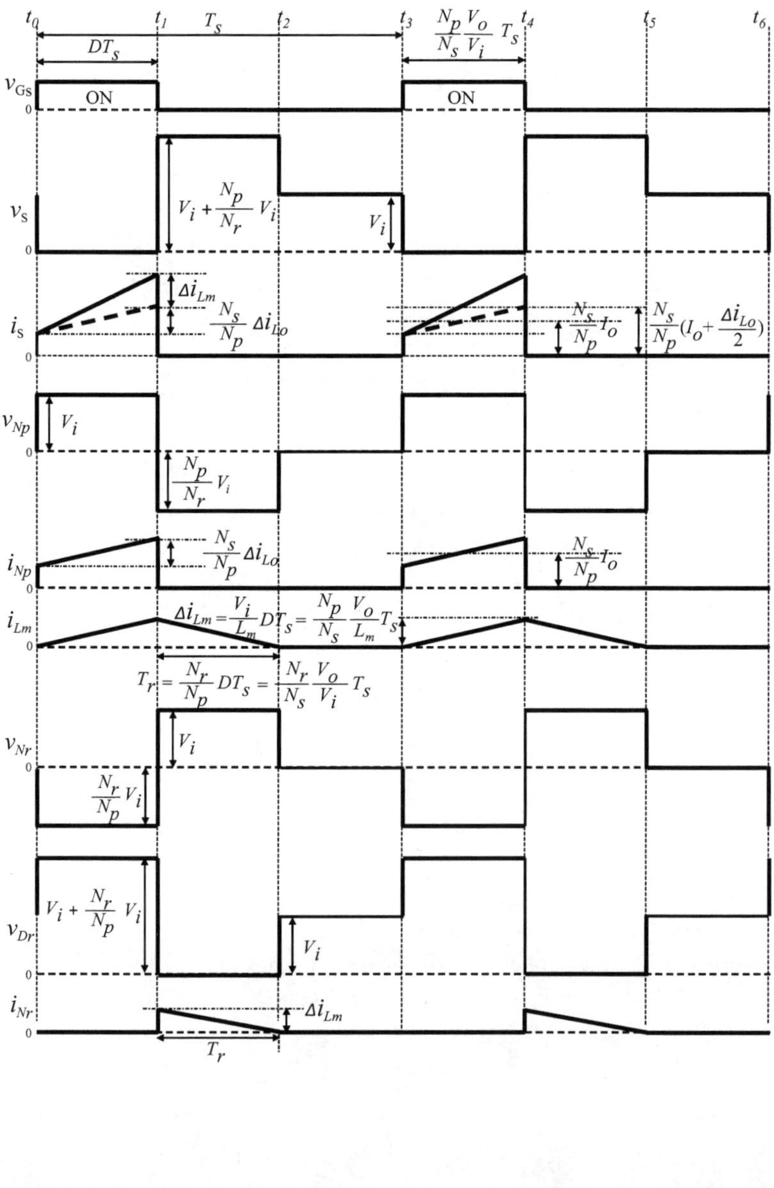

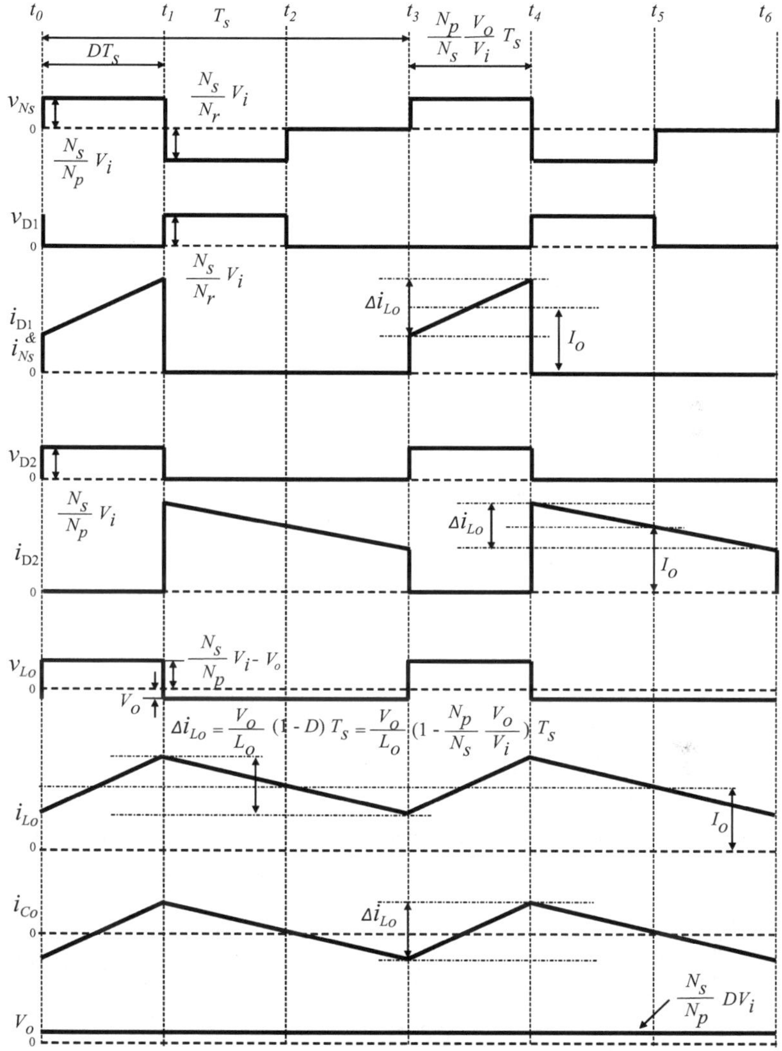

Figure 3.9 Typical waveforms of a single-switch forward converter.

$= (V_{Ns} - V_o)/L_o = [(N_s/N_p)V_i - V_o]/L_o$. At this time the energy is also stored in the magnetizing inductor, L_m, and the magnetizing inductor current is increased linearly with a slope of $di_{Lm}/dt = V_i/L_m$. Also, the transformer reset winding, N_r, generates a voltage of $V_{Nr} = (N_r/N_p)V_i$, and the reset diode D_r is off with a reverse bias of $V_{Dr} = V_{Nr} + V_i = (N_r/N_p)V_i + V_i$.

When S is open, the secondary winding voltage polarity is reversed and the rectifier diode D_1 is reverse biased. At this time, the magnetic energy stored in the output inductor is released to the output-smoothing capacitor, C_o, and the load through the rectifier diode D_2; thus the output inductor current, i_{Lo}, is decreased linearly with a slope of $di_{Lo}/dt = -V_o/L_o$.

From steady-state analysis, the voltage conversion ratio of the forward converter is given by

$$\frac{V_o}{V_i} = \frac{N_s}{N_p} D \tag{3.5}$$

During the switch's off period, the reset winding voltage polarity is also reversed, and the magnetic energy stored in the magnetizing inductor is released to the input through the reset winding N_r and the reset diode D_r. Because the primary and secondary winding voltages are $(N_p/N_r)V_i$ and $(N_s/N_r)V_i$, respectively, the magnetizing inductor current is decreased linearly with a slope of $di_{Lm}/dt = -(N_p/N_r)V_i/L_m$. In this state, the maximum voltage stresses of the switch S and the rectifier diode D_1 are $V_s = V_i + (N_p/N_r)V_i$ and $V_{D1} = (N_s/N_r)V_i$, respectively.

The magnetizing inductor current should be reset during the off period, T_{off}, in order to guard against transformer core saturation. The reset time, T_r, should be limited to an amount of time less than the off period. Since the magnetizing inductor slope of the on period is V_i/L_m and the slope during the off period is $-(N_p/N_r)V_i/L_m$, the following equations are obtained from steady-state analysis:

$$\frac{V_i}{L_m} T_{on} = \frac{N_r}{N_p} \frac{V_i}{L_m} T_r \tag{3.6}$$

$$T_r = \frac{N_r}{N_p} T_{on} < T_{off} \tag{3.7}$$

where T_{on} is the on period of the switch S, $T_{off} = T_s - T_{on}$. As a result, the maximum duty cycle, D_{max}, of the single-switch forward converter is derived by

$$D_{max} = \frac{1}{1 + N_r/N_p} \tag{3.8}$$

where the duty cycle $D = T_{on}/T_s$ and T_s is the switching period. If the number of turns of the primary winding is equal to that of the reset winding, the maximum duty cycle is limited to 0.5.

The characteristics of the forward converter are similar to those of the buck converter, the average output inductor current is equal to the output current I_o, and the output ripple current is sufficiently smaller than that of the flyback converter. The peak-to-peak current of the output inductor, that is, that of the output-smoothing capacitor, Δi_{Co}, is given by

$$\Delta i_{Co} = \Delta i_{Lo} = \frac{V_o}{L_o} T_{off} = \frac{V_o}{L_o}(1-D)T_s \tag{3.9}$$

From Eqs. 3.5 and 3.9

$$\Delta i_{Co} = \Delta i_{Lo} = \frac{V_o}{L_o}\left(1 - \frac{N_p}{N_s}\frac{V_o}{V_i}\right)T_s \tag{3.10}$$

The magnetizing inductor current only flows toward the positive direction during the on period, that is, the transformer core has a unipolar flux swing in the positive direction. Because of this, the core size of the forward converter will be larger than that of other converters with bipolar flux swing, as described later. However, the forward converter has a simple circuit structure using a single switch. As a result, this converter is a good candidate for middle-power applications such as AC adaptors for notebook PCs and ATX power supply units for desktop PCs.

To prevent the switching surge and noise, a snubber circuit is essential for the single-switch forward converter in the same manner as in the flyback converter.

3.3.2 Two-Switch Forward Converter

Figure 3.10 shows an ideal two-switch forward converter circuit, and Fig. 3.11 shows a two-switch forward converter with the transformer-equivalent circuit. This forward converter requires two switches, S_1 and S_2, and two reset diodes, D_{r1} and D_{r2}, in the primary side but does not require a reset transformer winding and a snubber circuit and can reduce the voltage stresses of the switches to half the value of the single-switch forward converter.

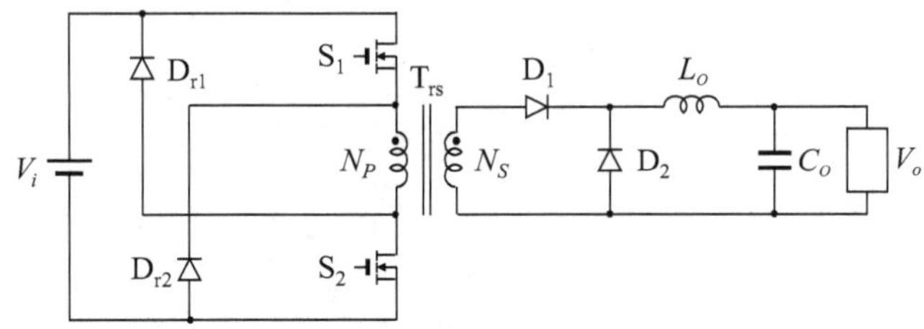

Figure 3.10 Ideal two-switch forward converter.

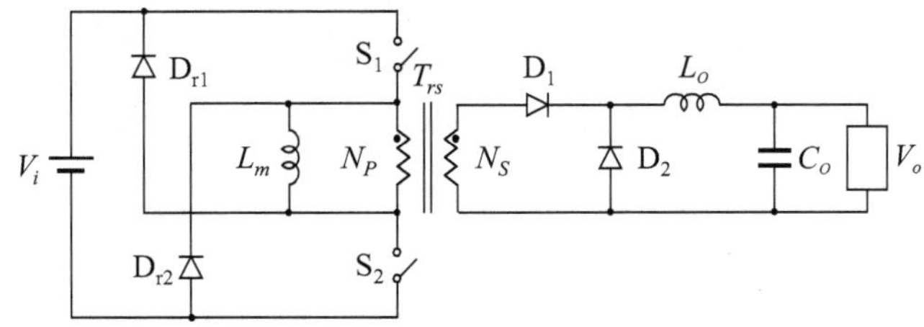

Figure 3.11 Two-switch forward converter with the transformer-equivalent circuit.

Figure 3.12 shows two equivalent circuits for each operation state, and Fig. 3.13 shows typical waveforms of the two-switch forward converter.

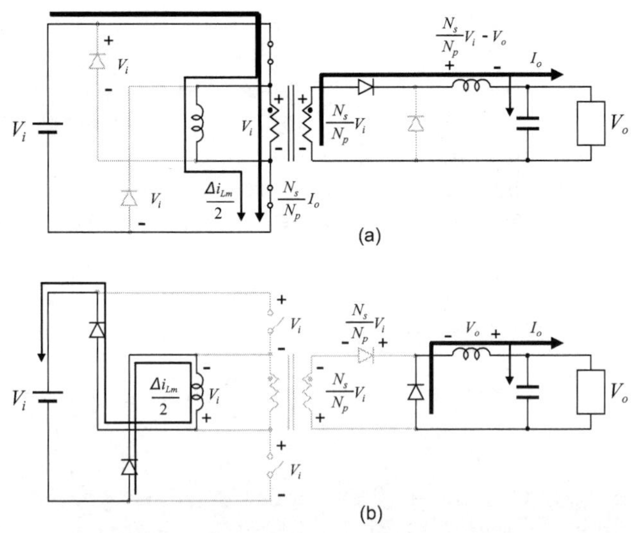

Figure 3.12 Equivalent circuits for each operation state. (a) $t_0 - t_1$ and (b) $t_1 - t_2$.

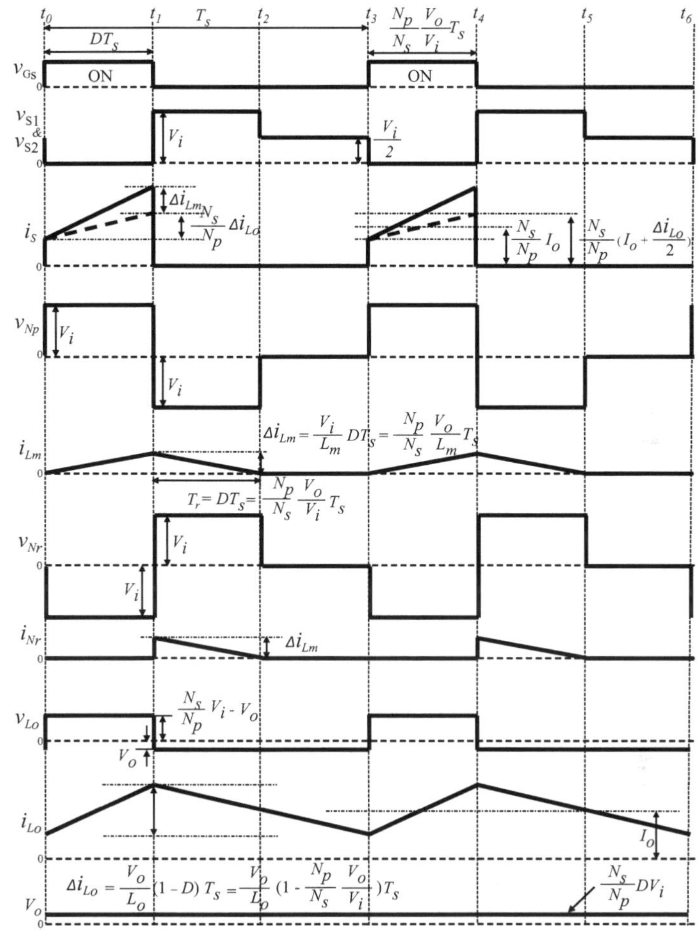

Figure 3.13 Typical waveforms of a two-switch forward converter.

Both switches are turned on and off synchronously. When the switches are closed, the input voltage V_i is directly supplied to the positive side of the primary winding, N_p, and energy is stored in the output inductor and the magnetizing inductor in the same manner as the single-switch forward converter. During this time, the magnetizing inductor current is increased linearly with a slope of $di_{Lm}/dt = V_i/L_m$.

When both switches are open, the magnetic energy stored in the output inductor L_o is released to the output through the rectifier diode D_2. At the same time, the energy stored in the magnetizing inductor L_m is released to the input through the reset diodes D_{r1}

and D_{r2}. Because the primary and secondary winding voltages are V_i and $(N_s/N_p)V_i$, respectively, the magnetizing inductor current is decreased linearly with a slope of $di_{Lm}/dt = -V_i/L_m$. In this state, the voltage stresses of the switches and the diode D_1 are $V_{s1} = V_{s2} = V_i$ and $V_{D1} = (N_s/N_p)V_i$, respectively.

Similar to the single-switch forward converter, the magnetizing inductor current should be reset during the switch's off period T_{off}, in order to guard against transformer core saturation. Since the slope of the magnetizing inductor current during the on period is V_i/L_m and that slope in the off period is $-V_i/L_m$, the maximum duty cycle D_{max} of the two-switch forward converter is 0.5.

Switching surge and noise can be suppressed because the energy stored in the leakage inductor of the primary side is recycled to the input through the reset diodes during the switch's off period.

3.4 Push–Pull Converter

Figure 3.14 shows an ideal push–pull converter circuit, and Fig. 3.15 shows a push–pull converter with the transformer-equivalent circuit. This converter can be regarded as a forward converter with two primary circuits: when either of the switches is closed, energy is stored in the output inductor; when both switches are open, the stored energy is released to the output through the rectifier diodes.

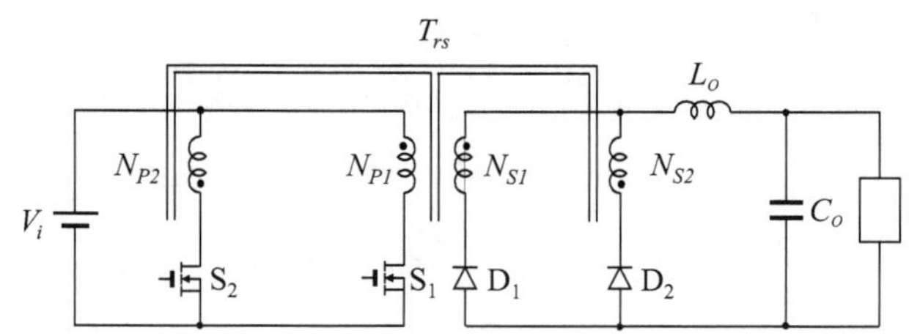

Figure 3.14 Ideal push–pull converter.

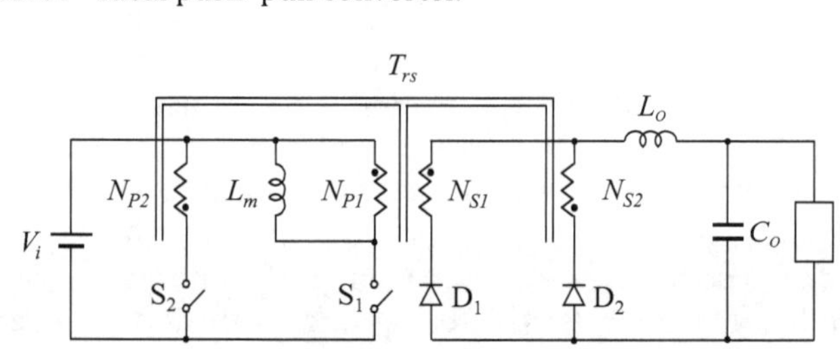

Figure 3.15 Push–pull converter with the transformer-equivalent circuit.

Figure 3.16 shows four equivalent circuits for each operation state, and Fig. 3.17 shows typical waveforms of the push–pull converter. The secondary circuit, which is called a center tap rectifier circuit, is constructed from two secondary windings with the same number of turns, two rectifier diodes, and an output filter. The two switches S_1 and S_2 are turned on and off 180° out of phase, and each switch operates during one half cycle ($T_s/2$). Therefore the duty cycles of switches should be less than 0.5.

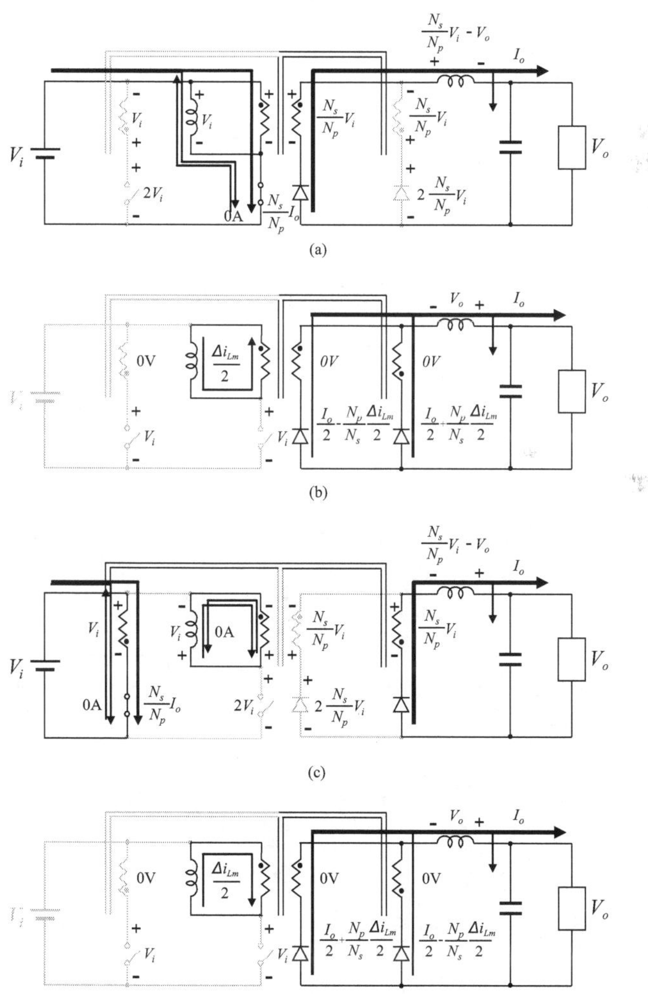

Figure 3.16 Equivalent circuits for each operation state. (a) $t_0 - t_1$, (b) $t_1 - t_2$, (c) $t_2 - t_3$, and (d) $t_3 - t_4$.

During the first half cycle, when the first switch, S_1, is closed, the input voltage, V_i, is directly supplied to the positive side of the first primary winding, N_{p1}, so the secondary windings, N_{s1} and N_{s2}, generate a voltage of $V_{Ns1} = V_{Ns2} = (N_s/N_p)V_i$. During this time, the diode D_1 is forward biased and the diode D_2 is reverse biased with a voltage of $V_{D2} = V_{Ns1} + V_{Ns2} = 2(N_s/N_p)V_i$. Current flows to the output-smoothing capacitor, C_o, and the load through the secondary winding N_{s1}, diode D_1, and output inductor L_o. Energy is stored in the output inductor, L_o, during this period, and the inductor current increases

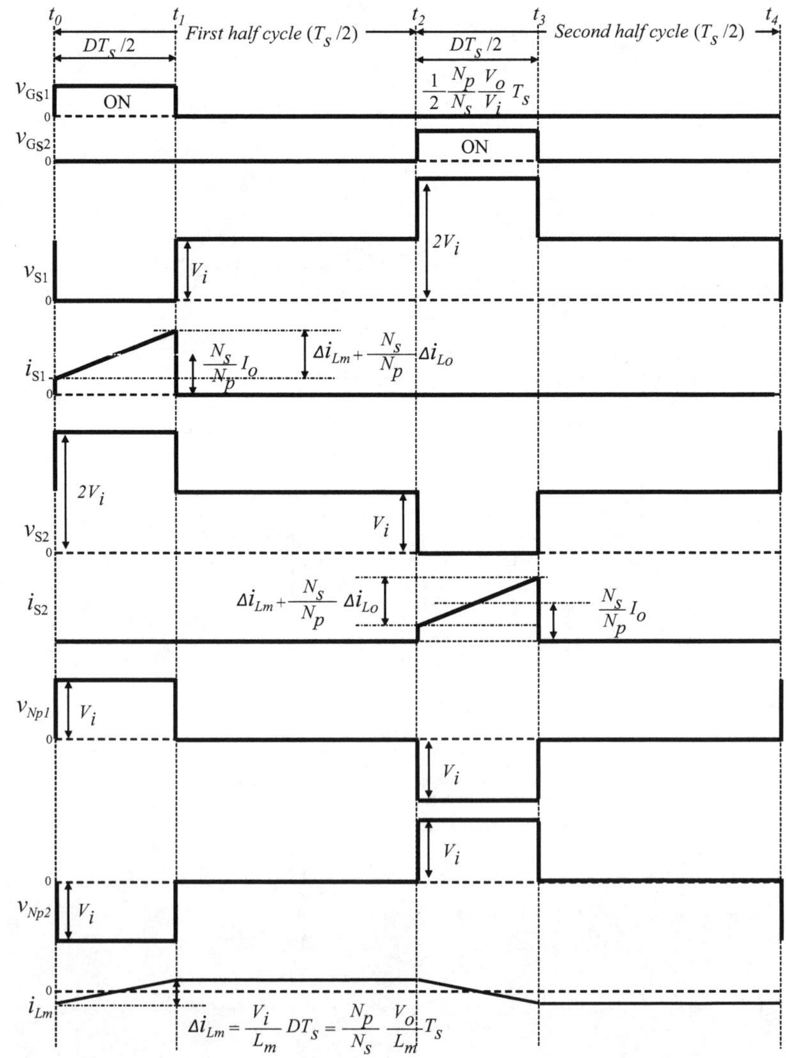

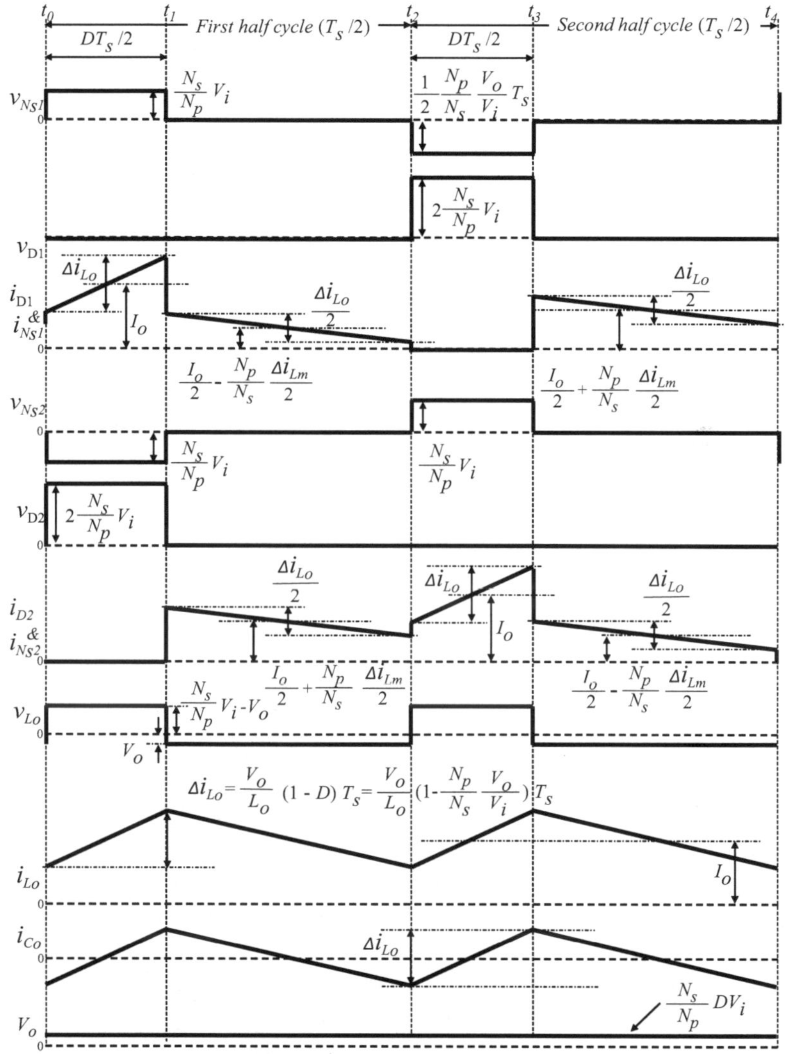

Figure 3.17 Typical waveforms of a push–pull converter.

linearly with a slope of $di_{Lo}/dt = (V_{Ns1} - V_o)/L_o = [(N_s/N_p) V_i - V_o]/L_o$. At the same time, the magnetizing inductor current has waveform of positive–negative symmetric with a slope of $di_{Lm}/dt = V_i/L_m$, and the average value is zero. The second primary winding, N_{p2}, generates a voltage of $V_{Np2} = V_i$, and hence the maximum voltage stress of the second switch, S_2, is $V_{s2} = V_i + V_{Np2} = 2V_i$.

During the first half cycle, when both switches are open, the voltage polarity of the output inductor, L_o, is reversed, and the energy stored in the output inductor is released to the output-smoothing capacitor, C_o, and the load through both the rectifier diodes and secondary windings. In this state, all winding voltages of the transformer are zero because the variation of the transformer's magnetic flux is cancelled by the secondary winding currents in opposite directions (positive and negative). Therefore, the output inductor current, i_{Lo}, is decreased linearly with a slope of $di_{Lo}/dt = -V_o/L_o$, and the magnetizing inductor current i_{Lm} remains at a constant value (half the peak-to-peak current = $\Delta i_{Lm}/2$). Accordingly, the average currents of the rectifier diodes are given as $I_{D1} = I_o/2 - \Delta i_{Lm}/2$ and $I_{D2} = I_o/2 + \Delta i_{Lm}/2$.

The voltage stresses of the switches are $V_{s1} = V_{s2} = V_i$ in the theoretical analysis.

During the second half cycle, the push–pull converter operates in the same manner as the first half cycle. When the second switch, S_2, is closed, voltage is supplied to the negative side of the transformer windings and energy is stored in the output inductor, L_o, through the windings N_{p2} and N_{s2} and the rectifier diode D_2. When both switches are opened, the stored energy is released to the output through both secondary windings and rectifier diodes.

By utilizing steady-state analysis, the voltage conversion ratio of the push–pull converter is given by

$$\frac{V_o}{V_i} = \frac{N_s}{N_p} D \tag{3.11}$$

where $D = 2T_{on}/T_s$ is the duty cycle of the switches. The duty cycle of the push–pull converter is two times of that of the forward converter because of the on periods during both half cycles.

The transformer core size of the push–pull converter can be smaller than that of the forward converter. In the case of the forward converter, the magnetizing inductor current always flows to the positive side, and the maximum value equal to the peak-to-peak of the inductor current (Δi_{Lm}). Therefore the magnetizing flux swing of the forward converter is an asymmetrical waveform. On the other hand in the case of the push–pull converter, the magnetizing inductor current flows to both positive and negative sides during the switch's on periods of the first and second half cycles, and hence the maximum value is equal to half the peak-to-peak inductor current ($\Delta i_{Lm}/2$). Therefore the magnetizing flux swing of the push–pull converter is a symmetrical waveform.

Because of the bidirectional flux swing, the push–pull converter can remove the additional reset transformer winding that is required in the forward converter topology. However, if the duty cycle of the first cycle is not equal to that of the second cycle, the transformer core will saturate by the imbalance of the increase and decrease in currents within the magnetizing inductor. This core saturation problem may also be caused by imbalances between the characteristics of the circuit components. Therefore the current sensing information must be used for duty control in order to prevent core saturation.

A snubber circuit is also essential for the push–pull converter in the same manner as the flyback and forward converters.

3.5 Half-Bridge Converter

Figure 3.18 shows an ideal half-bridge converter circuit, and Fig. 3.19 shows a half-bridge converter with the transformer-equivalent circuit. This converter can be regarded as a push–pull converter in which input voltages are divided into half by two input capacitors: when either of the switches is closed, energy is stored in the output inductor; when both switches are open, the stored energy is released to the output through the rectifier diodes.

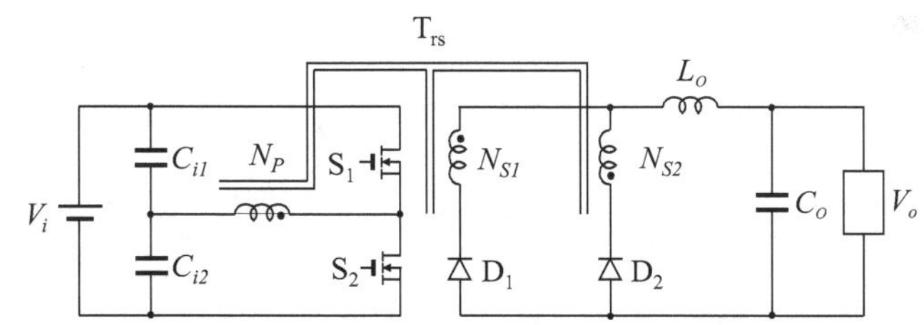

Figure 3.18 Ideal half-bridge converter.

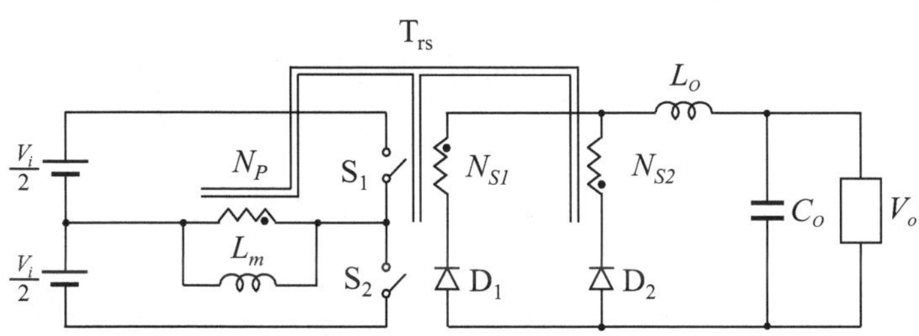

Figure 3.19 Half-bridge converter with the transformer-equivalent circuit.

Figure 3.20 shows four equivalent circuits for each operation state, and Fig. 3.21 shows typical waveforms of the half-bridge converter. In the same manner as the push–pull converter, the two switches S_1 and S_2 are turned on and off 180° out of phase, and each switch operates during one half cycle ($T_s/2$). The input voltage, V_i, is divided into half of the input voltage ($V_{Ci1} = V_{Ci2} = V_i/2$) by the input capacitors C_{i1} and C_{i2}.

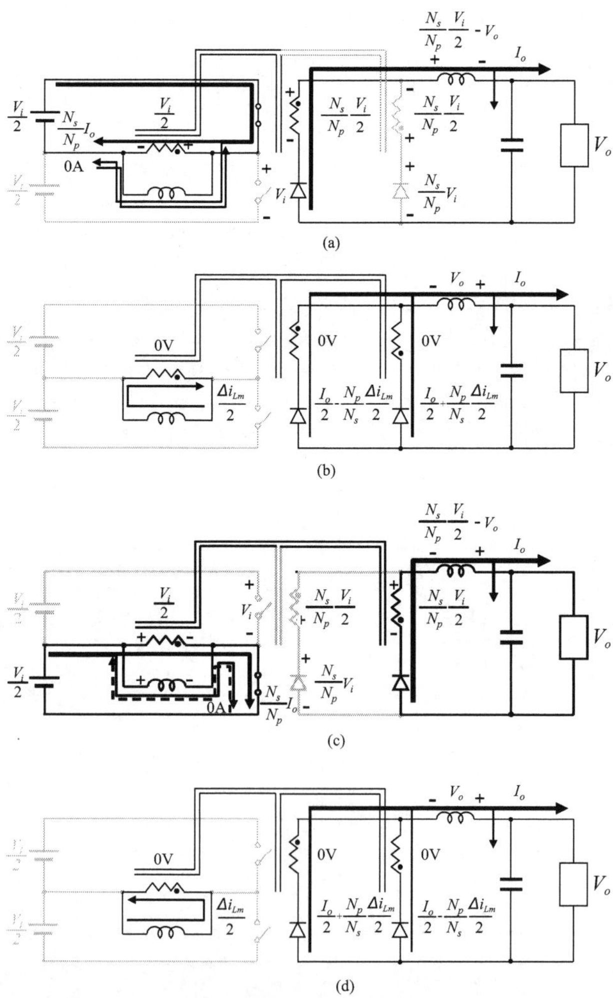

Figure 3.20 Equivalent circuits for each operation state. (a) $t_0 - t_1$, (b) $t_1 - t_2$, (c) $t_2 - t_3$, and (d) $t_3 - t_4$.

During the first half cycle, when the first switch, S_1, is closed, the capacitor voltage, V_{Ci1}, is directly applied to the positive side of the primary winding, N_p, so that the secondary windings, N_{s1} and N_{s2}, generate a voltage of $V_{Ns1} = V_{Ns2} = (N_s/N_p)V_i/2$. Since the rectifier diode D_1 is forward biased and the rectifier diode D_2 is reverse biased with a voltage of $V_{D2} = V_{Ns1} + V_{Ns2} = (N_s/N_p)V_i$, current flows to the output-smoothing capacitor, C_o, and the load through the secondary winding N_{s1}, rectifier diode D_1, and output inductor L_o. During this time, energy is stored in the output inductor, L_o, and the inductor current is increased linearly with a slope of $di_{Lo}/dt = (V_{Ns1} - V_o)/L_o = [(N_s/N_p)(V_i/2] - V_o)/L_o$. At the same time, the magnetizing inductor current has waveform of positive–negative symmetric with a slope of $di_{Lm}/dt = V_i/2L_m$, and the average value is zero. During this period, the maximum voltage stress of the other switch, S_2, is $V_{s2} = V_i$.

During the first half cycle, when both switches are open, the voltage polarity of the output inductor, L_o, is reversed, and the energy stored in the output inductor is released to the output-smoothing capacitor, C_o, and the load through both rectifier diodes and secondary windings. In this state, all winding voltages of the transformer are zero because the transformer's magnetic flux is cancelled by the secondary winding currents flowing in the opposite direction (positive and negative). Additionally, the output inductor current, i_{Lo}, is decreased linearly with a slope of $di_{Lo}/dt = -V_o/L_o$, and the magnetizing inductor current, i_{Lm}, is held at a constant value (half the peak-to-peak current = $\Delta i_{Lm}/2$). The voltage stresses of the switches during this period are $V_{s1} = V_{s2} = V_i/2$.

During the second half cycle, the converter operates in the same manner as the first half cycle. When the switch S_2 is closed, voltage is applied to the negative side of the transformer windings, and energy is stored in the output inductor, L_o, through the windings N_p and N_{s2} and the rectifier diode D_2. When both switches are open, the stored energy is released to the output through both secondary windings and rectifier diodes.

By using steady-state analysis, the voltage conversion ratio of the half-bridge converter is given by

$$\frac{V_o}{V_i} = \frac{1}{2}\frac{N_s}{N_p}D \tag{3.12}$$

where $D = 2T_{on}/T_s$ is the duty cycle of the switches. The duty cycle of the half-bridge converter is two times that of the forward converter because of the two on periods during both half cycles. The voltage conversion ratio of the half-bridge converter is half that of the push–pull converter because the primary winding, N_p, is supplied by half of the input voltage by using the input capacitors.

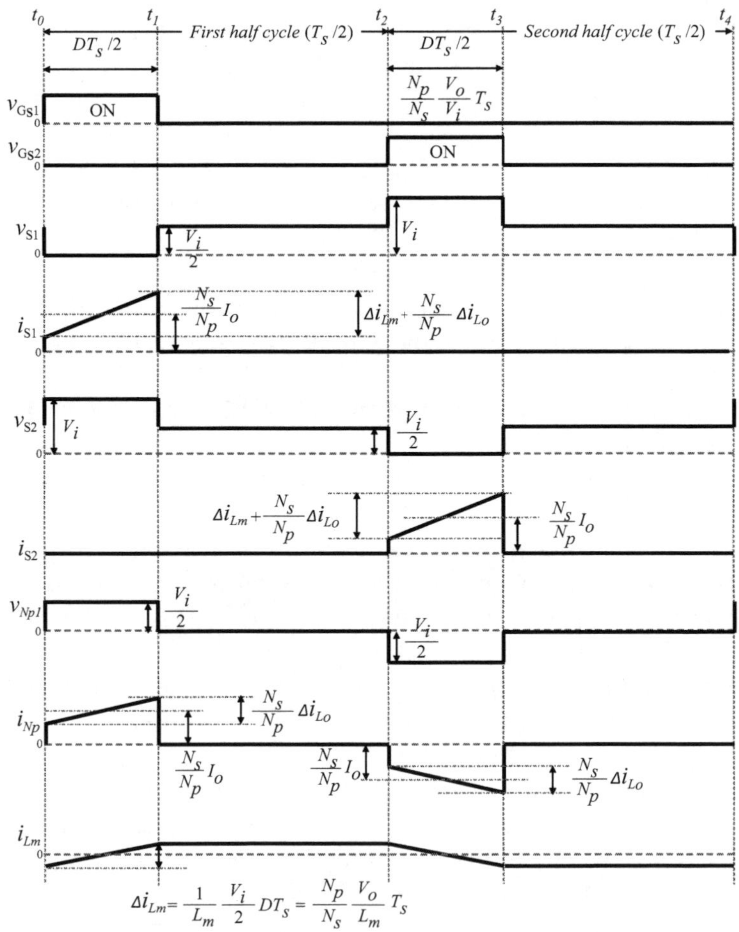

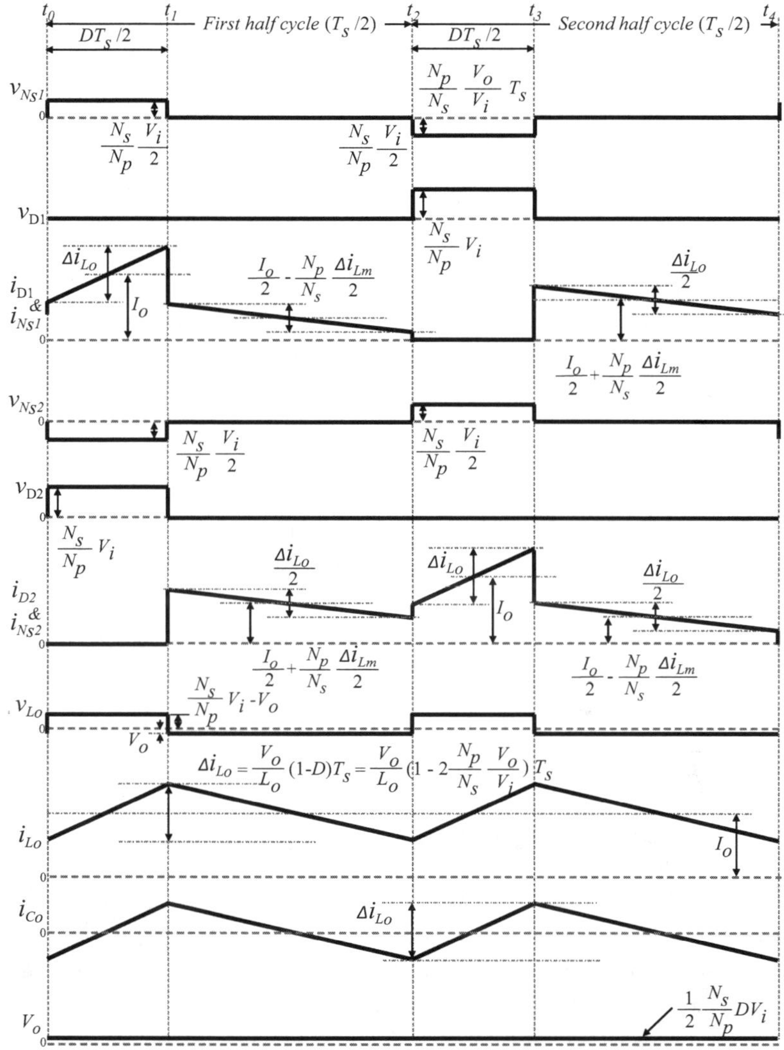

Figure 3.21 Typical waveforms of a half-bridge converter.

In the same manner as the push–pull converter, voltage is applied to both the positive and the negative side of the primary windings during both half cycles. Therefore, the magnetizing inductor current is a symmetrical waveform with a peak-to-peak current of $\Delta i_{Lm} = V_i DT_s/2L_m$.

Unlike the push–pull converter, the half-bridge converter has no risk of transformer core saturation, because the variations of the magnetizing current in the first and second cycles are balanced automatically. For example, if the variation amount of the magnetizing inductor current in the first cycle is larger than that in the second cycle, the discharge amount of the input capacitor C_{i1} is larger than that of the input capacitor C_{i2}. This situation causes the capacitor voltage V_{Ci1} to decrease to less than $V_i/2$ and the capacitor voltage V_{Ci2} to increase to more than $V_i/2$. Accordingly, the variation amount of the magnetizing current in the first half cycle will be reduced and that in the second half cycle will be increased so that the unbalance of the magnetizing inductor currents is removed without high-accuracy current detection and control.

Switching surge and noise can be suppressed incorporating additional diodes connected between the drain and source terminals of both switches so that the magnetic energy stored in the leakage inductor of the primary side is recycled to the input through the diodes during the switch's off time. If MOSFETs are used for the switches, the leakage energy is recycled through the body diodes built into the MOSFETs without a need for additional diodes.

3.6 Full-Bridge Converter

3.6.1 PWM-Controlled Full-Bridge Converter

Figure 3.22 shows an ideal full-bridge converter circuit, which is an isolated-type buck converter composed of one primary winding and four switches in the primary circuit. Figure 3.23 shows a full-bridge converter with the transformer-equivalent circuit.

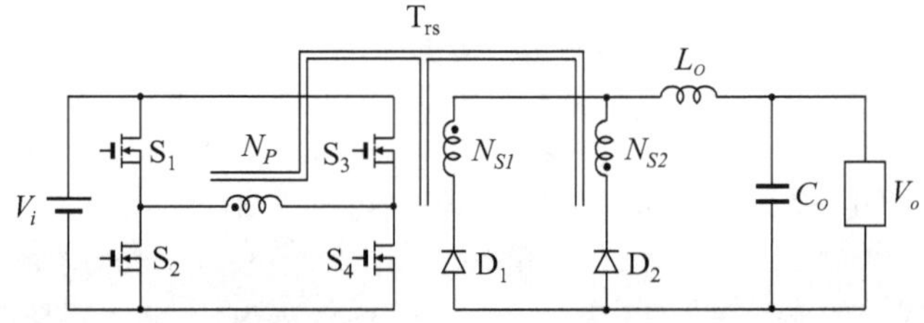

Figure 3.22 Ideal pulse width modulation (PWM)-controlled full-bridge converter circuit.

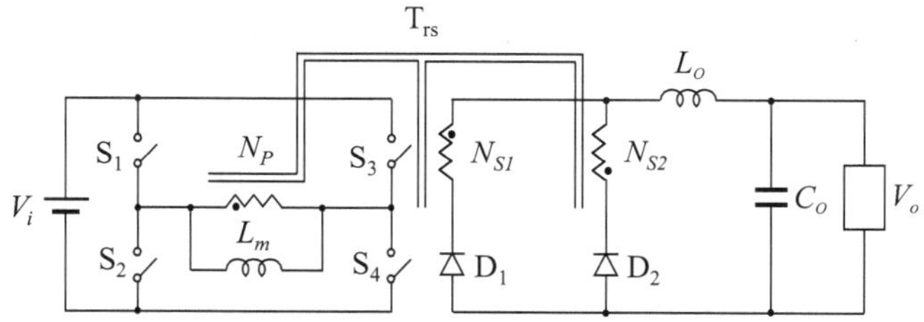

Figure 3.23 PWM-controlled full-bridge converter with the transformer-equivalent circuit.

Figure 3.24 shows four equivalent circuits for each operation state, and Fig. 3.25 shows typical waveforms of the full-bridge converter. In the same manner as the push–pull converter, the two switches S_1 and S_3 are turned on and off 180° out of phase, and each operates during one half cycle ($T_s/2$). Each pair of switches S_1 and S_2 (or S_3 and S_4) is switched complementarily, meaning that they perform the same function but are switched on at opposite times.

During the first half cycle, when the first switch, S_1, is closed, the input voltage, V_i, is directly supplied to the positive side of the primary winding, N_p, through the switches S_1 and S_4, so the secondary windings, N_{s1} and N_{s2}, generate a voltage of $V_{Ns1} = V_{Ns2} = (N_s/N_p)V_i$. Since the rectifier diode D_1 is forward biased and the rectifier diode D_2 is reverse biased with a voltage of $V_{D2} = V_{Ns1} + V_{Ns2} = 2(N_s/N_p)V_i$, the current flows to the output-smoothing capacitor, C_o, and the load through the secondary winding N_{s1}, rectifier diode D_1, and output inductor L_o. During this time, energy is stored in the output inductor, L_o, and the inductor current is increased linearly with a slope of $di_{Lo}/dt = (V_{Ns1} - V_o)/L_o = [(N_s/N_p)V_i - V_o]/L_o$. At the same time, the magnetizing inductor current has waveform of positive–negative symmetric with a slope of $di_{Lm}/dt = V_i/L_m$, and the average value is zero. The maximum voltage stresses of the other switches S_2 and S_3 are $V_{s2} = V_i$.

During the first half cycle, when the switch S_1 is open, all winding voltages of the transformer are zero because the primary winding, N_p, is shorted through switches S_2 and S_4. This causes the voltage polarity of the output inductor, L_o, to reverse, and the stored energy in the output inductor is released to the output-smoothing capacitor, C_o, and the load. In this state, the output inductor current, i_{Lo}, is decreased linearly with a slope of $di_{Lo}/dt = -V_o/L_o$, and the

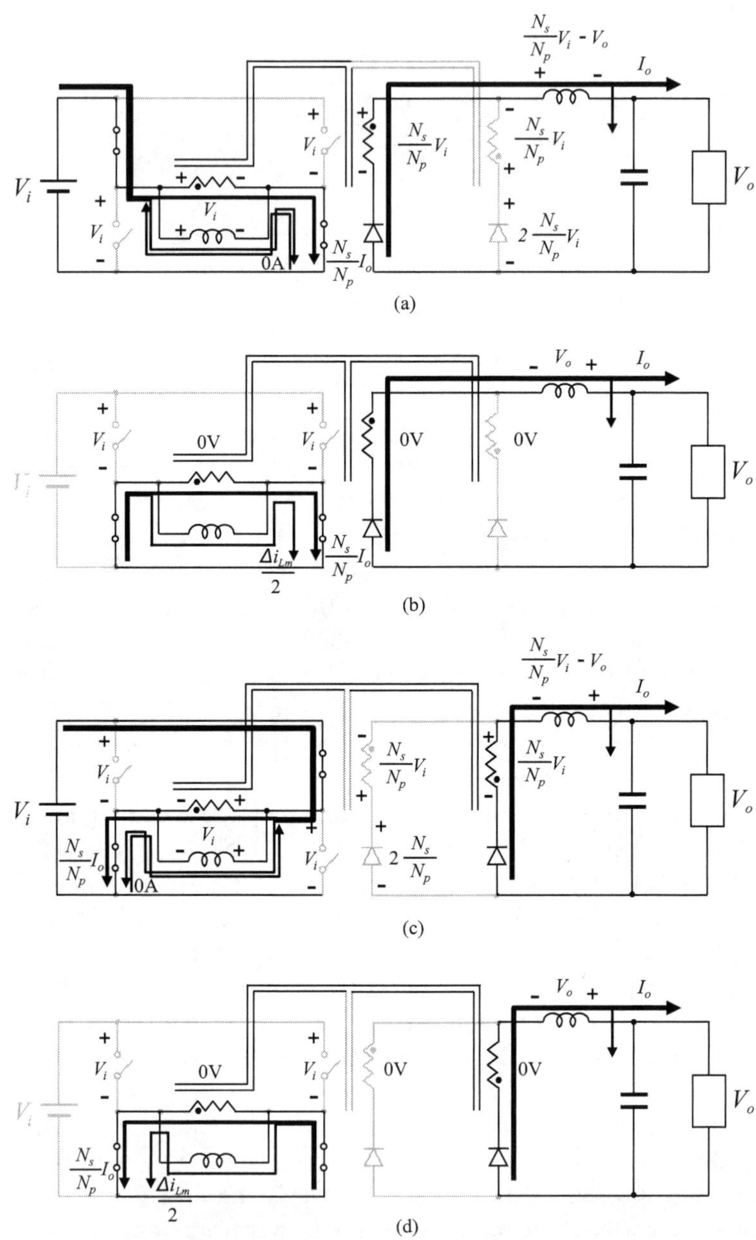

Figure 3.24 Equivalent circuits for each operation state with PWM control. (a) $t_0 - t_1$, (b) $t_1 - t_2$, (c) $t_2 - t_3$, and (d) $t_3 - t_4$.

magnetizing inductor current, i_{Lm}, is held at a constant value (half of the inductor peak-to-peak current = $\Delta i_{Lm}/2$). Because of leakage inductance of the transformer, the output inductor current, i_{Lo}, will continuously flow through the diode D_1 and the secondary winding N_{s1}. Additionally, a current with a value of $(N_s/N_p)i_{Lo} + \Delta i_{Lm}/2$ will continuously flow through the switches S_2 and S_4.

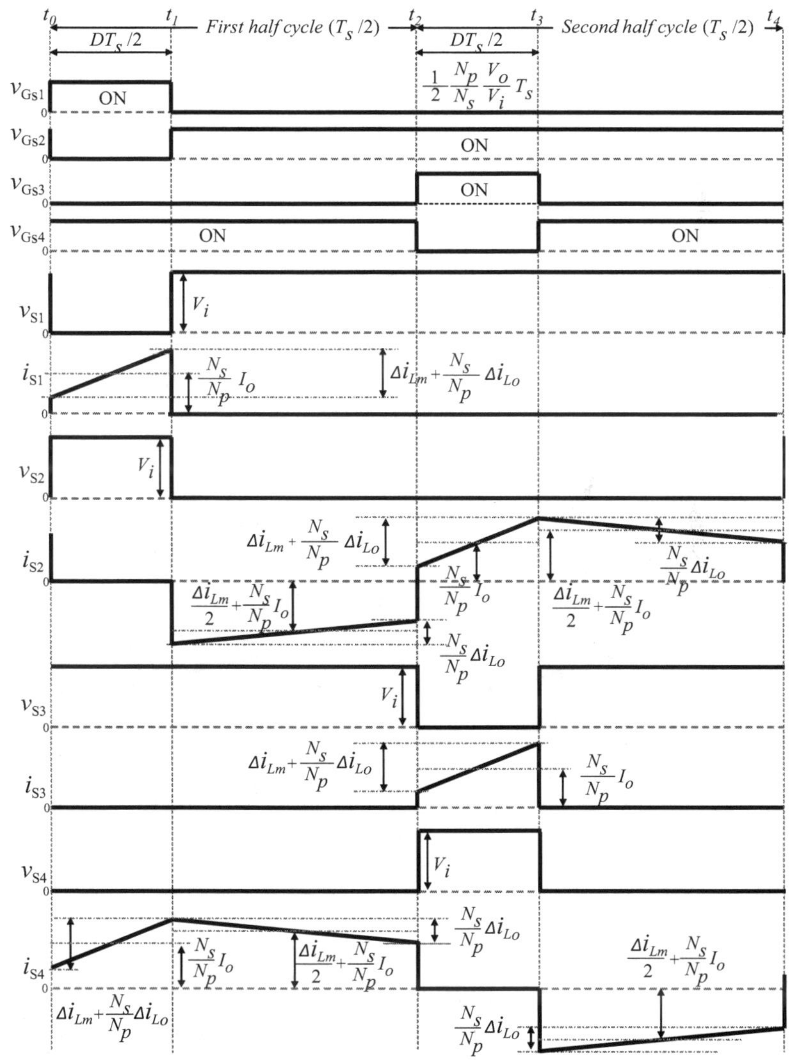

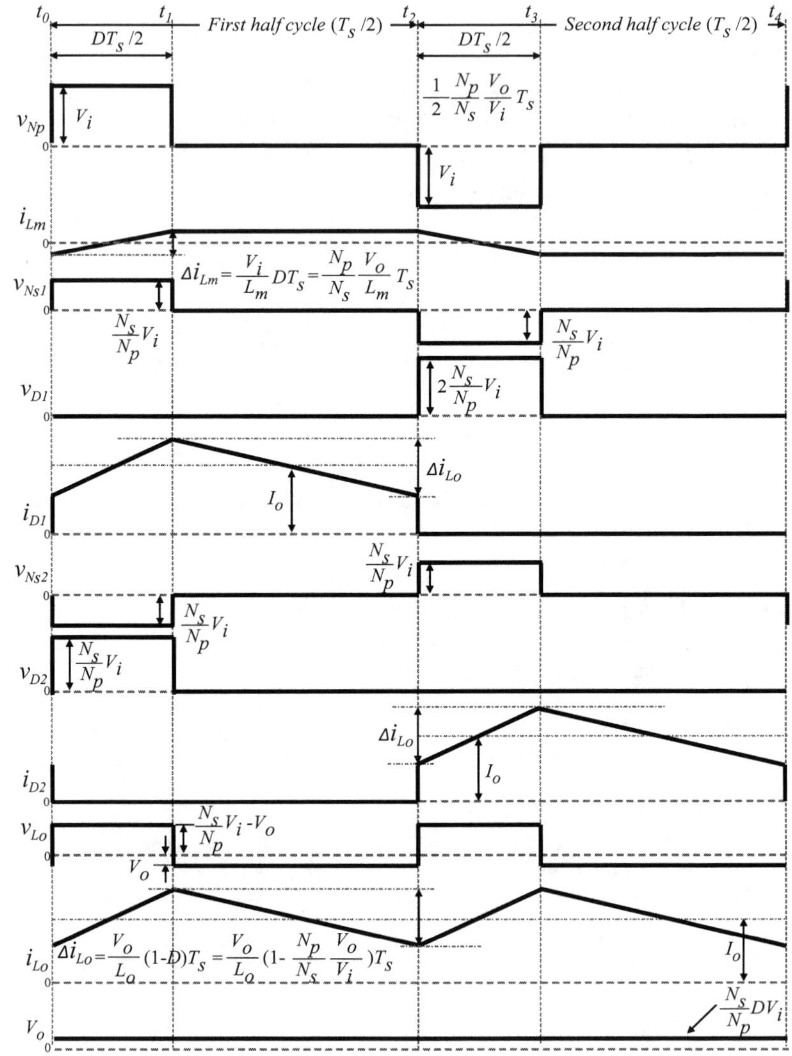

Figure 3.25 Typical waveforms of a PWM-controlled full-bridge converter.

During the second half cycle, the converter operates in the same manner as the first half cycle. When the switch S_3 is closed, voltage is applied to the negative side of the transformer windings and energy is stored in the output inductor through the windings N_p and N_{s2} and the rectifier diode D_2. When S_3 is open, the transformer winding voltages are zero with the primary short circuited, and the stored energy is released to the output through the secondary winding N_{s2} and the rectifier diode D_2.

By using steady-state analysis, the voltage conversion ratio of the full-bridge converter is given by

$$\frac{V_o}{V_i} = \frac{N_s}{N_p} D \qquad (3.13)$$

where $D = 2T_{on}/T_s$ is the duty cycle of the switches. The duty cycle of the full-bridge converter is two times that of the forward converter because of the on periods during both half cycles. The voltage conversion ratio of the full-bridge converter is the same as that of the push–pull converter.

In the same manner as the push–pull converter, voltage is applied to the positive side of the magnetizing inductor during the first half cycle and to the negative side during the second half cycle. Therefore the magnetizing inductor current is a symmetrical waveform with a peak-to-peak current of $\Delta i_{Lm} = V_i D T_s / L_m$.

The full-bridge converter experiences transformer core saturation similar to that of the push–pull converter. One solution to remove this core saturation effect is to add a DC cut capacitor connected in series with the primary winding, N_p. In a similar manner to the half-bridge converter, if the charge and discharge amounts of the DC-cut capacitor during the first and second half cycles are not equal, the voltage of the DC-cut capacitor automatically changes to reduce the unbalance of the magnetizing inductor currents.

The full-bridge converter requires four switches in the primary side, but the maximum voltage stress of all switches is half of the push–pull converter (V_{s1}, V_{s2}, V_{s3} and $V_{s4} = V_i$.)

The circulating currents led by the leakage inductance increase the conduction losses of the switches and transformer windings. However, the switching loss and surge current are suppressed by the continuously circulating currents using the zero-voltage-switching (ZVS) technique described in Section 3.6.4. Switching losses are especially important in converters operating at high input voltage, because turn-on and turn-off losses are proportional to the input voltage. Additionally, stored energy that dissipates into parasitic capacitances between switching device terminals is proportional to the square of the input voltage.

Because of these benefits, the full-bridge converter is widely used for high-input-voltage and high-power applications such as internet servers and EV (electric vehicle) chargers.

3.6.2 Phase-Shift-Controlled Full-Bridge Converter

Phase-shift control is possible with full-bridge converters. Figure 3.26 shows two equivalent circuits for each operation state, and Fig. 3.27 shows typical waveforms of a full-bridge converter with phase-shift control. Using this control method, each pair of switches is operated with a fixed duty cycle of $D = 0.5$, but they are controlled by varying the phase shift between their commutation timings.

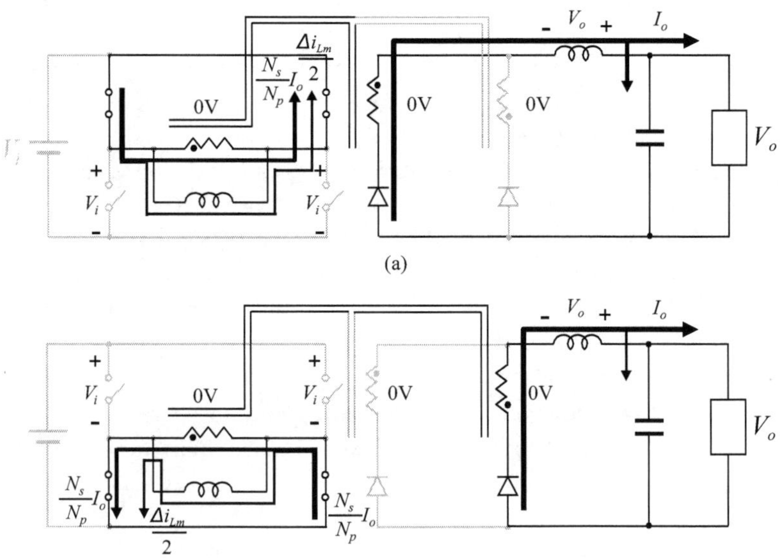

Figure 3.26 Equivalent circuits for each operation state with phase-shift control. (a) $t_1 - t_2$ and (b) $t_3 - t_4$.

During the first half cycle, when switches S_1 and S_4 are conducting, the input voltage is applied to the positive side of the transformer windings, and hence energy is stored in the output inductor, L_o, through the windings N_p and N_{s1} and the rectifier diode D_1. After the switch S_3 turns off and S_1 and S_4 are conducting, the primary side becomes a short circuit and the stored energy is released to the output. Because of leakage inductance, the secondary winding current continuously flows through the rectifier diode D_1 and the secondary winding N_{s1}, while the primary winding current continuously flows through the switches S_1 and S_4.

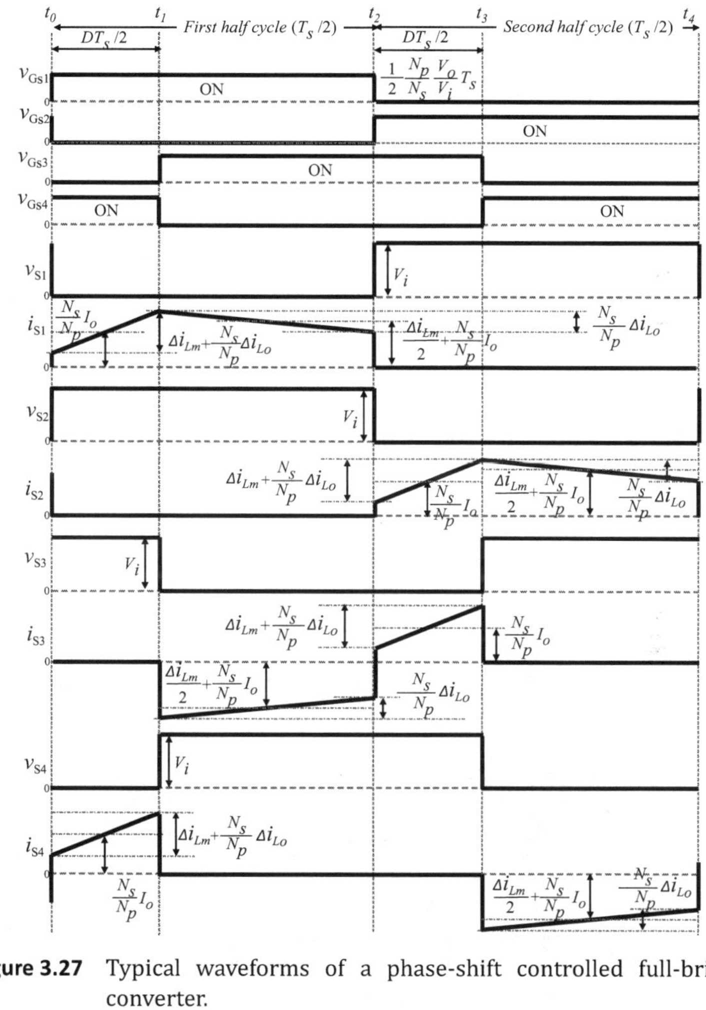

Figure 3.27 Typical waveforms of a phase-shift controlled full-bridge converter.

During the second half cycle, the converter operates in the same manner as the first half cycle. When S_2 and S_3 are closed, the voltage is supplied to the negative side of the transformer windings, so the energy is stored in the output inductor through the windings N_p and N_{s2} and the rectifier diode D_2. After S_3 turns off, when S_2 and S_4 are closed, the magnetic energy stored in the output inductor is released to the output through the secondary winding N_{s2} and the diode D_2.

The voltage conversion ratio of a phase-shift-controlled full-bridge converter is identical to that of a PWM-controlled full-bridge circuit.

In the case of phase-shift control, the same effective current flows through all the switches independent of the phase-shift amount.

3.6.3 Full-Bridge Converter with a Current-Doubler Rectifier

Figure 3.28 shows an equivalent circuit of a full-bridge converter with a current-doubler rectifier circuit. The current doubler is constructed by utilizing one secondary winding, two output inductors, and two rectifier diodes. Figure 3.29 shows four equivalent circuits for each operation state, and Fig. 3.30 shows typical waveforms of a full-bridge converter with current-doubler control.

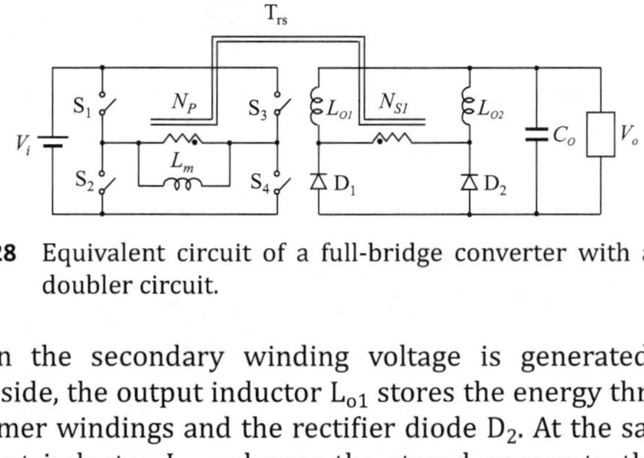

Figure 3.28 Equivalent circuit of a full-bridge converter with a current-doubler circuit.

When the secondary winding voltage is generated on the positive side, the output inductor L_{o1} stores the energy through the transformer windings and the rectifier diode D_2. At the same time, the output inductor L_{o2} releases the stored energy to the output through the rectifier diode D_2.

When the secondary winding voltage is zero by a short circuit in the primary side, both output inductors, L_{o1} and L_{o2}, release their stored energy to the output. Because of leakage inductance of the transformer, the output inductor currents i_{Lo1} and i_{Lo2} will continuously flow through the rectifier diode D_2.

When a secondary winding voltage is generated in the negative direction, the output inductor L_{o2} stores the energy through the transformer windings and the rectifier diode D_1 and the output inductor L_{o1} releases the stored energy to the output through the rectifier diode D_1.

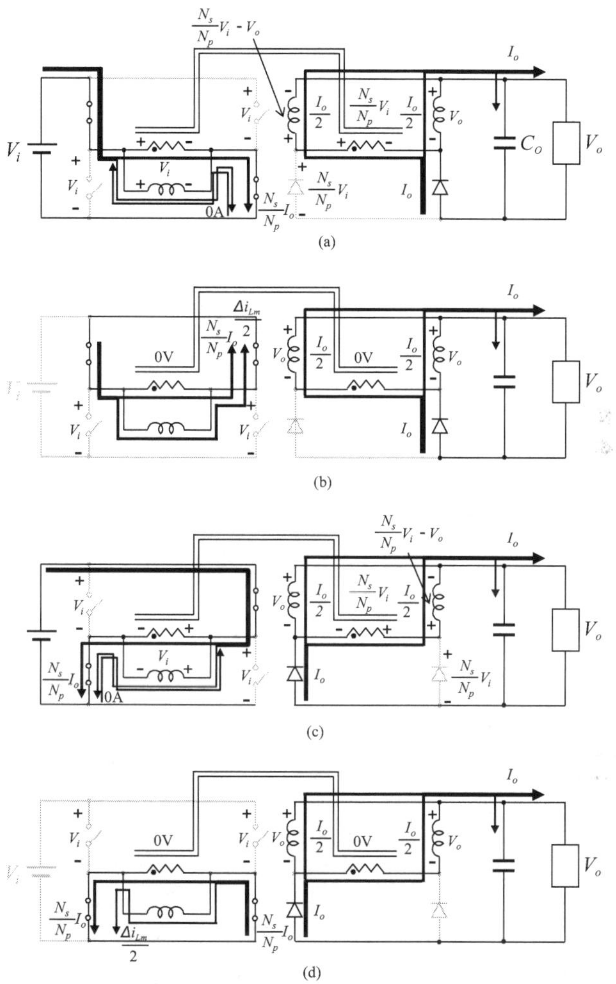

Figure 3.29 Equivalent circuits for each operation state with a current-doubler circuit. (a) $t_0 - t_1$, (b) $t_1 - t_2$, (c) $t_2 - t_3$, and (d) $t_3 - t_4$.

When the secondary winding voltage is zero by the primary short circuit, both output inductors, L_{o1} and L_{o2}, release their stored energy to the output through the rectifier diode D_2.

3.6.4 Full-Bridge Converter with Zero Voltage Switching

During switching commutations between turn-on and turn-off states, switching losses occur in switching elements caused by both voltage and current overlap and by dissipation of the stored energy into parasitic capacitances between switch terminals.

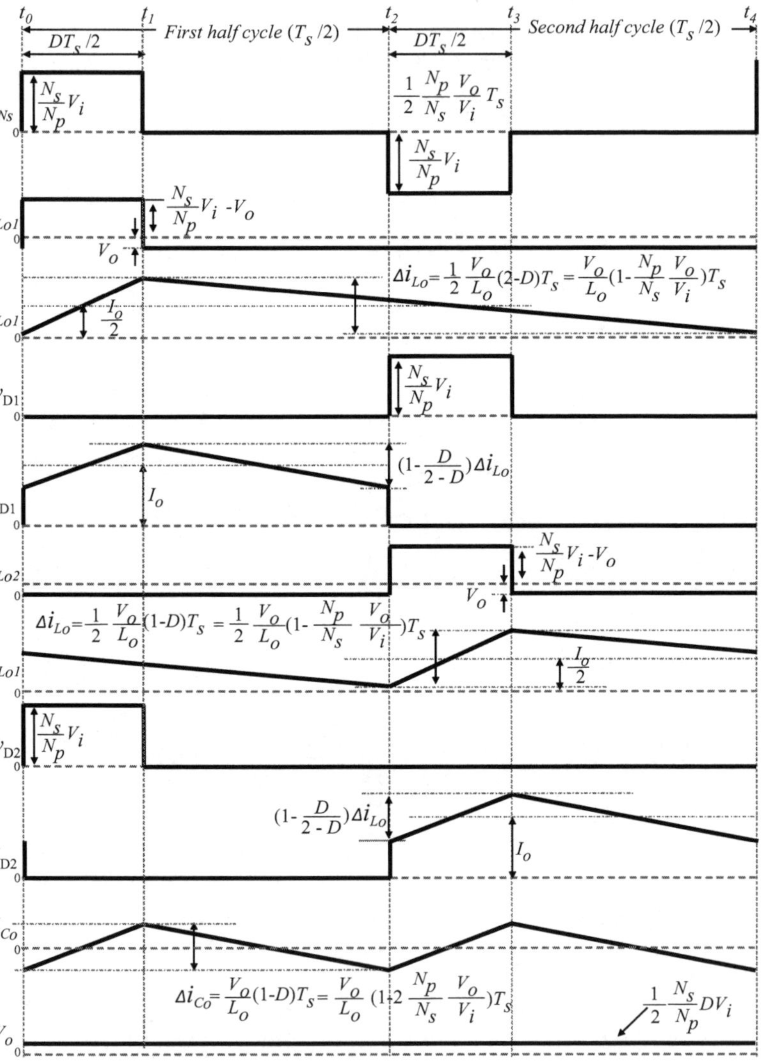

Figure 3.30 Typical waveforms of a current-doubler circuit.

A full-bridge converter is capable of removing switching losses by a technique called ZVS. In the case of a full-bridge converter, ZVS is achieved by making a short interval (dead time) where both switches are off during switching commutations.

Figure 3.31 shows two equivalent circuits for dead time after turn-off of the switch S_4 in a phase-shift-controlled full-bridge converter.

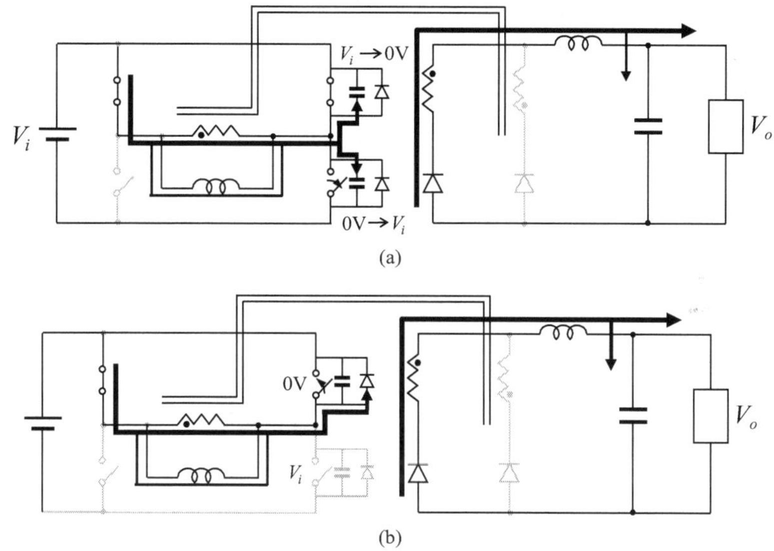

Figure 3.31 Equivalent circuits for dead time with zero voltage switching (a) before reaching zero voltage and (b) after reaching zero voltage.

In this state, as the magnetizing and leakage inductor currents are discharging, the output parasitic capacitor of the switch S_4 is charged and the output parasitic capacitor of the switch S_2 is discharged. Accordingly, the drain-to-source voltage of S_2 is reduced to zero. After reaching zero voltage, the current continuously flows through the body diode of S_2. Next, S_2 can be turned on at zero voltage and without switching losses. Similarly, ZVS is easily achieved at all switching commutations by adding dead time.

Chapter 4

Modeling and Analysis of Switching Converters

Terukazu Sato
Department of Electrical and Electronics Engineering,
Oita University, 700 Dannoharu, Oita 870-1192, Japan
tsato@oita-u.ac.jp

4.1 Introduction

Steady-state analysis describes static characteristics such as the voltage conversion ratio, the voltage, and current ripples, while dynamic or small-signal analysis predicts the load and line regulation characteristics, the transient response for the load current or the input voltage changes, and the margin of stability. It is not so difficult to determine the steady state of switching converters. It can be easily performed by dividing the operation of the switching converter into several states, deriving the equivalent

Handbook of Power Management Circuits
Edited by Haruo Kobayashi and Takashi Nabeshima

ISBN 978-981-4613-15-6 (Hardcover), 978-981-4613-16-3 (eBook)
www.panstanford.com

circuits and equations for each state and solving them under steady-state condition such that integrals of the voltages across the inductors and the currents through the capacitors along the switching period are zero. On the other hand analysis of the dynamic characteristics is not so easy because of the nonlinear operation of switching converters. So far several analysis methods have been developed, including state-space averaging, which is well known for the analysis of pulse-width-modulated switching converters. Unfortunately, it cannot be applied to frequency-modulated converters like resonant converters. The extended state-space averaging method is a method that revises state-space averaging to make it applicable to frequency-modulated converters. However, it is not easy to comprehend and apply this method, especially for students or engineers who have started to learn about switching converters. For example, it requires the distinction between low-frequency elements and high-frequency ones, which is so difficult for not only beginners but also skilled engineers.

This chapter describes a new unified approach to analyzing switching converters, which is applicable to both pulse-width-modulated and frequency-modulated converters. The method is quite simple. Commonly, when analyzing linear circuits, we derive circuit equations using Kirchhoff's voltage law (KVL) and Kirchhoff's current law (KCL) and the relation between the voltage and current of the circuit element, like Ohm's law. Then we find unknown voltages and currents by solving them. Actually, a method like this can be applicable for averaged voltages and currents of circuit elements that are even if nonlinear or time invariant. After demonstrating that KVL and KCL holds for the averaged voltages and currents of nonlinear circuit elements, equivalent devices for the averaged model are derived and the analysis procedure is described. Then the steady-state and dynamic characteristics of a buck converter in continuous conduction mode (CCM) and discontinuous conduction mode (DCM) are analyzed.

4.2 Switching Converter Analysis Using the Averaged Device Model

4.2.1 Kirchhoff's Law for Averaged Voltage and Current

Even for nonlinear circuits like switching converters, Kirchhoff's law is valid for averaged voltages and currents. Suppose one switching cycle is divided into n intervals and the circuit is linear in each interval. When the circuit comprises p branches, at node q in an interval m, KCL is expressed as

$$\sum_{k=1}^{p} a_k i_{k,m} = 0 \tag{4.1}$$

Where $i_{\mathrm{k,m}}$ is the current of branch k in an interval m and a_{k} is one of 1, 0, or −1. Integrating Eq. 4.1 over one switching cycle, we have

$$\frac{1}{T_s}\sum_{m=1}^{n}\left(\int_{T_m}\left(\sum_{k=1}^{p} a_k i_{k,m}\right)dt\right) = 0 \tag{4.2}$$

where T_s is the switching period and $\int_{T_m} dt$ represents the integral over the interval m.

Equation 4.2 is rewritten as

$$\sum_{k=1}^{p} a_k \left(\frac{1}{T_s}\sum_{m=1}^{n}\int_{T_m} i_{k,m}dt\right) = 0 \tag{4.3}$$

The average current of i_{k} is defined as

$$\bar{i}_k = \frac{1}{T_s}\sum_{m=1}^{n}\int_{T_m} i_{k,m}dt \tag{4.4}$$

From Eqs. 4.3 and Eq. 4.4, we have

$$\sum_{k=1}^{p} a_k \bar{i}_k = 0 \tag{4.5}$$

Equation 4.5 is exactly KCL for averaged voltages and currents. A similar discussion can be made for KVL. Since Eq. 4.5 is valid for the very large number of states in one switching cycle, it is valid for nonlinear circuits.

4.2.2 The Equivalent Device Model

Faraday's law of induction is given by

$$v_{\mathrm{L}} = N\frac{d}{dt}\phi \tag{4.6}$$

Where v_{L} is the voltage across the inductor, ϕ is magnetic flux, and N is the number of windings. Averaging Eq. 4.6 over one switching cycle gives

$$\frac{1}{T_{\mathrm{s}}}\int_{T_{\mathrm{s}}} v_{\mathrm{L}} dt = \frac{1}{T_{\mathrm{s}}}\int_{T_{\mathrm{s}}}\left(N\frac{d}{dt}\phi\right)dt \tag{4.7}$$

where $\int_{T_{\mathrm{s}}} dt$ represents the integral over one switching cycle. Equation 4.7 can be expressed as

$$\frac{1}{T_{\mathrm{s}}}\int_{T_{\mathrm{s}}} v_{\mathrm{L}} dt = \frac{d}{dt}\left(\frac{1}{T_{\mathrm{s}}}\int_{T_{\mathrm{s}}} N\phi dt\right) \tag{4.8}$$

The average values of the voltage across the inductor, $\overline{v}_{\mathrm{L}}$, the average value of the magnetic flux of the inductor, $\overline{\phi}$, and the average value of the current in the inductor, $\overline{i}_{\mathrm{L}}$, are defined as

$$\overline{v}_{\mathrm{L}} = \frac{1}{T_{\mathrm{s}}}\int_{T_{\mathrm{s}}} v_{\mathrm{L}} dt \tag{4.9}$$

$$\overline{\phi} = \frac{1}{T_{\mathrm{s}}}\int_{T_{\mathrm{s}}} \phi dt \tag{4.10}$$

$$\overline{i}_L = \frac{1}{T_{\mathrm{s}}}\int_{T_{\mathrm{s}}} i_{\mathrm{L}} dt \tag{4.11}$$

From Eqs. 4.8–4.11 we get

$$\overline{v}_L = \overline{L}\frac{d}{dt}\overline{i}_L \tag{4.12}$$

$$\overline{L} = \frac{N\int_{T_{\mathrm{s}}} \phi dt}{\int_{T_{\mathrm{s}}} i_{\mathrm{L}} dt} \tag{4.13}$$

where $\overline{L}$ is the equivalent inductance for the averaged model. Equations 4.12 and 4.13 are valid for the averaged voltage and current in the inductor even if the inductor is nonlinear and time invariant. Similar discussions can be applied to the capacitor and the resistor, including a switching device. The equations defining the relation between the averaged voltage across and the average current through the capacitor is

$$\bar{i}_C = \bar{C}\frac{d}{dt}\bar{v}_C \tag{4.14}$$

$$\bar{C} = \frac{\int_{T_s} q dt}{\int_{T_s} v_C dt} \tag{4.15}$$

where $\bar{i}_C$, $\bar{v}_C$, q, v_C, and $\bar{C}$ are the averaged current in the capacitor, the averaged voltage across the capacitor, the electric charge of the capacitor, the instantaneous value of the voltage across the capacitor, and the equivalent capacitance for the averaged model, respectively. As for resistive elements, including switching elements, we have

$$\bar{i}_R = \bar{R}\bar{v}_R \tag{4.16}$$

$$\bar{R} = \frac{\int_{T_s} v_R dt}{\int_{T_s} i_R dt} \tag{4.17}$$

where $\bar{i}_R$, $\bar{v}_R$, i_R, v_R, and $\bar{R}$ are the averaged current in the resistor, the averaged voltage across the resistor, the instantaneous value of the current in the resistor, the instantaneous value of the voltage across the resistor, and the equivalent resistance for the averaged model, respectively.

Note that the averaged equivalent value of the element is not the average of the instantaneous value of the element but the quotient of the integral of the voltage divided by the integral of the current. Commonly inductors and capacitors are supposed to be linear elements throughout one switching period; in that case the averaged inductance and capacitance are identical to the their own values.

As for power losses, the following expressions must be used:

$$P_{\text{inductor}} = \frac{1}{T_s}\int_{T_s} v_L i_L dt \tag{4.18}$$

$$P_{\text{capacitor}} = \frac{1}{T_s}\int_{T_s} v_C i_C dt \tag{4.19}$$

$$P_{\text{resistor}} = \frac{1}{T_s}\int_{T_s} v_R i_R dt \tag{4.20}$$

These are not products of the averaged voltage and current but the averaged value of products of the instantaneous voltage and current.

4.2.3 Analysis Procedure Using the Averaged Device Model

The analysis procedure using the averaged device model is described to apply the theory mentioned in the previous section practically.

STEP 1: Derivation of waveforms of voltages and currents
According to states of switching devices, one switching cycle should be divided into some stages. After deriving equivalent circuits and corresponding circuit equations for each stage, derive waveforms of voltages and currents by solving them. At this point, let currents in inductors and voltages across capacitors be unknown values.

STEP 2: Derivation of the averaged device model
Integrate the voltages and currents over one switching cycle and derive equivalent device models using Eqs. 4.13, 4.15, and 4.17. Using the derived equivalent device models, draw an equivalent circuit for the averaged model. Switching devices may be replaced as resistors, voltage sources, or current sources.

STEP 3: Steady-state characteristics
Short-circuiting inductors and open-circuiting capacitors in the equivalent circuit derived in the previous step, steady-state characteristics are given.

STEP 4: Dynamic characteristics
Small signal AC characteristics are obtained by considering small perturbations around the DC operating point in the equivalent circuit for the averaged model.

4.3 Buck Converter in Continuous Conduction Mode

The circuit configuration of a buck converter is shown in Fig. 4.1. In the figure, V_i is an input DC voltage source, S is a switch, D is a freewheel diode, L is a smoothing inductor, C is a smoothing capacitor, and R_L is a load resistor. The operation mode of the buck converter is divided into two modes, CCM and DCM, according to the continuity of the current in the output inductor. In this section, we analyze the characteristics of the buck converter in CCM. Switching states of the converter in CCM are shown in Fig. 4.2. S is on and D is off in state I, while S is off and D is on in state II.

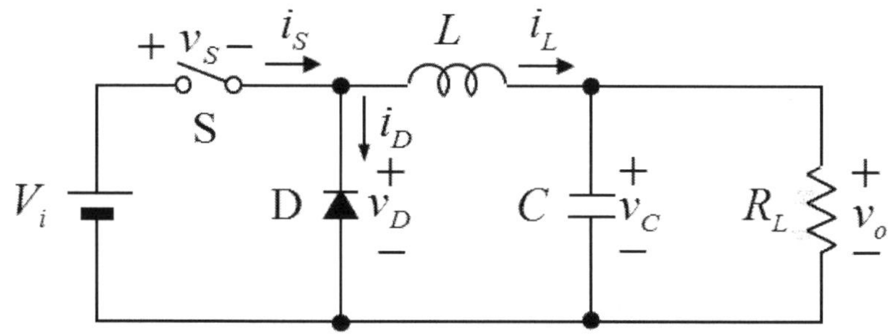

Figure 4.1 Buck converter.

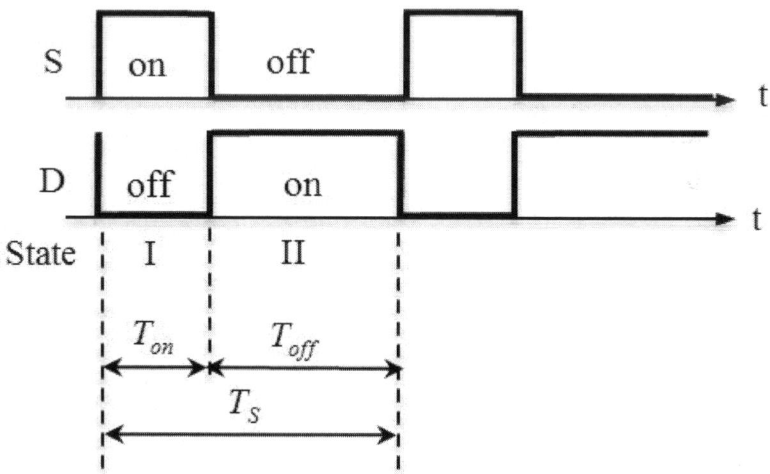

Figure 4.2 Switching state.

4.3.1 Derivation of Waveforms of Currents and Voltages

Suppose S and D have a small resistance in the on state and an infinite resistance in the off state. Then the equivalent circuit in each state is shown in Fig. 4.3.

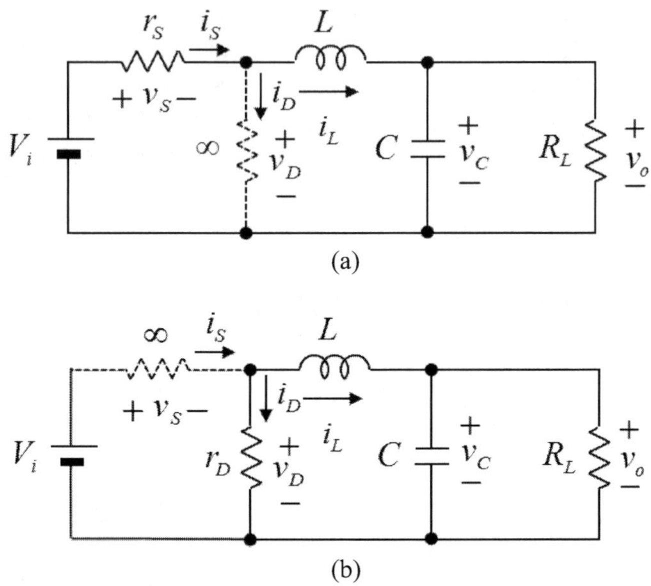

Figure 4.3 Equivalent circuits in continuous conduction mode. (a) State I and (b) state II.

In Fig. 4.3, r_s and r_D represent on resistances of S and D, respectively. Then the following equations are obtained for state I:

$$\text{KVL: } V_i = r_S i_L + L\frac{di_L}{dt} + v_C \tag{4.21}$$

$$\text{KCL: } i_L = C\frac{dv_C}{dt} + \frac{1}{R_L}v_C \tag{4.22}$$

Assuming that the switching period is sufficiently shorter than the time constant of the converter, the voltage across the capacitor can be approximated as constant as

$$v_C = V_C = V_o \tag{4.23}$$

and the current in the inductor rises almost linearly during state I. Let $i_L = I_L$ in Eq. 4.21; using Eq. 4.23, the gradient of the inductor current can be approximated as

$$\frac{di_L}{dt} = \frac{V_i - V_o - r_S I_L}{L} \tag{4.24}$$

Where I_L is the averaged value of the current in the inductor. Then the increased inductor current during state I is

$$I_{L,inc} = \frac{V_i - V_o - r_S I_L}{2L} T_{on} \tag{4.25}$$

Next, the circuit equations for state II are

$$\text{KVL: } 0 = r_D i_L + L\frac{di_L}{dt} + v_C \tag{4.26}$$

$$\text{KCL: } i_L = C\frac{dv_C}{dt} + \frac{1}{R_L} v_C \tag{4.27}$$

Similarly, the voltage across the capacitor can be approximated as constant and the current in the inductor falls almost linearly during state II. The gradient of the inductor current is approximated as

$$\frac{di_L}{dt} = -\frac{V_o + r_D I_L}{L} \tag{4.28}$$

Then the decreased inductor current during state II is

$$I_{L,dec} = \frac{V_o + r_D I_L}{2L} T_{off} \tag{4.29}$$

In the steady state, since the increased inductor current during state I is equal to the decreased inductor current during state II, the peak-to-peak current ripple of the inductor is

$$I_{L,ripple} = I_{L,inc} = I_{L,dec} \tag{4.30}$$

From the above analysis, the key waveforms of the currents and voltages in CCM are obtained, as shown in Fig. 4.4.

4.3.2 Derivation of the Averaged Device Model

From Fig. 4.4, integration over one switching cycle yields

$$V_s = (1 - \mathrm{D})V_i + rI_L \tag{4.31}$$

$$V_D = \mathrm{D}V_i - rI_L \tag{4.32}$$

$$I_S = DI_L \tag{4.33}$$

$$I_D = -(1 - \mathrm{D})I_L \tag{4.34}$$

where

$$D = \frac{T_{on}}{T_S} \tag{4.35}$$

$$r = Dr_s + (1 - \mathrm{D})r_D \tag{4.36}$$

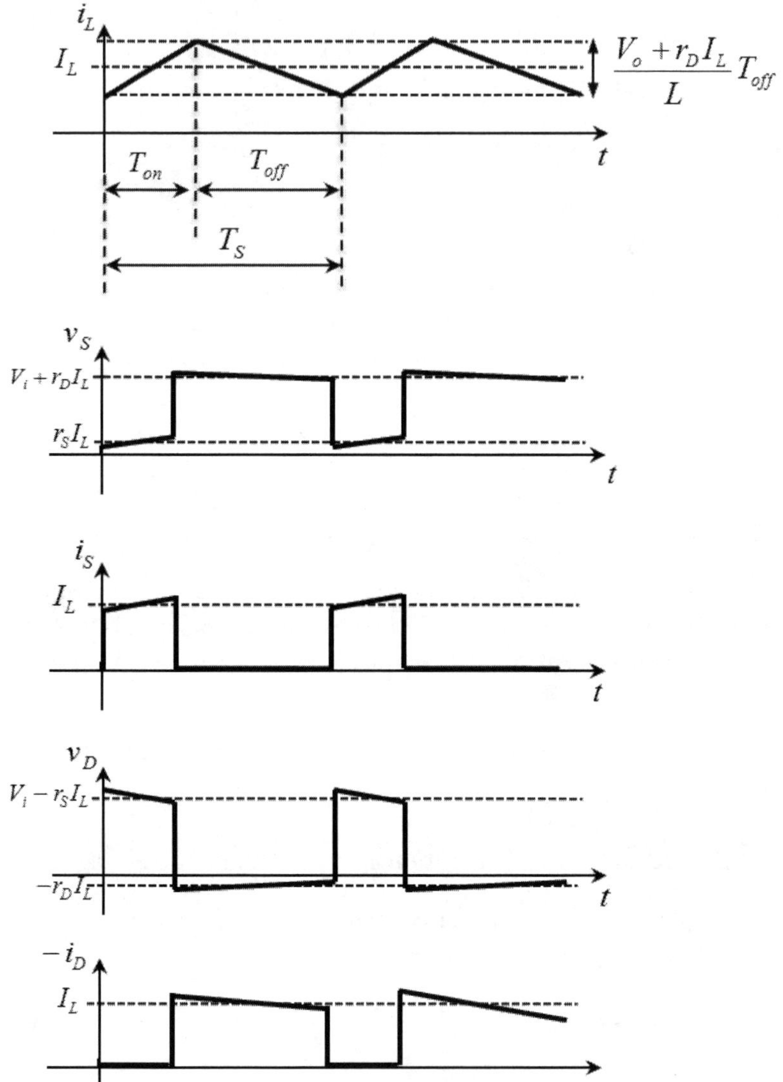

Figure 4.4 Key waveforms of the currents and voltages in CCM.

And V_S and V_D, are the averaged voltages across S and D, I_S and I_D the averaged currents in S and D, and D the duty cycle.

The equivalent device model of the buck converter in CCM is shown in Fig. 4.5. In the figure, R_S and R_D represent the equivalent resistances of the switch and the diode, respectively.

$$R_S = \frac{(1-D)V_i}{DI_L} + \frac{r}{D} \quad (4.37)$$

$$R_D = \frac{DV_i}{(1-D)I_L} - \frac{r}{1-D} \quad (4.38)$$

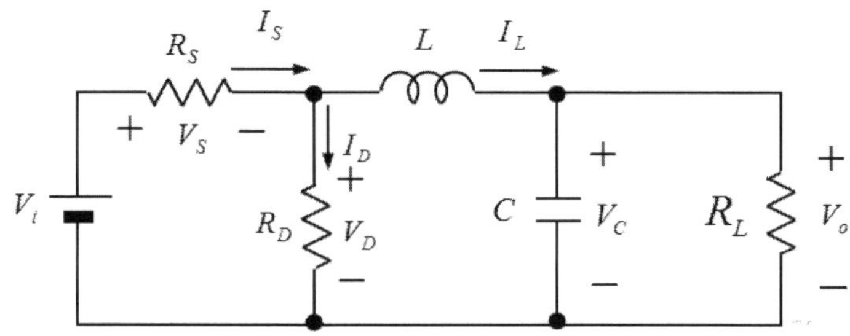

Figure 4.5 The equivalent device model.

4.3.3 Steady-State Characteristics

In the steady state, since the averaged voltage across the inductor and the averaged current in the capacitor are zero, short-circuiting the inductor and open-circuiting the capacitor in Fig. 4.5 yields

$$V_o = V_D = DV_i - rI_o \quad (4.39)$$

$$V_C = V_o \quad (4.40)$$

$$I_L = I_o = \frac{V_o}{R_L} \quad (4.41)$$

Using Eqs. 4.39 and 4.40, the voltage conversion ratio, M, is

$$M = \frac{V_o}{V_i} = D\frac{1}{1+r/R_L} \quad (4.42)$$

4.3.4 Small-Signal AC Analysis

4.3.4.1 Control to the output transfer function

The small perturbation of the duty cycle is assumed as

$$D \rightarrow D + \Delta D \quad (4.43)$$

where D represents a steady-state term and ΔD a small-signal term. By this variation, suppose the following small changes occur:

$$I_L \rightarrow I_L + \Delta I_L \tag{4.44}$$

$$V_C \rightarrow V_C + \Delta V_C \tag{4.45}$$

Substituting Eqs. 4.43–4.45 into Eq. 4.32, the voltage, V_D, changes as

$$V_D \rightarrow V_D + \Delta V_D = (D + \Delta D)V_i - (r + \Delta r)(I_L + \Delta I_L) \tag{4.46}$$

where

$$\Delta r = (r_S - r_D)\Delta D \tag{4.47}$$

Using Eqs. 4.46, 4.47, 4.41, and 4.42, considering only variations from the steady state, and neglecting the product of any AC terms, we have

$$\Delta V_D = \left(\frac{V_o}{D}\left(1+\frac{r}{R_L}\right) - (r_S + r_D)\frac{V_o}{R_L}\right)\Delta D - r\Delta I_L \tag{4.48}$$

From Fig. 4.5 and Eq. 4.48, taking the Laplace transform, we have the equivalent circuit for small-signal analysis, as shown in Fig. 4.6.

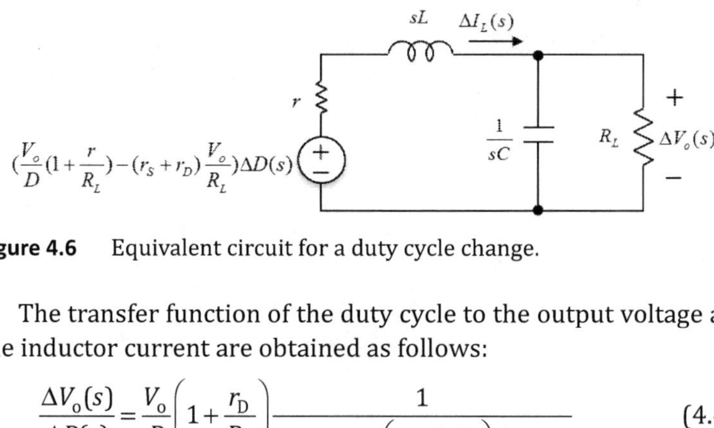

Figure 4.6 Equivalent circuit for a duty cycle change.

The transfer function of the duty cycle to the output voltage and the inductor current are obtained as follows:

$$\frac{\Delta V_o(s)}{\Delta D(s)} = \frac{V_o}{D}\left(1+\frac{r_D}{R_L}\right)\frac{1}{1+\dfrac{r}{R_L}+s\left(Cr+\dfrac{L}{R_L}\right)+s^2LC} \tag{4.49}$$

$$\frac{\Delta I_L(s)}{\Delta D(s)} = \frac{V_o}{DR_L}\left(1+\frac{r_D}{R_L}\right)\frac{1+sCR_L}{1+\dfrac{r}{R_L}+s\left(Cr+\dfrac{L}{R_L}\right)+s^2LC} \tag{4.50}$$

Equations 4.49 and 4.50 can be expressed as

$$\frac{\Delta V_o(s)}{\Delta D(s)} = \frac{G_{vd0}}{P(s)} \tag{4.51}$$

$$\frac{\Delta I_{\mathrm{L}}(s)}{\Delta D(s)}=\frac{G_{\mathrm{id0}}}{P(s)}(1+s/\omega_{\mathrm{idz}}) \tag{4.52}$$

where

$$G_{\mathrm{vd0}}=\frac{V_{\mathrm{o}}}{D}\frac{1+r_{\mathrm{D}}/R_{\mathrm{L}}}{1+rR_{\mathrm{L}}} \tag{4.53}$$

$$P(s)=1+2\delta\frac{s}{\omega_0}+\left(\frac{s}{\omega_0}\right)^2 \tag{4.54}$$

$$\omega_0=\sqrt{\frac{1+r/R_{\mathrm{L}}}{LC}} \tag{4.55}$$

$$\delta=\frac{1}{2}\frac{\frac{\sqrt{L/C}}{R_{\mathrm{L}}}+\frac{r}{\sqrt{L/C}}}{\sqrt{1+r/R_{\mathrm{L}}}} \tag{4.56}$$

$$G_{\mathrm{id0}}=\frac{V_{\mathrm{o}}}{DR_{\mathrm{L}}}\frac{1+r_{\mathrm{D}}/R_{\mathrm{L}}}{1+rR_{\mathrm{L}}} \tag{4.57}$$

$$\omega_{\mathrm{idz}}=\frac{1}{R_{\mathrm{L}}C} \tag{4.58}$$

4.3.4.2 Input to the output transfer function

For the small perturbation of the input voltage from the steady state

$$V_{\mathrm{i}}\rightarrow V_{\mathrm{i}}+\Delta V_{\mathrm{i}} \tag{4.59}$$

From the similar procedure described above, we have the equivalent circuit in the s-domain, as shown in Fig. 4.7.

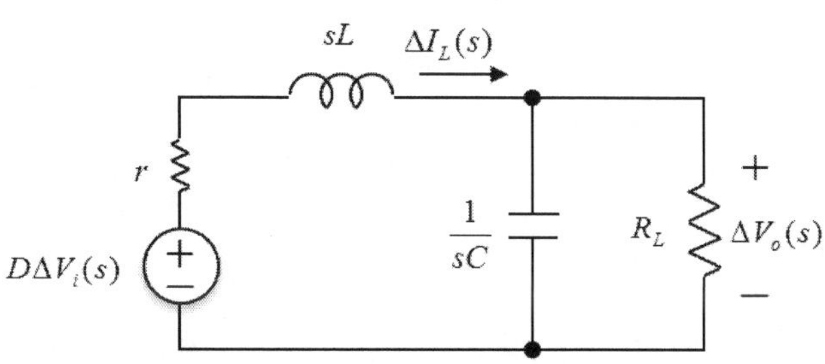

Figure 4.7 Equivalent circuit for an input voltage change.

Then the transfer functions of the input voltage to the output voltage and the inductor current are

$$\frac{\Delta V_{\mathrm{o}}(s)}{\Delta V_{\mathrm{i}}(s)} = \frac{G_{\mathrm{vv0}}}{P(s)} \tag{4.60}$$

$$\frac{\Delta I_{\mathrm{L}}(s)}{\Delta V_{\mathrm{i}}(s)} = \frac{G_{\mathrm{iv0}}}{P(s)}(1 + s/\omega_{\mathrm{ivz}}) \tag{4.61}$$

where

$$G_{\mathrm{vv0}} = D\frac{1}{1 + r/R_{\mathrm{L}}} \tag{4.62}$$

$$G_{\mathrm{iv0}} = \frac{D}{R_{\mathrm{L}}}\frac{1}{1 + r/R_{\mathrm{L}}} \tag{4.63}$$

$$\omega_{\mathrm{ivz}} = \frac{1}{R_{\mathrm{L}}C} \tag{4.64}$$

4.3.4.3 Load to the output transfer function

From the equivalent circuit shown in Fig. 4.5, we have the circuit equations as follows:

$$I_{\mathrm{o}}R_{\mathrm{L}} = V_{\mathrm{o}} \tag{4.65}$$

$$I_{\mathrm{o}} = I_{\mathrm{L}} - C\frac{d}{dt}V_{\mathrm{o}} \tag{4.66}$$

$$rI_{\mathrm{L}} + L\frac{d}{dt}I_{\mathrm{L}} = V_{\mathrm{D}} - V_{\mathrm{o}} \tag{4.67}$$

For the small perturbation of the load resistance from the steady state

$$R_{\mathrm{L}} \rightarrow R_{\mathrm{L}} + \Delta R_{\mathrm{L}} \tag{4.68}$$

From Eqs. 4.65–4.68, considering variations from the steady state, taking the Laplace transform, we have

$$R_{\mathrm{L}}\,\Delta I_{\mathrm{o}}\,(s) + \mathrm{I_o}\Delta I_{\mathrm{L}}\,(s)\,\Delta V_{\mathrm{o}}\,(\mathrm{s}) \tag{4.69}$$

$$\Delta I_{\mathrm{o}}\,(s) = \Delta I_{\mathrm{L}}\,(s) - s\mathrm{C}\Delta V_{\mathrm{o}}\,(\mathrm{s}) \tag{4.70}$$

$$(\mathrm{r} + \mathrm{sL})\Delta I_{\mathrm{L}}\,(s) = -\Delta V_{\mathrm{o}}(s) \tag{4.71}$$

Then the equivalent circuit in the s-domain is as shown in Fig. 4.8.

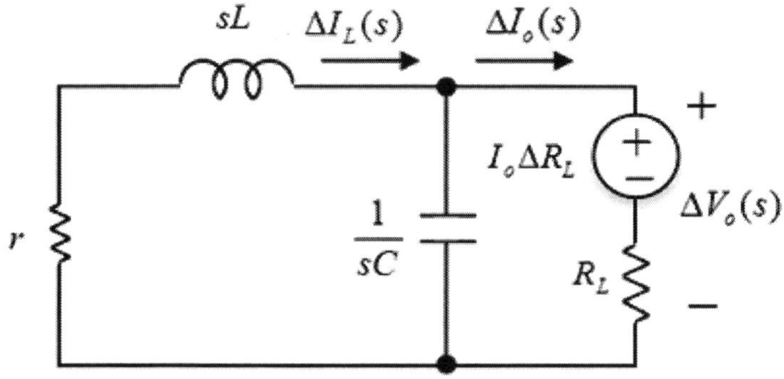

Figure 4.8 Equivalent circuit for load resistance change.

From the above equations and Fig. 4.8, the transfer functions of the load resistance to the output voltage and the inductor current are

$$\frac{\Delta V_{\mathrm{o}}(s)}{\Delta R_{\mathrm{L}}(s)} = \frac{G_{\mathrm{vr0}}}{P(s)}(1 + s/\omega_{\mathrm{vrz}}) \tag{4.72}$$

$$\frac{\Delta I_{\mathrm{L}}(s)}{\Delta R_{\mathrm{L}}(s)} = \frac{-G_{\mathrm{ir0}}}{P(s)} \tag{4.73}$$

where

$$G_{\mathrm{vr0}} = \frac{r}{R_{\mathrm{L}}^{\,2}} V_{\mathrm{o}} \frac{1}{1 + r/R_{\mathrm{L}}} \tag{4.74}$$

$$G_{\mathrm{ir0}} = \frac{V_{\mathrm{o}}}{R_{\mathrm{L}}^{\,2}} \frac{1}{1 + r/R_{\mathrm{L}}} \tag{4.75}$$

$$\omega_{\mathrm{vrz}} = \frac{r}{L} \tag{4.76}$$

4.4 Buck Converter in Discontinuous Conduction Mode

Switching states of a converter in DCM are shown in Fig. 4.9.

State I: S is on and D is off.

State II: S is off and D is on.

State III: Both S and D are off.

The equivalent circuit in each state is shown in Fig. 4.10.

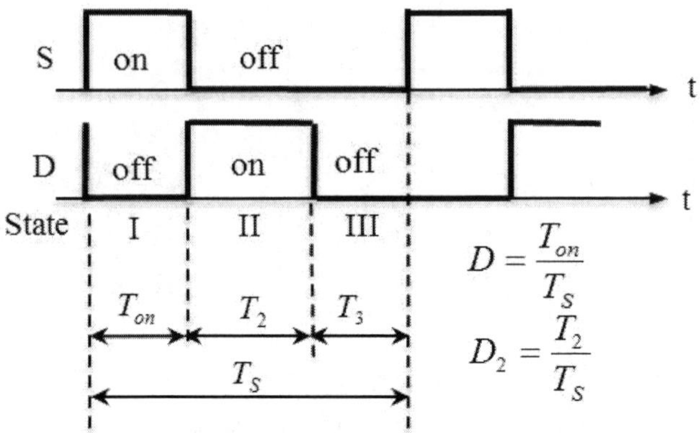

Figure 4.9 Switching state in discontinuous conduction mode.

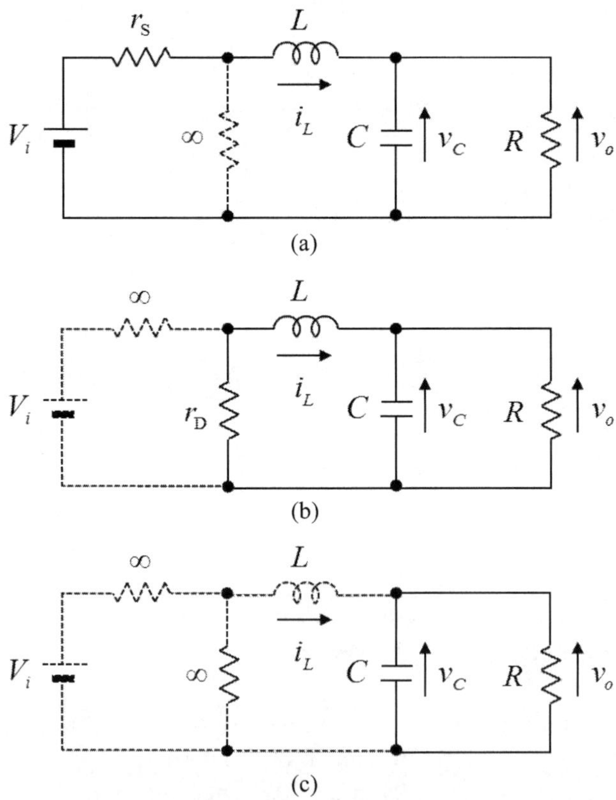

Figure 4.10 Equivalent circuits in discontinuous conduction mode. (a) State I, (b) state II, and (c) state III.

4.4.1 Derivation of Waveforms of Currents and Voltages

Assuming that the switching period is sufficiently shorter than the time constant of the converter, the voltage across the capacitor can be approximated as constant as follows:

$$v_C = V_C = V_o \tag{4.77}$$

The current in the inductor rises almost linearly during state I. The gradient of the inductor current can be approximated as

$$\frac{di_L}{dt} = \frac{V_i - V_o}{L} \tag{4.78}$$

Then the increased inductor current during state I is

$$I_{L,inc} = \frac{V_i - V_o}{2L} T_{on} \tag{4.79}$$

Similarly, the current in the inductor falls almost linearly during state II, and the gradient of the inductor current is approximated as

$$\frac{di_L}{dt} = -\frac{V_o}{L} \tag{4.80}$$

Then the decreased inductor current during state II is

$$I_{L,dec} = \frac{V_o}{2L} T_{off} \tag{4.81}$$

The inductor current is zero in state III, and the voltage across the diode is equal to that across the load resistor.

Figure 4.11 shows waveforms of the currents and the voltages of the switch and the diode.

4.4.2 Derivation of the Averaged Device Model

Integrating over one switching cycle, the averaged value of the voltages and currents are

$$V_S = D_2 V_i + D_3 (V_i - V_o) + (D r_S + D_2 r_D) \frac{V_o}{2L} D_2 T_S \tag{4.82}$$

$$V_D = D V_i + D_3 V_o - (D r_S + D_2 r_D) \frac{V_o}{2L} D_2 T_S \tag{4.83}$$

$$I_S = \frac{V_i - V_o}{2L} D^2 T_S \tag{4.84}$$

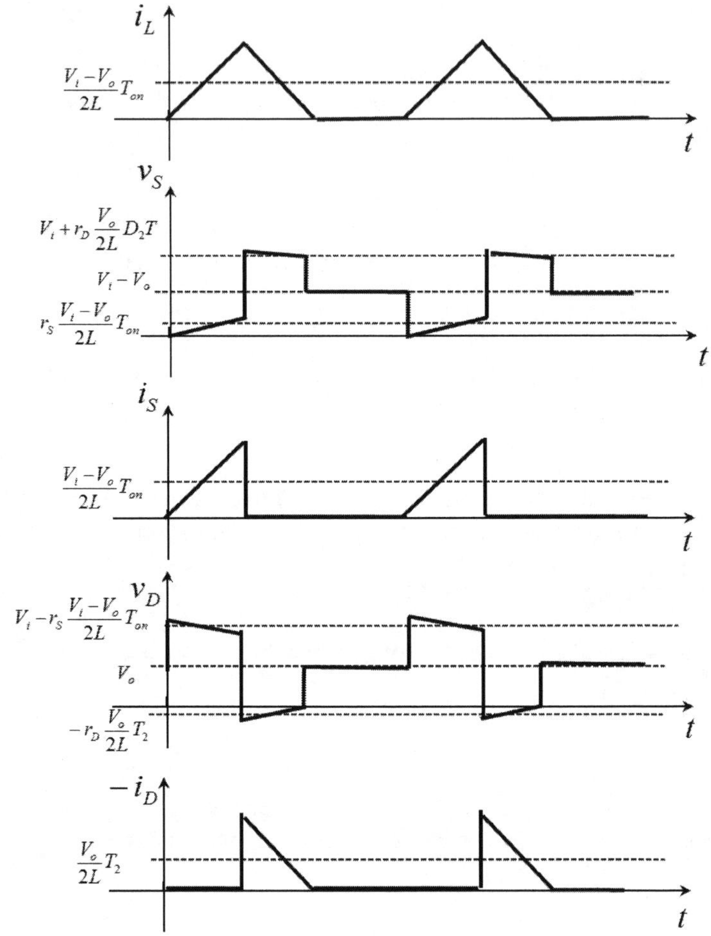

Figure 4.11 Waveforms of currents and voltages of S and D.

$$I_D = -\frac{V_o}{2L} D_2^2 T_S \tag{4.85}$$

where

$$D = \frac{T_{on}}{T_S},\ D_2 = \frac{T_2}{T_S},\ D_3 = 1 - D_1 - D_2 \tag{4.86}$$

$$I_L = I_S - I_D = \frac{V_i - V_o}{2L} D^2 T_S + \frac{V_o}{2L} D_2^2 T_S \tag{4.87}$$

Since currents in S and D at t = T_{on} are equal, we have

$$\frac{V_{\text{i}} - V_{\text{o}}}{L} DT_{\text{S}} = \frac{V_{\text{o}}}{L} D_2 T_{\text{S}} \tag{4.88}$$

The equivalent device model of the buck converter in DCM is shown in Fig. 4.12. In the figure, R_{s} and R_{D} represent the equivalent resistances of the switch and the diode.

$$R_{\text{S}} = \frac{D + D_2}{D}(D_2 + D_3)\frac{V_{\text{i}}}{I_{\text{L}}} - \frac{D + D_2}{D} D_3 \frac{V_{\text{o}}}{I_{\text{L}}} + \frac{Dr_{\text{S}} + D_2 r_{\text{D}}}{D} \tag{4.89}$$

$$R_{\text{D}} = \frac{D + D_2}{D_2} D \frac{V_{\text{i}}}{I_{\text{L}}} + \frac{D + D_2}{D_2} D_3 \frac{V_{\text{o}}}{I_{\text{L}}} - \frac{Dr_{\text{S}} + D_2 r_{\text{D}}}{D_2} \tag{4.90}$$

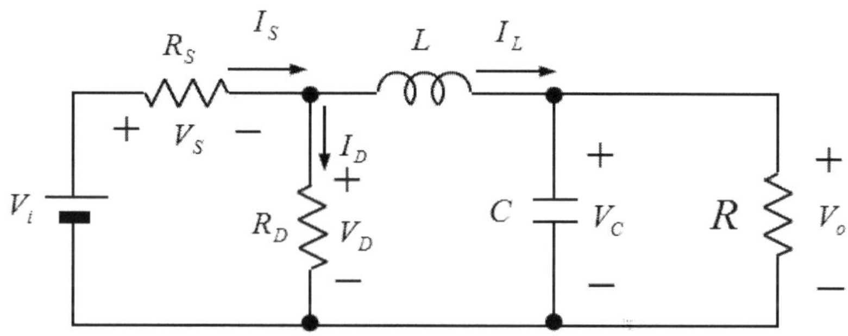

Figure 4.12 The equivalent device model for discontinuous conduction mode.

4.4.3 Steady-State Characteristics

From Eq. 4.88 we have

$$V_{\text{o}} = \frac{D}{D + D_2} V_{\text{i}} \tag{4.91}$$

In the steady state, the averaged current in the capacitor is zero, so we get

$$I_{\text{L}} = I_{\text{o}} = \frac{V_{\text{o}}}{R_{\text{L}}} \tag{4.92}$$

Substituting Eqs. 4.87 and 4.88 into Eq. 4.92, we have

$$\frac{V_{\text{o}}}{2L} DD_2 T_{\text{S}} + \frac{V_{\text{o}}}{2L} D_2^{\,2} T_{\text{S}} = \frac{V_{\text{o}}}{R_{\text{L}}} \tag{4.93}$$

Solving the above equation, we obtain

$$D_2 = \frac{F}{D}\frac{2}{1+\sqrt{1+4F/D^2}} \quad (4.94)$$

where we define

$$F = \frac{2L}{R_L T_S} = \frac{2f_S L}{R_L} \quad (4.95)$$

The condition of DCM is

$$D_2 \le 1 - D \quad (4.96)$$

Then the criterion of DCM is

$$D_2 = 1 - D \quad (4.97)$$

Equation 4.97 is rewritten as

$$F_{crit} = (1 - D)^2 \quad (4.98)$$

$$R_{L,\,crit} = \frac{2L}{(1-D)^2 T_S} \quad (4.99)$$

$$I_{o,\,crit} = \frac{D(1-D)T_S V_i}{2L} \quad (4.100)$$

4.4.4 Small-Signal AC Characteristics

From Fig. 4.12, we get the following equations:

$$\text{KVL: } V_D = L\frac{d}{dt}I_L + V_o \quad (4.101)$$

$$\text{KCL: } I_L = C\frac{d}{dt}V_o + \frac{V_o}{R_L} \quad (4.102)$$

4.4.4.1 Control to the output transfer function

For small-signal analysis, taking the Laplace transform, Eqs. 4.101, 4.102, 4.87, and 4.88 become

$$\Delta V_D(s) = sL\Delta I_{L(s)} + \Delta V_o(s) \quad (4.103)$$

$$\Delta I_L(s) = (sC + \frac{1}{R_L})\Delta V_o(s) \quad (4.104)$$

$$\Delta I_L(s) = \frac{V_i - V_o}{L}DT_S\Delta D(s) + \frac{V_o}{L}D_2T_S\Delta D_2(s) + \frac{T_S}{2L}(D_2{}^2 - D^2)\Delta V_o(s)$$

(4.105)

$$V_o\Delta D_2(s) = (V_i - V_o)\,\Delta D(s) - (D + D_2)\,\Delta V_o(s) \tag{4.106}$$

From Eqs. 4.103–4.106 we have the transfer function of the duty cycle to the output voltage as

$$\frac{\Delta V_o(s)}{\Delta D(s)} = \frac{2\dfrac{V_i - V_o}{F}(D + D_2)}{sCR_L + 1 + \dfrac{1}{F}(D + D_2)^2} \tag{4.107}$$

The above equation is rewritten as

$$\frac{\Delta V_o(s)}{\Delta D(s)} = \frac{G_{vd0}}{1 + \dfrac{s}{\omega_0}} \tag{4.108}$$

where

$$G_{vd0} = \frac{V_o}{D}\frac{2D_2}{D + 2D_2} \tag{4.109}$$

$$\omega_0 = \frac{1}{CR}\frac{D + 2D_2}{D_2} \tag{4.110}$$

4.4.4.2 Input to the output transfer function

For small-signal analysis, taking the Laplace transform, Eqs. 4.101, 4.102, and 4.87 become

$$\Delta I_L(s) = \left(sC + \frac{1}{R_L}\right)\Delta V_o(s) \tag{4.111}$$

$$V_o\Delta D_2(s) = D\Delta V_i(s) - (D + D_2)\,\Delta V_o(s) \tag{4.112}$$

$$\Delta I_L(s) = \frac{D^2 T_S}{2L}\Delta V_i(s) + \frac{V_o}{L}D_2 T_S \Delta D_2(s) + \frac{T_S}{2L}(D_2{}^2 - D^2)\Delta V_o(s) \tag{4.113}$$

From Eqs. 4.111–4.113 the transfer function of the input voltage to the output voltage is given as

$$\frac{\Delta V_o(s)}{\Delta V_i(s)} = \frac{\dfrac{D(D + 2D_2)}{F}}{sCR_L + 1 + \dfrac{1}{F}(D + D_2)^2} \tag{4.114}$$

The above equation is rewritten as

$$\frac{\Delta V_o(s)}{\Delta V_i(s)} = \frac{G_{vv0}}{1 + \dfrac{s}{\omega_0}} \tag{4.115}$$

where

$$G_{\mathrm{vv0}} = \frac{D}{D+D_2} \tag{4.116}$$

$$\omega_0 = \frac{1}{CR_\mathrm{L}}\frac{D+2D_2}{D_2} \tag{4.117}$$

4.4.4.3 Load to the output transfer function

From the equivalent circuit shown in Fig. 4.12, we get the circuit equations as follows:

$$I_\mathrm{o}R_\mathrm{L} = V_\mathrm{o} \tag{4.118}$$

$$I_\mathrm{o} = I_\mathrm{L} - C\frac{d}{dt}V_\mathrm{o} \tag{4.119}$$

For the small perturbation of the load resistance from the steady state

$$R_\mathrm{L} \rightarrow R_\mathrm{L} + \Delta R_\mathrm{L} \tag{4.120}$$

From Eqs. 4.118, 4.119, 4.87, and 4.88, considering variations from the steady state and taking the Laplace transform, we have

$$R_\mathrm{L}\Delta I_\mathrm{o}(s) + I_\mathrm{o}\Delta R_\mathrm{L}(s) = \Delta V_\mathrm{o} \tag{4.121}$$

$$\Delta I_\mathrm{L}(s) = \Delta I_\mathrm{o}(s) + sC\Delta V_\mathrm{o}(s) \tag{4.122}$$

$$\Delta I_\mathrm{L}(s) = \frac{V_\mathrm{o}}{L}D_2T_\mathrm{S}\Delta D_2(s) + \frac{T_\mathrm{S}}{2L}(D_2{}^2 - D^2)\Delta V_\mathrm{o}(s) \tag{4.123}$$

$$V_\mathrm{o}\Delta D_2(s) = -(D + D_2)\Delta V_\mathrm{o}(s) \tag{4.124}$$

From Eqs. 4.120–4.123, we have the load to the output voltage transfer function as follows:

$$\frac{\Delta V_\mathrm{o}(s)}{\Delta R_\mathrm{L}(s)} = \frac{\dfrac{V_\mathrm{o}}{R_\mathrm{L}}\dfrac{F}{F+(D+D_2)^2}}{sCR_\mathrm{L} + 1 + \dfrac{1}{F}(D+D_2)^2} \tag{4.125}$$

The above equation is rewritten as

$$\frac{\Delta V_\mathrm{o}(s)}{\Delta R_\mathrm{L}(s)} = \frac{G_{\mathrm{vr0}}}{1+\dfrac{s}{\omega_0}} \tag{4.126}$$

where

$$G_{\mathrm{vr0}} = \frac{V_{\mathrm{o}}}{R_{\mathrm{L}}} \frac{D_2}{D+2D_2} \tag{4.127}$$

$$\omega_0 = \frac{1}{CR_{\mathrm{L}}} \frac{D+2D_2}{D_2} \tag{4.128}$$

4.5 Summary of Steady-State and Dynamic Characteristics of Basic Converters

This section summarizes the characteristics of three basic converters. A similar procedure as in the previous section is performed for buck, boost, and buck–boost converters shown in Fig. 4.13. Block diagrams of the small-signal model in CCM and DCM are shown in Figs. 4.14 and 4.15, respectively. Tables 4.1 and 4.2 show static characteristics of the three basic converters in CCM and DCM, respectively. Parameters for the small-signal model in CCM and DCM are listed in Tables 4.3 and 4.4, respectively.

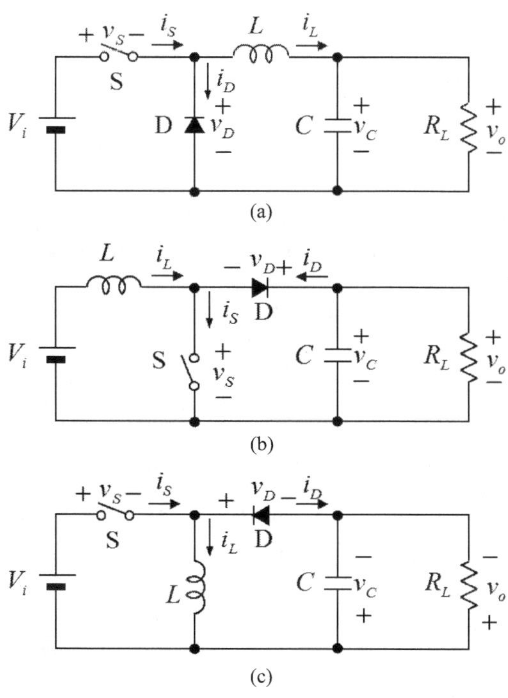

Figure 4.13 Basic converters: (a) Buck, (b) boost, and (c) buck–boost.

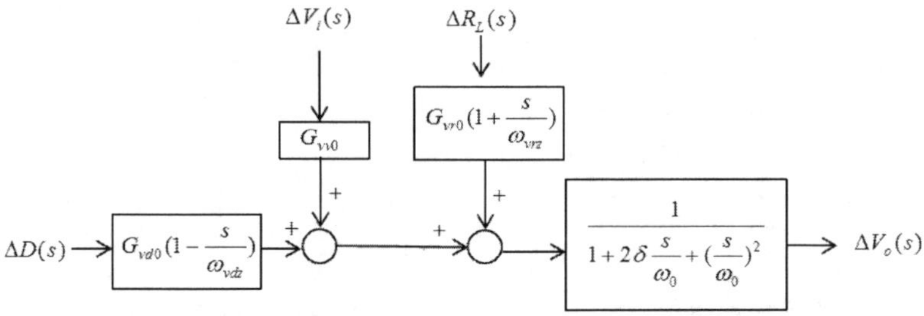

Figure 4.14 Small-signal model for CCM.

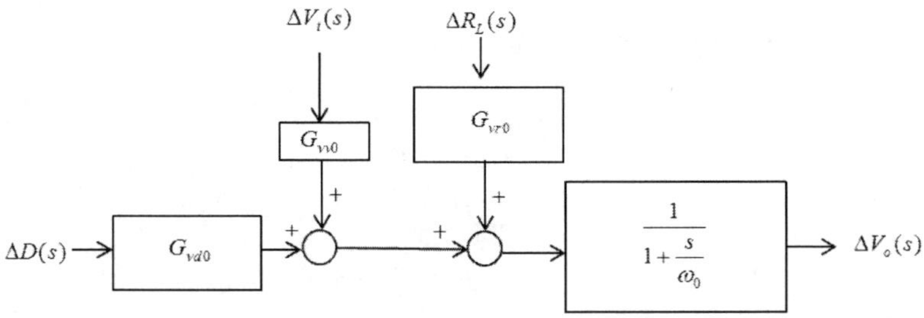

Figure 4.15 Small-signal model for DCM.

Table 4.1 Steady-state characteristics in CCM

	Buck	Boost	Buck–boost
M	$D\dfrac{1}{1+Z_o/R_L}$	$\dfrac{1}{1-D}\dfrac{1}{1+Z_o/R_L}$	$\dfrac{D}{1-D}\dfrac{1}{1+Z_o/R_L}$
Z_o	$Dr_S+(1-D)r_D$	$\dfrac{Dr_S+(1-D)r_D}{(1-D)^2}$	$\dfrac{Dr_S+(1-D)r_D}{(1-D)^2}$
$I_{L,\ ripple}$	$\dfrac{V_o+rI_L}{L}(1-D)T_S$	$\dfrac{V_i-r_SI_L}{L}DT_S$	$\dfrac{V_i-r_SI_L}{L}DT_S$
r	$Dr_S+(1-D)r_D$	$Dr_S+(1-D)r_D$	$Dr_S+(1-D)r_D$

Table 4.2 Steady-state characteristics in DCM

	Buck	Boost	Buck–boost
M	$\dfrac{D}{D+D_2}$	$\dfrac{D+D_2}{D_2}$	$\dfrac{D}{D_2}$
D_2	$\dfrac{F}{D}\dfrac{2}{1+\sqrt{1+4F/D^2}}$	$\dfrac{F}{D}\dfrac{1+\sqrt{1+4F/D^2}}{2}$	$\sqrt{F}$

$I_{L, \text{ripple}}$	$\frac{V_i-V_o}{L}DT_S$	$\frac{V_i}{L}DT_S$	$\frac{V_i}{L}DT_S$
F_{crit}	$1-D$	$D(1-D)^2$	$(1-D)^2$
F	$\frac{2L}{R_L T_S}=\frac{2L}{R_L}f_S$	$\frac{2L}{R_L T_S}=\frac{2L}{R_L}f_S$	$\frac{2L}{R_L T_S}=\frac{2L}{R_L}f_S$

Table 4.3 Parameters for a small-signal model in CCM

	Buck	**Boost**	**Buck–boost**
δ	$\frac{\frac{\sqrt{L/C}}{R_L}+\frac{Z_o}{\sqrt{L/C}}}{2\sqrt{1+Z_o/R_L}}$	$\frac{\frac{\sqrt{L/C}}{(1-D)R_L}+\frac{(1-D)Z_o}{\sqrt{L/C}}}{2\sqrt{1+Z_o/R_L}}$	$\frac{\frac{\sqrt{L/C}}{(1-D)R_L}+\frac{(1-D)Z_o}{\sqrt{L/C}}}{2\sqrt{1+Z_o/R_L}}$
ω_0	$\frac{1}{\sqrt{L/C}}\sqrt{1+Z_o/R_L}$	$\frac{1-D}{\sqrt{L/C}}\sqrt{1+Z_o/R_L}$	$\frac{1-D}{\sqrt{L/C}}\sqrt{1+Z_o/R_L}$
G_{vv0}	$D\frac{1}{1+Z_o/R_L}$	$\frac{1}{1-D}\frac{1}{1+Z_o/R_L}$	$\frac{D}{1-D}\frac{1}{1+Z_o/R_L}$
G_{vd0}	$\frac{V_o}{D}\frac{1+r_D/R_L}{1+Z_o/R_L}$	$\frac{V_o}{1-D}\frac{1-\frac{r_S}{(1-D)^2R_L}}{1+Z_o/R_L}$	$\frac{V_o}{D(1-D)}\frac{1-\frac{Dr-(1-D)r_D}{(1-D)^2R_L}}{1+Z_o/R_L}$
ω_{vdz}	∞	$\frac{(1-D)^2R_L}{L}\left(1-\frac{r_S}{(1-D)^2R_L}\right)$	$\frac{(1-D)^2R_L}{DL}\left(1-\frac{Dr-(1-D)r_D}{(1-D)^2R_L}\right)$
G_{vr0}	$\frac{Z_oV_o}{R_L{}^2}\frac{1}{1+Z_o/R_L}$	$\frac{Z_oV_o}{R_L{}^2}\frac{1}{1+Z_o/R_L}$	$\frac{Z_oV_o}{R_L{}^2}\frac{1}{1+Z_o/R_L}$
ω_{vrz}	$\frac{Z_o}{L}$	$\frac{(1-D)^2Z_o}{L}$	$\frac{(1-D)^2Z_o}{L}$

Table 4.4 Parameters for a small-signal model in DCM

	Buck	**Boost**	**Buck–boost**
ω_0	$\frac{1}{R_LC}\frac{D+2D_2}{D_2}$	$\frac{1}{R_LC}\frac{2D+D_2}{D}$	$\frac{2}{R_LC}$
G_{vv0}	$\frac{D}{D+D_2}$	$\frac{D+D_2}{D_2}$	$\frac{D}{D_2}$
G_{vd0}	$\frac{V_o}{D}\frac{2D_2}{D+2D_2}$	$\frac{V_o}{D}\frac{2D}{2D+D_2}$	$\frac{V_o}{D}$
G_{vr0}	$\frac{V_o}{R_L}\frac{D_2}{D+2D_2}$	$\frac{V_o}{R_L}\frac{D}{2D+D_2}$	$\frac{V_o}{R_L}\frac{1}{2}$

Chapter 5

Control Schemes of Switching Converters

Terukazu Sato
Department of Electrical and Electronics Engineering, Oita University,
700 Dannoharu, Oita 870-1192, Japan
tsato@oita-u.ac.jp

5.1 Introduction

An ideal switching power supply maintains a constant output voltage, even though the load current or the input voltage changes for some events. However, real switching power supplies have transient variations and steady-state errors for them. It is very important for power supplies to regulate the output voltage for stable and safe operation of electronic circuits. To regulate the output of a switching converter, a negative feedback control is generally used. This chapter presents basic control techniques for DC–DC power converters, including hysteretic pulse width modulation (PWM) control and current-mode control.

5.2 Voltage-Mode PWM Control

Figure 5.1 shows a circuit diagram of a buck converter with a fixed switching frequency voltage-mode PWM controller. The feedback

Handbook of Power Management Circuits
Edited by Haruo Kobayashi and Takashi Nabeshima

ISBN 978-981-4613-15-6 (Hardcover), 978-981-4613-16-3 (eBook)
www.panstanford.com

circuit consists of an error amplifier, a PWM generator, a sawtooth wave generator and a reference voltage. Consider the small-signal component of v_{fb} and v_{ref} around the steady-state operating point as

$$V_{fb} \rightarrow V_{fb} + \Delta V_{fb} \tag{5.1}$$

$$V_{ref} \rightarrow V_{ref} + \Delta V_{ref} \tag{5.2}$$

where V_{fb} and V_{ref} represent steady-state terms and ΔV_{fb} and ΔV_{ref} are small-signal terms. By those variations, suppose the following small changes occur:

$$V_a \rightarrow V_a + \Delta V_a \tag{5.3}$$

$$V_{PWM} \rightarrow V_{PWM} + \Delta V_{PWM} \tag{5.4}$$

$$D \rightarrow D + \Delta D \tag{5.5}$$

where V_a, V_{PWM}, and D represent steady-state terms of v_a, v_{PWM}, and the duty cycle, respectively. Similarly, ΔV_a, ΔV_{PWM}, and ΔD represent small-signal terms of v_a, v_{PWM}, and the duty cycle, respectively.

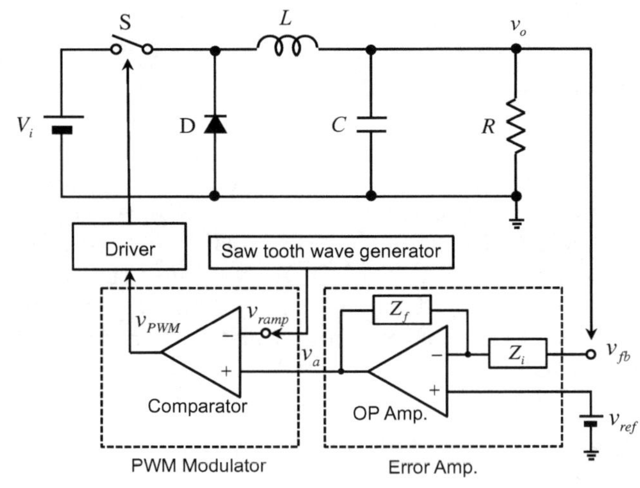

Figure 5.1 A buck converter with a fixed switching frequency voltage-mode PWM controller.

5.2.1 Transfer Function of an Error Amplifier

If OP amplifier in Fig. 5.1 has ideal characteristics, the small-signal component of the output voltage of the error amplifier is given as

$$\Delta V_{\mathrm{a}}(s) = \left(1 + \frac{Z_{\mathrm{f}}(s)}{Z_{\mathrm{i}}(s)}\right)\Delta V_{\mathrm{ref}}(s) - \frac{Z_{\mathrm{f}}(s)}{Z_{\mathrm{i}}(s)}\Delta V_{\mathrm{fb}}(s) \tag{5.6}$$

5.2.2 Transfer Function of a PWM Generator

The output voltage of the error amplifier is the control signal and is fed to the noninverting input of the comparator of the PWM generator. A sawtooth signal is fed to the inverting input of the comparator, which compares the control signal with the sawtooth signal and generates a PWM signal at the comparator output. The operating waveforms of the comparator are shown in Fig. 5.2.

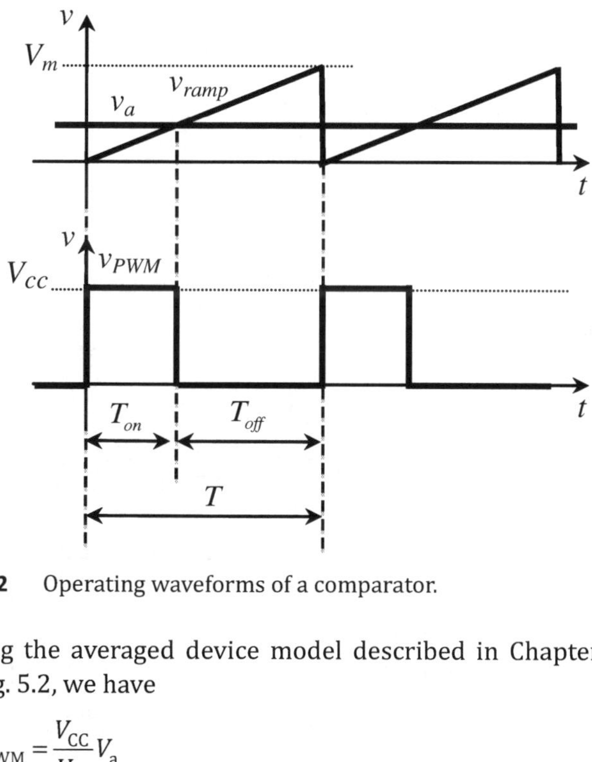

Figure 5.2 Operating waveforms of a comparator.

Using the averaged device model described in Chapter 5, and from Fig. 5.2, we have

$$V_{\mathrm{PWM}} = \frac{V_{\mathrm{CC}}}{V_{\mathrm{m}}} V_{\mathrm{a}} \tag{5.7}$$

The relation between the duty cycle and the PWM output is

$$V_{\mathrm{PWM}} = DV_{\mathrm{CC}} \tag{5.8}$$

Considering only variations from the steady state, neglecting the product of any AC terms, and taking the Laplace transform, we have

$$\Delta V_{\mathrm{PWM}}(s) = \frac{V_{\mathrm{CC}}}{V_{\mathrm{m}}}\Delta V_{\mathrm{a}}(s) \tag{5.9}$$

$$\Delta V_{\mathrm{PWM}}(s) = V_{\mathrm{CC}}\Delta D(s) \tag{5.10}$$

Finally the following transfer function is obtained:

$$\Delta D(s) = \frac{1}{V_{\mathrm{m}}}\left(1 + \frac{Z_{\mathrm{f}}(s)}{Z_{\mathrm{i}}(s)}\right)\Delta V_{\mathrm{ref}}(s) - \frac{1}{V_{\mathrm{m}}}\frac{Z_{\mathrm{f}}(s)}{Z_{\mathrm{i}}(s)}\Delta V_{\mathrm{fb}}(s) \tag{5.11}$$

5.3 Self-Oscillating Hysteretic PWM Control

Hysteretic PWM control is often referred to as ripple-based control or bang-bang control and has excellent dynamic performance for a load current transient. A circuit diagram of a buck converter with a hysteretic PWM controller is shown in Fig. 5.3. This type of hysteretic controller operates in self-oscillating mode. A method of synchronizing with an external clock signal is described in the next section. The feedback circuit consists of a hysteretic comparator, an RC network, and a reference voltage. The operating waveforms of the comparator are shown in Fig. 5.4. The voltage of the negative

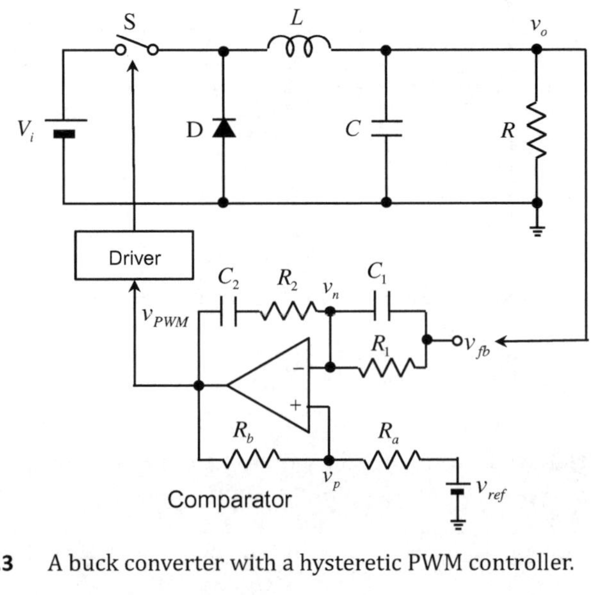

Figure 5.3 A buck converter with a hysteretic PWM controller.

input of the comparator, v_n, rises when the output of the comparator is at a high level, while it falls when the output is at a low level. The positive input of the comparator, v_p, has two values as

$$v_p = V_H = \frac{R_a}{R_a + R_b} V_{OH} + \frac{R_b}{R_a + R_b} V_{ref} \quad \text{for} \quad v_{PWN} = V_{OH} \tag{5.12}$$

$$v_p = V_L = \frac{R_a}{R_a + R_b} V_{OL} + \frac{R_b}{R_a + R_b} V_{ref} \quad \text{for} \quad v_{PWM} = V_{OL} \tag{5.13}$$

where V_{OH} and V_{OL} represent the high-level and low-level output voltage of the comparator.

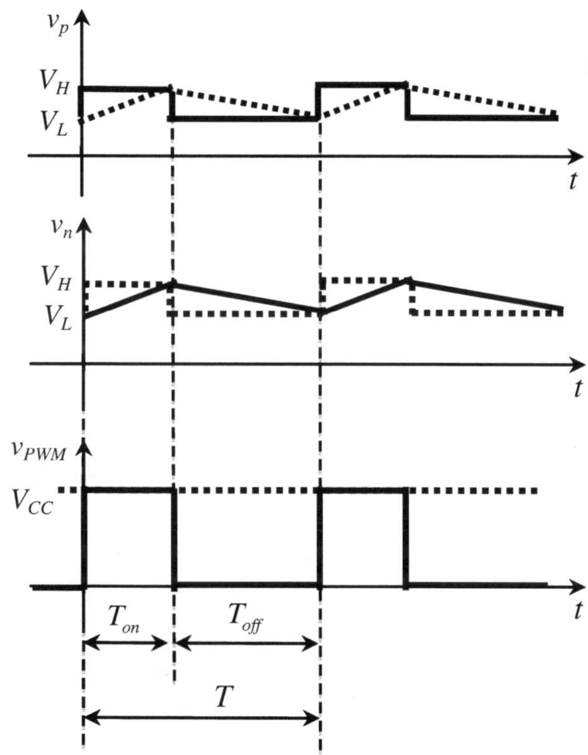

Figure 5.4 Operating waveforms of a comparator.

5.3.1 Transfer Function of a Hysteretic PWM Generator

As seen from Fig. 5.4, the averaged value of a comparator's negative input is equal to that of the positive input. That is

$$V_{\mathrm{p}} - V_{\mathrm{n}} = \frac{R_{\mathrm{a}}}{R_{\mathrm{a}} + R_{\mathrm{b}}} V_{\mathrm{PWM}} - \frac{R_{\mathrm{a}}}{R_{\mathrm{a}} + R_{\mathrm{b}}} \frac{V_{\mathrm{OH}} - V_{\mathrm{OL}}}{2} \tag{5.14}$$

Considering only small-signal terms, and taking the Laplace transform, we have

$$\Delta V_{\mathrm{p}}(s) - \Delta V_{\mathrm{n}}(s) = \frac{R_{\mathrm{a}}}{R_{\mathrm{a}} + R_{\mathrm{b}}} \Delta V_{\mathrm{PWM}}(s) \tag{5.15}$$

Likewise, we have

$$\Delta V_{\mathrm{n}}(s) = \frac{Z_2(s)}{Z_1(s) + Z_2(s)} \Delta V_{\mathrm{fb}}(s) + \frac{Z_1(s)}{Z_1(s) + Z_2(s)} \Delta V_{\mathrm{PWM}}(s) \tag{5.16}$$

$$\Delta V_{\mathrm{p}}(s) = \frac{R_{\mathrm{b}}}{R_{\mathrm{a}} + R_{\mathrm{b}}} \Delta V_{\mathrm{ref}}(s) + \frac{R_{\mathrm{a}}}{R_{\mathrm{a}} + R_{\mathrm{b}}} \Delta V_{\mathrm{PWM}}(s) \tag{5.17}$$

where

$$Z_1(s) = \frac{R_1}{1 + sR_1C_1} \tag{5.18}$$

$$Z_2(s) = R_2 + \frac{1}{sC_2} \tag{5.19}$$

From (5.15) to (5.17), the following transfer function is obtained.

$$\Delta V_{\mathrm{PWM}}(s) = \frac{R_{\mathrm{b}}}{R_{\mathrm{a}} + R_{\mathrm{b}}} \frac{Z_1(s) + Z_2(s)}{Z_1(s)} \Delta V_{\mathrm{ref}}(s) - \frac{Z_2(s)}{Z_1(s)} \Delta V_{\mathrm{fb}}(s) \tag{5.20}$$

For $R_{\mathrm{a}} << R_{\mathrm{b}}$, we have

$$\Delta V_{\mathrm{PWM}}(s) = \left(1 + \frac{Z_2(s)}{Z_1(s)}\right) \Delta V_{\mathrm{ref}}(s) - \frac{Z_2(s)}{Z_1(s)} \Delta V_{\mathrm{fb}}(s) \tag{5.21}$$

Then we get

$$\Delta D(s) = \frac{1}{V_{\mathrm{CC}}} \left(1 + \frac{Z_2(s)}{Z_1(s)}\right) \Delta V_{\mathrm{ref}}(s) - \frac{1}{V_{\mathrm{CC}}} \frac{Z_2(s)}{Z_1(s)} \Delta V_{\mathrm{fb}}(s) \tag{5.22}$$

5.3.2 Constant-Frequency Operation of a Hysteretic PWM Generator

It is very easy to synchronize a hysteretic PWM generator with an external clock signal. This can be achieved by inputting the clock signal with a frequency higher than the free running frequency to the positive input of the comparator, as shown in Fig. 5.5. The operating waveforms are shown in Fig. 5.6. A clock signal with a 50% duty cycle is differentiated by R_3 and C_3 and fed to the positive input of the comparator. When the duty cycle is less than 50%, the comparator output is set to a high level by the clock signal, the voltage of the negative input of the comparator rises, and at the time it reaches the higher threshold level, the comparator is inverted to a low level. On the other hand, when the duty cycle is larger than 50%, the comparator output is set to a low level by the clock signal, the voltage of the negative input of the comparator falls, and at the time it reaches the lower threshold level, the comparator is inverted to a low level.

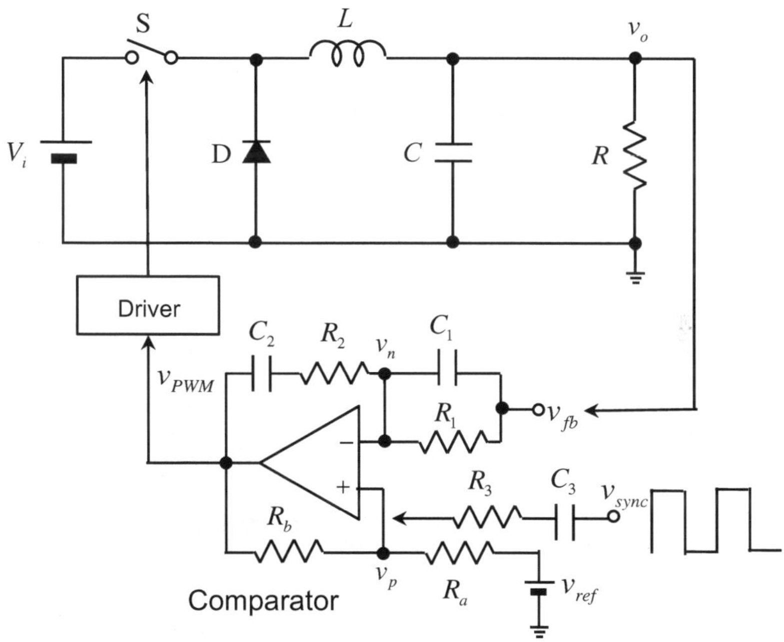

Figure 5.5 Constant-frequency hysteretic PWM generator.

5.4 Current-Mode Control

In Fig. 5.1, instead of the sawtooth wave, the voltage converted from the inductor current is input to the negative input of the comparator, and the hysteresis characteristic is produced by the resisters. Shown

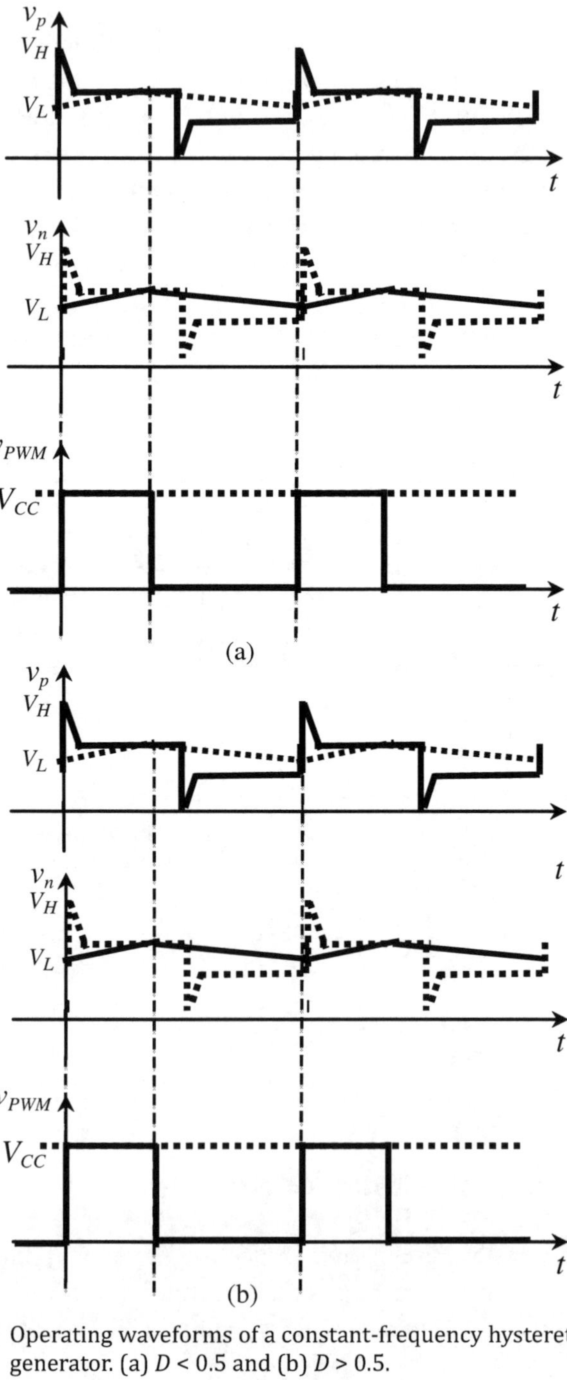

Figure 5.6 Operating waveforms of a constant-frequency hysteretic PWM generator. (a) $D < 0.5$ and (b) $D > 0.5$.

in Fig. 5.7 is a current-mode PWM controller. Since current-mode control is a first-order delay system, it has good system stability. Furthermore it can easily achieve pulse-by-pulse overcurrent protection.

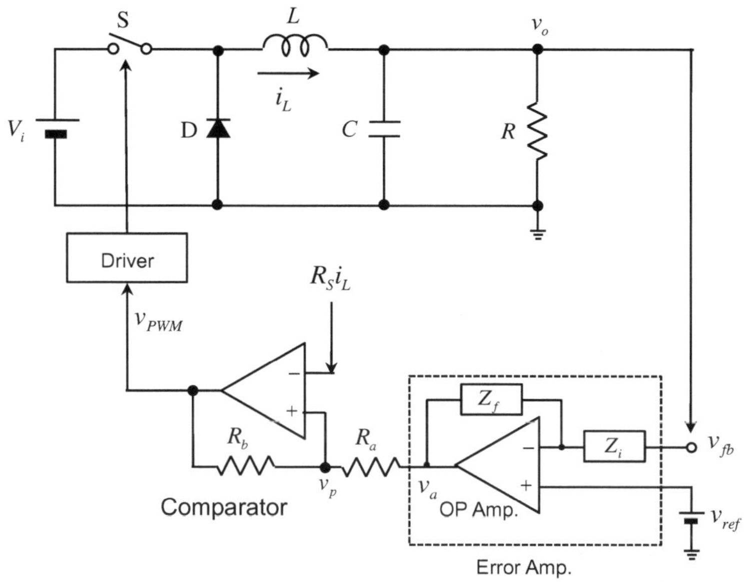

Figure 5.7 A buck converter with a current-mode PWM controller.

5.4.1 Transfer Function of Current-Mode Control

Using the same procedure as in Section 5.3.1, we have

$$\Delta V_{\mathrm{n}}(s) = R_{\mathrm{s}} \Delta I_{\mathrm{L}}(s) \tag{5.23}$$

$$\Delta V_{\mathrm{p}}(s) = \frac{R_{\mathrm{b}}}{R_{\mathrm{a}} + R_{\mathrm{b}}} \Delta V_{\mathrm{a}}(s) + \frac{R_{\mathrm{a}}}{R_{\mathrm{a}} + R_{\mathrm{b}}} \Delta V_{\mathrm{PWM}}(s) \tag{5.24}$$

$$\Delta V_{\mathrm{n}}(s) = \Delta V_{\mathrm{p}}(s) \tag{5.25}$$

$$\Delta D(s) = \frac{1}{V_{\mathrm{CC}}} \Delta V_{\mathrm{PWM}}(s) \tag{5.26}$$

From Eqs. 5.23–5.26, the following transfer function is obtained:

$$\Delta D(s) = \frac{1}{V_{\mathrm{CC}}} \frac{R_{\mathrm{S}}(R_{\mathrm{a}} + R_{\mathrm{b}})}{R_{\mathrm{a}}} \Delta I_{\mathrm{L}}(s) - \frac{1}{V_{\mathrm{CC}}} \frac{R_{\mathrm{b}}}{R_{\mathrm{a}}} \Delta V_{\mathrm{a}}(s) \tag{5.27}$$

Assuming the transfer function of $\Delta D(s)$ to $\Delta I_L(s)$ is expressed as

$$\Delta I_L(s) = G_{id}(s)\Delta D(s) \tag{5.28}$$

Eq. 5.27 becomes

$$\Delta D(s) = -\frac{1}{V_{CC} - \frac{R_S(R_a + R_b)}{R_a}G_{id}(s)}\frac{R_b}{R_a}\Delta V_a(s) \tag{5.29}$$

For $R_a << R_b$ and $V_{CC} << \frac{R_S(R_a + R_b)}{R_a}G_{id}(s)$, we have

$$\Delta D(s) = \frac{1}{R_S G_{id}(s)}\Delta V_a(s) \tag{5.30}$$

Substituting Eq. 5.30 into 5.28 gives

$$\Delta I_L(s) = \frac{1}{R_S}\Delta V_a(s) \tag{5.31}$$

Equation 5.31 is the well-known result for current-mode control.

5.4.2 Constant-Frequency Operation of a Current-Mode PWM Generator

It is very easy to synchronize a current-mode PWM generator with an external clock signal. The same method described in Section 5.3.2 can be applied for current-mode control. The method introduced here has no instable phenomenon known as subharmonic oscillation.

Chapter 6

Passive Components

Yuya Tamai[a] **and Yoshiyuki Ishihara**[b]
[a]*Solution R & D Department, R & D Headquarters, Nippon Chemi-Con Corporation, 6-4, Osaki 5-Chome, Shinagawa-ku, Tokyo 141-8605, Japan*
[b]*Electrical Engineering Department, Faculty of Science and Engineering, Doshisha University, Kyotanabe 610-0321, Japan*
tamaii@nippon.chemi-con.co.jp, yishihar@mail.doshisha.ac.jp

6.1 Inductors and Transformers

6.1.1 Inductors

6.1.1.1 Definition of an inductor

An inductor is a circuit element that is widely used in electronic circuits. Current flow through a conductive wire produces magnetic flux around the wire, following the Maxwell corkscrew (right-handed screw) rule. The main characteristic of an inductor is the magnetic flux that it produces when a current is passed through it. Although wire conductors also possess inductance components, the value is negligibly small.

Handbook of Power Management Circuits
Edited by Haruo Kobayashi and Takashi Nabeshima

ISBN 978-981-4613-15-6 (Hardcover), 978-981-4613-16-3 (eBook)
www.panstanford.com

Inductance L (H) is the proportionality constant between the current flow I (A) through the wire and the linkage flux Φ, and the relation is expressed as follows:

$$\Phi = LI \quad \text{(Wb)} \tag{6.1}$$

Here L is the value of inductance.

When an inductor is used as a circuit component, it is usually constructed by winding a wire into a coil. This case, inductance is expressed as follows:

$$L = n/R_{\mathrm{m}} \quad \text{(H)} \tag{6.2}$$

Here, n is the number of turns and R_{m} is the magnetic reluctance (A/Wb).

Magnetic reluctance, R_{m}, is given by

$$R_{\mathrm{m}} = l/(\mu S) \quad \text{(A/Wb)} \tag{6.3}$$

Here, l is the average magnetic path length (m), μ is the permeability (H/m), and S is the cross-sectional area of the magnetic path (m^2).

Using Eqs. 6.2 and 6.3, the value of magnetic reluctance is determined from the required inductance; the reluctance determines the required permeability, which in turn determines the required characteristics and shape of the core material.

In an ideal inductor, the phase of the current flowing through the inductor lags 90° behind the voltage across its terminals.

The energy stored in the inductor due to a current flow of I (A) is given by

$$W = \frac{1}{2}LI^2 \tag{6.4}$$

6.1.1.2 Construction of inductors

Inductors are constructed by winding a wire on an iron core. The material and form of the iron core are determined by the required inductor characteristics and inductance. Usually small inductors have ring-shaped cores; medium-sized inductors use two C-shaped cores, E-I shaped cores, or E-E shaped cores. Multiple block-type cores are used to construct large inductors. Here we will restrict this discussion to inductors used in electronic circuits and omit discussion of the various types of laminated-core inductors used in power systems.

For cores other than ring-shaped cores, when there are multiple cores in the magnetic path, there will be a gap where the cores join. Sometimes a variable gap is used to adjust the inductance value. Since the opposite sides of the gap may be magnetically attracted, this may cause audible noise.

For a ring-core inductor whose dimensions are as shown in Fig 6.1, the inductance can be determined as follows:

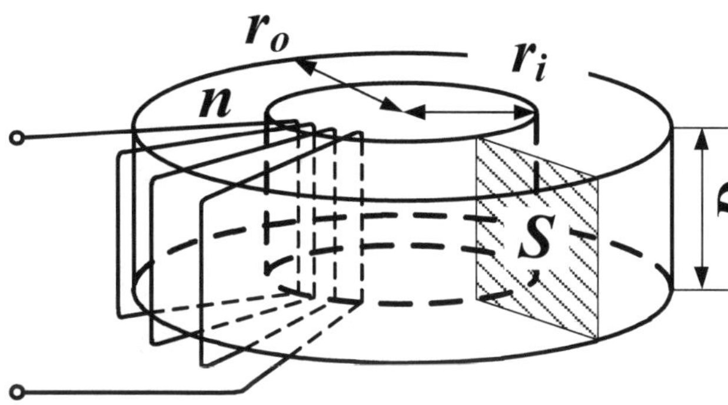

Figure 6.1 Ring-shaped core.

First, it should be determined that the proposed core material can provide suitable magnetic flux density at the operating frequency. Also, current density and skin effect considerations will determine the cross-sectional area of the wire used in the windings.

Here, the average magnetic path is determined as follows:

$$l = \pi(r_0 + r_i) \quad \text{(m)} \tag{6.5}$$

Here, r_i is the inner diameter and r_o is the outer diameter.

When current I (A) flows through the coil, the magnetic flux density B (T) can be determined as follows:

$$B = \mu H = \mu nI/l \quad \text{(T)} \tag{6.6}$$

Here μ is the core permeability, H is the magnetic field intensity, and n is the number of turns on the coil.

The magnetic flux through the core is given by

$$\phi = BS \quad \text{(Wb)} \tag{6.7}$$

As the number of turns of the coil is given by n, the magnetic flux Φ that interlinks the coil is as follows:

$$\Phi = n\phi \quad \text{(Wb)} \tag{6.8}$$

From Eq. 6.1, the value of inductance is as follows:

$$L = \Phi/I = n\phi/I \quad \text{(H)} \tag{6.9}$$

In an air-core inductor, the magnetic flux is proportional to the current, and—as a result—the inductance value remains constant. However, in iron-core inductors, the magnetic flux density is determined by the amount of magnetic flux through the core, which is determined by the magnetic permeability—but, as shown in Fig. 6.4, the relationship between magnetic flux density and magnetic reluctance is nonlinear: as a result the inductance value varies with the current flowing through the core.

6.1.2 Transformers

6.1.2.1 Principles of transformers

The ideal transformer transforms voltage and current without any energy loss between the primary and secondary windings and/or electrically isolates primary and secondary circuits. The magnetic flux links the primary and secondary windings, and the magnetic material passing through the primary and secondary windings is used to carry this flux. Real transformers have "copper loss" due to current flow through the windings and "core loss" due to eddy current flow in the iron core. Therefore the output power is less than the input power. An equivalent circuit to represent this is shown in Fig. 6.2.

x_1 and x_2 are leakage reactances (Ω), r_1 and r_2 are the winding resistances of primary and secondary windings, respectively (Ω), and g_0 and b_0 are the exciting conductance and exciting susceptance, respectively, and are usually represented by ℧. Magnetic conductance corresponds to core loss, while exciting susceptance corresponds to the inductor that generates the magnetic flux.

The number of turns on the primary and secondary windings is usually represented by the turns ratio of the transformer.

$$a = n_1/n_2 \tag{6.10}$$

Using the turns ratio, the theoretical relationship between primary and secondary voltages can be expressed as shown in the following equation:

$$v_1 = av_2 \tag{6.11}$$

In practical transformers, due to magnetic flux leakage, the secondary voltage is slightly smaller than this.

The principles of transformers—the transfer of energy from the primary winding to the secondary winding—are described below, considering only ideal theoretical conditions.

1. A power source is connected to the primary side.
2. Magnetization current flows through the primary winding.
3. Magnetic flux interlinks the secondary winding.
4. Voltage is induced into the secondary winding.
5. Load current, whose value is determined by the size of the load, flows through the secondary winding.
6. This load current flowing through the winding produces magnetic flux.
7. The primary load current flows through the primary winding to cancel the magnetic flux produced by the secondary winding. This magnetic flux created by the excitation current maintains the magnetic flux in the iron core.

As mentioned in point 7, in theory the flux in the core is constant regardless of the size of the load. Thus, in theory, the full-load and no-load iron core loss are the same.

In a practical transformer, magnetic flux leaks outside the iron core, rather than just flowing through it. There is magnetic flux leakage due to both primary and secondary winding currents, and in the equivalent circuit it is expressed as a leakage reactance (L_m).

Furthermore, windings are usually made of copper wire, which has finite resistance. These primary and secondary winding resistances are shown as circuit elements in the equivalent circuit of Fig. 6.2.

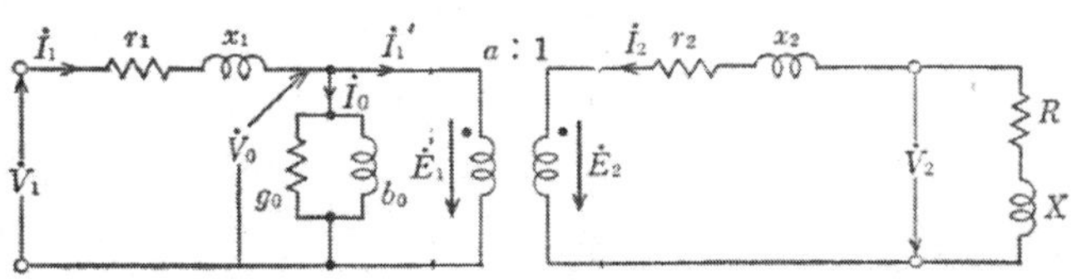

Figure 6.2 Equivalent circuit of a transformer. V_1: input voltage, E_1: primary induced electromotive force, E_2: secondary induced electromotive force, V_2: output voltage, I_1: input current, I_1': primary load current, I_0: exciting current, I_2: secondary current, r_1: primary winding resistance, r_2: secondary winding resistance, x_1: leakage reactance of primary winding, x_2: leakage reactance of secondary winding, g_0: exciting conductance, b_0: exciting susceptance, and $R + jX$: load.

The losses caused by current flow through the winding are copper loss, and the losses caused by alternating magnetic flux passing through the core are iron loss (core loss). Core loss is also referred to as no-load loss, while copper loss is referred to as load loss. In large transformers, stray losses in the transformer, such as eddy current loss generated by leakage flux in the transformer case, may become too large to be ignored. It is also very important to consider the effects of magnetic flux leakage from large transformers on any electronic circuits that are close to the transformers. Transformer efficiency is determined by the size of these losses.

Iron loss depends on the material used, the magnetic flux density in the core, the magnetic flux waveform, and the frequency. Copper loss is determined by the size of the current flowing through the resistance of the winding. These items must be considered in designing a transformer.

6.1.2.2 Structure of transformers

Transformers usually consist of a primary winding, a secondary winding, and a core. The cores of power transformers usually consist of a stack of precut electromagnetic steel sheet laminations or steel strips fabricated in a toroidal shape. Some transformers use amorphous magnetic-core materials.

Transformer cores in electronic circuits operating at very high frequencies are usually made of ferrite because it has very low core loss over a very wide frequency range. Since these cores are usually produced by compression molding or sintering, this limits their size.

Usually, transformers used in electronic circuits consist of a single primary and a single secondary winding to produce single-phase power output. However, in some special cases—to get two or more outputs or for various other reasons—multiple secondary windings are used. To reduce leakage flux, the primary and secondary windings are usually overlapped. However, to actively use the leakage inductance for circuit elements, sometimes the windings are separated to greatly increase leakage magnetic flux.

6.1.2.3 Basics of transformer design

When designing a transformer, the operating frequency largely determines the magnetic-core material. The core material's saturation magnetic flux density and core loss characteristic at the operating frequency determine the required core cross-sectional area.

When designing transformers for electronic circuits, the required transformer inductance is the core consideration, and this determines the number of turns in the windings. Voltage—determined by the circuit design—is another important consideration, as it is important to stay within the maximum permissible voltage per turn. The current is determined by the power and voltage. The cross-sectional area of the wire is determined by the current it must carry. The cross-sectional area of the wire and the number of turns of the winding determine the space needed for the winding. Thus, the core size of the transformer is determined.

It is also a good idea to consider the effect of using given core and winding specifications but changing the operating frequency.

The relationship between the maximum flux density in the iron core cross section and the sinusoidal voltage applied to the winding of the transformer is as given below.

$$V = 4.44 fnsB_{\mathrm{m}} \quad \text{(V)} \tag{6.12}$$

Here, f is the frequency (Hz), n is the number of turns, s is the cross-sectional area (m^2), B_{m} is the maximum flux density in the core (T).

According to this equation, for a given magnetic flux density and number of turns, increasing the frequency results in an increase in voltage. When this is adjusted for a given load current flow in the secondary side, the equation becomes as follows:

$$P = VI\cos\theta \tag{6.13}$$

From this equation, for a given load power factor, increasing the voltage increases the power that the transformer can handle. From Eq. 6.12, the power that can be handled by the transformer is proportional to the frequency. However, for practical high-frequency power transformer design, it is important to consider core loss. For this reason, it is necessary to reduce the magnetic flux density at higher frequencies or (for a given flux) to use a larger core cross-sectional area at high frequencies.

For a given core and magnetic flux, the necessary magnetization current is independent of frequency. For materials with higher eddy current, increasing the frequency will cause a slight increase in leakage current. As magnetization current is small compared to load current, the cross-sectional area of the copper wire is determined by the size of the load current. However, the magnetization current of transformers with a gap in their magnetic path is high. Thus, this should be remembered when calculating the required wire cross-sectional area.

6.1.3 Materials Used in Inductors and Transformers

The characteristics of electrical machines are highly dependent on the shape of their magnetic components (inductors and transformers), as well as the composition and characteristics of the materials used. The magnetic materials, conductors, and insulation used in inductors and transformers have the most effect on system performance. Ideally, materials should be chosen such that magnetic materials have suitable permeability, conductors have low resistance, and insulators provide high resistance between conductive parts such as wires. The most important characteristics of conductors and magnetic materials used in inductors and transformers are described in more detail below.

6.1.3.1 Magnetic materials

Magnetic materials are classified as "hard magnetic materials" (permanent magnets) or "soft magnetic materials."

Soft magnetic materials are used as cores of magnetic devices, such as inductors and transformers in a variety of electrical equipment. Soft magnetic materials are selected according to the magnetic properties required in the working environment. The characteristics of many widely used soft magnetic materials are shown in Table 6.1.

Table 6.1 Properties of soft magnetic materials

	Saturation flux density (T)	Resistivity (Ω·m)	Density (kg/dm³)	Core loss (W/kg)	Relative permeability
Electrical steel sheet (JIS 50A300)	2.03	59×10^{-8}	7.65	3.00 (50 Hz, 1.5 T)	13,000
Ferrite	0.54	8.0	4.85	130 (100 kHz, 0.2 T)	
	0.39	0.1	4.9		
Permalloy	0.78	50×10^{-6}	8.62		160,000
	1.03	55×10^{-8}	8.72		100,000
Amorphous (Fe based)	1.56	1.37×10^{-6}	7.18	0.05 (50 Hz, 1.0 T)	950,000
Powder magnetic core	1.36		7.36	0.89 (100 kHz, 0.1 T)	77

Soft magnetic materials used as cores are primarily chosen to provide minimal core losses and optimal magnetic characteristics. Core losses cause the temperature of the core to rise and are essentially determined by the core material, the magnetic flux density, the frequency, and the waveform. Core losses are generally measured as the loss per unit weight (W/kg) for a given magnetic flux density (T) at a given frequency. The magnetic characteristics are represented by the magnetic flux density (T) when a magnetic field strength (H/m) is applied. The core losses of typical materials are shown in Fig. 6.3, while the magnetic characteristics are shown in Fig. 6.4. There are big differences in the electrical and magnetic characteristics of different materials. In addition, as shown in Fig. 6.3b, there are big differences in the loss versus frequency characteristics of different materials at different operating frequencies. This variation in frequency response is determined by the ratio of the eddy current loss and hysteresis loss components of core loss. The reason for this is that materials with low eddy current loss, such as ferrite, have a smaller increase in core loss with frequency than materials with higher eddy current loss—such as electrical steel sheets, which are electrically conductive.

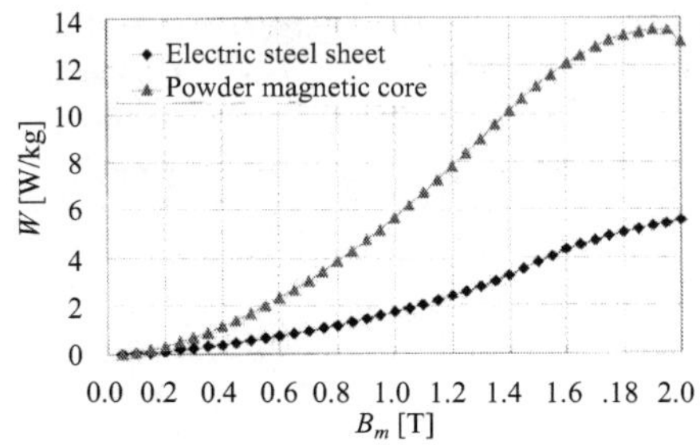

(a)

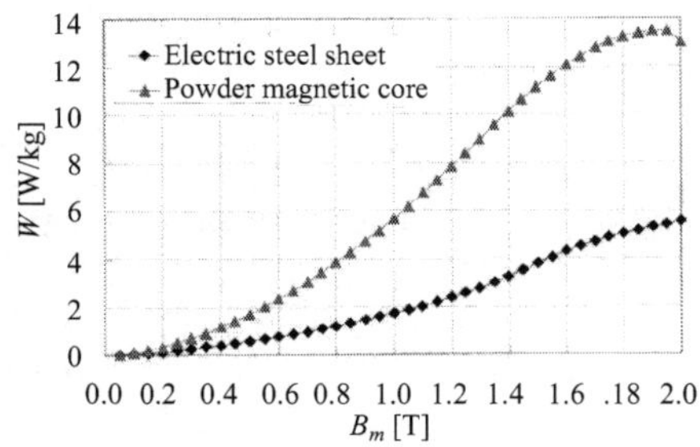

(b)

Figure 6.3 Core loss characteristics of various magnetic materials. (a) Magnetic flux density versus core loss characteristics. (b) Frequency versus core loss characteristics. Maximum flux density = 0.1T.

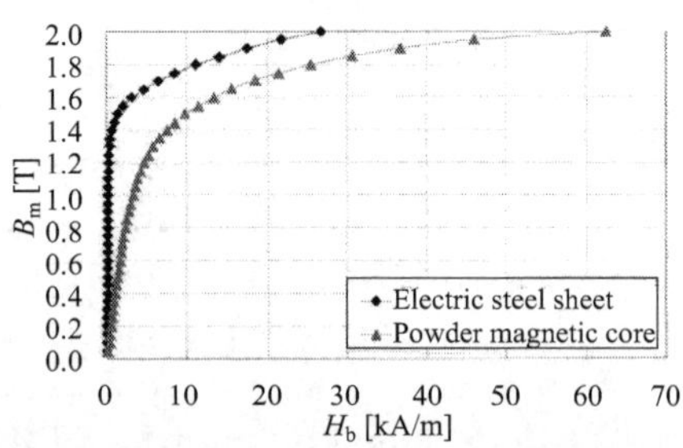

Figure 6.4 Magnetization characteristics of various magnetic materials (electrical steel sheet and powder magnetic core). Exciting frequency = 50 Hz.

Figure 6.5 shows frequency characteristics of hysteresis loops at a maximum flux density of 1.0T for an electrical steel sheet and a powder magnetic core. As we can see from Fig. 6.5a, all the hysteresis loops between 50 Hz and 1 kHz for the powder magnetic core overlap and together they appear to be a singular curve. While keeping that under consideration, if we look at Fig. 6.5b we can see that for an increase in the frequency of the electrical steel sheet, the shape of the hysteresis loops has expanded hugely. This is because the surface area of a hysteresis loop is equivalent to one cycle of iron loss. Iron loss is usually expressed by the following equation:

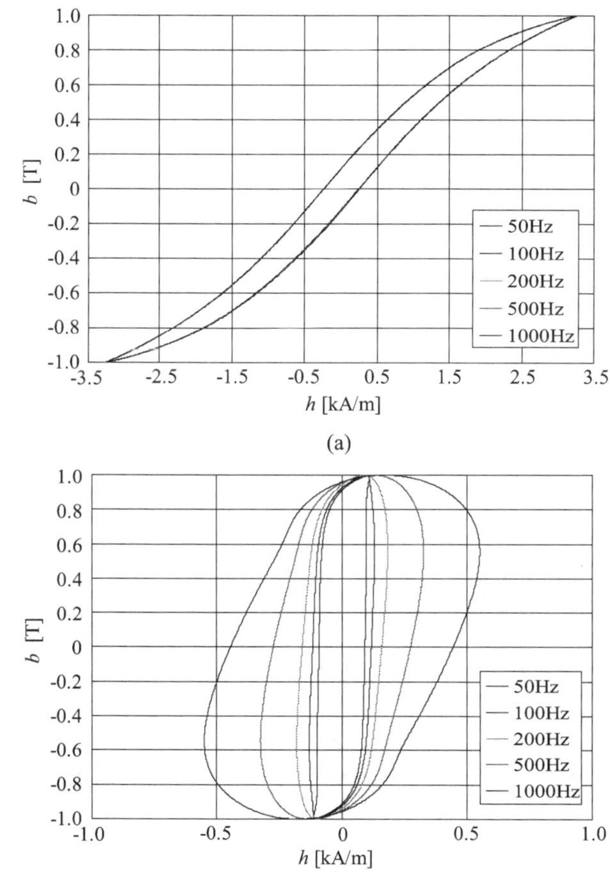

Figure 6.5 Frequency characteristics of hysteresis loops. (a) Powder magnetic core. (b) Nonoriented electrical steel sheet. Maximum flux density = 1.0 T.

$$P = af + bf^2 \tag{6.14}$$

Here, a and b are constants determined by the materials and f is the frequency. The equation shows that af corresponds to hysteresis loss proportional to the frequency, whereas bf^2 corresponds to eddy current loss proportional to square of the frequency. For the powder magnetic core, eddy current loss is only a small percentage of the core loss. For that reason, the shape of the hysteresis loop does not change much for frequency variation, while the magnetic flux density is constant, as shown in Fig. 6.5a. However, for materials like electrical steel sheets, the percentage of eddy current is very high. Therefore, the shape of the hysteresis loop varies greatly depending on the frequency, as shown in Fig. 6.5b.

The type of iron core used in electrical equipment is selected according to the purpose and operating conditions. For example, electrical power transformers typically operate at frequencies of 50 Hz or 60 Hz and handle very large electric power. So the magnetic flux through the core is very high, and the core needs to be large enough to handle this; therefore, a high-saturation-flux-density material such as electrical steel sheets is the most commonly used core material. Electronic circuits like inverters and DC–DC converters, on the other hand, operate at frequencies of a few hundred hertz or higher and need high efficiency. Electrical steel sheets cannot be used as the core material because of its high loss at such frequencies. Magnetic materials that are suitable for high-frequency use—such as ferrite—are used.

Transformers require high magnetic permeability because as much as possible of the magnetic flux induced by the primary winding needs to interlink the secondary winding. For inductors used in electronic circuits, the magnetic permeability of the core material must be determined accurately as the inductance depends on the permeability. Magnetic properties are usually measured by applying a sine wave. However, inductors in real circuits also carry DC as well as AC, so other magnetic properties such as linearity and core saturation need to be considered. Figure 6.6 shows typical magnetic permeability characteristics of an electrical steel sheet core with both AC and DC components of flux. The figure shows that permeability varies depending on the value of the DC component, that is, the magnitude of the magnetic flux. Figure 6.6c shows a nonoriented electrical steel sheet hysteresis loop when the superimposed DC magnetic flux density is varied from 0 T to 1.4 T, while the AC component is 0.1 T. However, in the figure the DC magnetic flux density has not been included. As we can understand from the figure, the area of the hysteresis loop increases when the amount of the superimposed DC magnetic flux density is increased. It eventually increases iron loss, as iron loss is proportional to the area of the hysteresis loop.

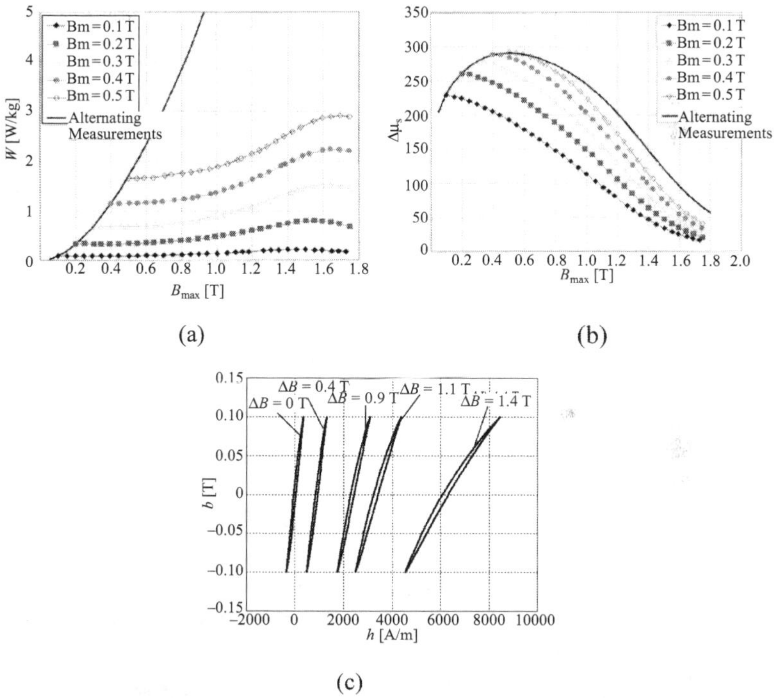

Figure 6.6 DC bias characteristics of a nonoriented electromagnetic sheet. (a) Core loss characteristics. (b) Increment permeability characteristics. (c) Hysteresis loops with DC bias. Exciting frequency = 50 Hz.

6.1.3.2 Conductors

The most important property of conductors is low resistance to electrical current flow. For pure metals, silver has the lowest resistivity (i.e., highest conductivity) of 1.62×10^{-2} μΩ·m (at 20°C), and copper has the next lowest resistivity of 1.72×10^{-2} μΩ·m (at 20°C). Relative to the resistivity of copper, the resistivity of silver is 0.94. As silver is a precious metal in limited supply, and thus expensive, it cannot be used for conductors in electrical equipment. Electrical equipment generally uses processed copper wire conductors. When weight is important, aluminum is also sometimes used. An aluminum wire with comparable resistance to a copper wire will have a larger cross-sectional area but weigh much less: aluminum has a resistivity of 2.84×10^{-2} μΩ·m (at 20°C), 1.65 times that of copper, but its specific gravity is only 30% that of copper.

The loss that occurs in the windings of transformers or inductors is called copper loss or load loss. Electrical machines such as transformers operate at optimum efficiency when their core loss (no-load loss) equals their copper loss (load loss). Therefore, when considering transformer efficiency or inductor loss, it is essential to take winding resistance into account.

Appropriate insulation of copper windings on a transformer or inductor is determined by the voltage that will be applied to it.

Resistance of a wire conductor R (Ω) is determined as follows:

$$R = \rho \frac{\ell}{S} \quad (\Omega) \tag{6.15}$$

Here, ρ is the resistivity (Ω·m), ℓ is the length (m), and S is the cross-sectional area of the copper wire (m^2).

The ρ of copper has a physical value of 1.72×10^{-2} μΩ·m at a temperature of 20°C. Depending on the conditions of use, the temperature coefficient of resistance of copper, 3.93×10^{-3} K^{-1} at 20°C, may also need to be taken into account. Equation 6.15 shows the DC resistance.

For electronic circuits, the DC resistance derived from Eq. 6.15 is not an accurate measure of the resistance at the high frequencies used in electronic circuits due to the skin effect. Therefore, when determining the winding resistance of transformers and inductors used at high frequencies, the skin effect at the operating frequency must be considered in determining the resistance value. In addition, the apparent resistance may vary due to the effect of magnetic flux leakage along the winding on the current distribution, especially when a gap is placed on the core.

The skin depth of the electric current caused by the skin effect is expressed as follows:

$$\delta = \sqrt{\frac{2}{\omega\mu\sigma}} \quad \text{(m)} \tag{6.16}$$

Here ω is the angular velocity (rad/s) ($2\pi f$), μ is permeability (for copper, $\mu \approx \mu_0$; μ_0 is the magnetic constant (H/m) ($4\pi \times 10^{-7}$), and σ is conductivity (S/m) (for copper $\sigma = 5.8 \times 10^7$).

When current of frequency 100 kHz flows through a copper wire, the skin depth is 0.2 mm. If the wire diameter is larger than 0.4 mm, the effective resistance of a single wire becomes quite high. As a result, it is necessary to use multiple wires in parallel (e.g., braided wire such as Litz wire) to reduce loss.

Another important criterion affecting the optimum wire diameter is current density. If the current density is too high, copper loss might cause high temperatures. On the other hand, if the current density is too small, the volume and weight of the winding will be larger than optimum.

It will probably be necessary to use Litz wires for inductors and transformers that carry a large current and operate at frequencies above several kilohertz.

The current density of wires in electrical equipment should be 10 A/mm^2 or less. Usually the design value of current density is set to 3–5 A/mm^2, and the required wire diameter is calculated on the basis of the current that will flow. In this case, the effective cross section of copper wire is determined by considering the skin effect.

6.1.4 Design Example

6.1.4.1 Inductor design example

Here we show a design example of an inductor using the E-I core. The target inductance value of the designed inductor is set to 3 μH, and the target DC bias current is 3 A, while the ripple amplitude is small.

Basic equations necessary for designing the inductor are shown below.

$$L = \frac{N^2}{R_\mathrm{m}} \tag{6.17}$$

$$B = \frac{NI}{R_\mathrm{m} S} \tag{6.18}$$

$$R_m = \frac{lc}{\mu_s \mu_0 S} + \frac{lg}{\mu_0 S} \tag{6.19}$$

Here, N is the number of turns, I is the maximum value of current flowing through the coil (A), B is the magnetic flux density (T), R_m is the magnetic resistance (A / Wb), S is the cross-sectional area of the core (m^2), ℓ_c is the magnetic path length of the core (m), ℓ_g is the gap length (m), μ_0 is the relative permeability of the core, and μ_0 the magnetic constant ($4\pi \times 10^{-7}$).

The shape of the core is shown in Fig. 6.7. The gap length is set to 1 mm, and the relative permeability is set to 1000. Leakage flux from the core is considered to be zero, and no spread of the magnetic flux in the gap and geometric path length is used as the magnetic path length. To have a core of symmetric magnetic flux with respect to the center axis of the central leg, we consider a core equivalent to the C-I core, whereas the cross-sectional area of the center leg is considered as the cross-sectional area of the core.

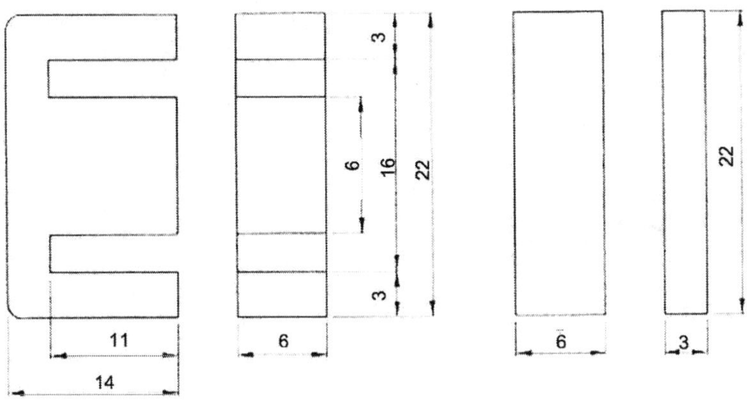

Figure 6.7 Core dimensions of the inductor (unit is mm).

The magnetic path length of the core ℓ_c and the gap length ℓ_g are determined as follows:

$$\ell_c = (11 + 3 + 5 + 3) \times 2 = 44 \quad \text{(mm)} \tag{6.20}$$

$$\ell_g = 1 + 1 = 2 \quad \text{(mm)} \tag{6.21}$$

Therefore, the magnetic resistance R_m is determined as

$$R_m = \left(\frac{lc}{\mu_s \mu_0} + \frac{lg}{\mu_0}\right)\frac{1}{S} = \left(\frac{0.044}{10^3 \times 4\pi \times 10^{-7}} + \frac{0.002}{4\pi \times 10^{-7}}\right)\frac{1}{6 \times 6 \times 10^{-6}}$$

$$= 4.52 \times 10^7 \quad (6.22)$$

If the value of R_m is substituted in Eq. 6.17 we have

$$L = \frac{N^2}{R_m} = \frac{N^2}{4.52 \times 10^7} \quad (6.23)$$

So we can deduce $N^2 = 4.52 \times 10^7 \times L$. As the target value of inductance L is 3 μH the number of turn N is

$$N = 11.6 \quad (6.24)$$

To make N a round figure we consider it as 11 or 12. For each of the cases, we calculate the inductance design value as 2.67 μH or 3.19 μH.

When considering the number of turns N = 12, magnetic flux density is

$$B = \frac{NI}{R_m S} = \frac{12 \times 3}{4.52 \times 10^7 \times 6 \times 6 \times 10^{-6}} = 0.022 \quad \text{(T)} \quad (6.25)$$

It is a reasonable value of magnetic flux density for an inductor.

Assuming that the current density is 3 A/mm^2, we calculate the cross-sectional area of the coil's winding, S_w, as follows:

$$S_w = \frac{3}{3} = 1 \quad (\text{mm}^2) \quad (6.26)$$

If we select a round wire with a cross-sectional area of 1 mm^2, from the standard table we have, the diameter equals 1.1 mm^2 and the cross-sectional area equals 0.9503 mm^2. If current density is recalculated, we find 3.16 A/mm^2, which is within the allowed range. In addition, the core window will have sufficient space because the winding has a total of 12 turns.

If there is any high-frequency component present in the current, it will be necessary to consider the skin effect and use a Litz wire.

6.1.4.2 Design example of a high-frequency transformer

Let us consider the transformer design condition as follows:

Primary voltage 140 V, capacity 100 W, frequency 100 (kHz), turn ratio 10 : 1

There are four transformer design procedures:

1. Assumption of voltage per turn: We assume the voltage per turn as 2 V. Therefore, the number of turns of the primary winding is 70 and secondary is 7.

2. Calculation of magnetic flux Φ:

$$\Phi_m = \frac{v_1}{4.443 f n_1} = \frac{140}{4.443 \times 100 \times 10^3 \times 70} = 4.50 \times 10^{-6} \text{ (Wb)} \tag{6.27}$$

Here, v_1 is the primary voltage, f is the frequency, and n_1 is the number of turns of the primary winding.

3. Calculation of the cross-sectional area of the core, assuming flux density of the core B_m = 100 mT:

$$S = \frac{\Phi_m}{B_m} = \frac{4.5 \times 10^{-6}}{0.1} = 45 \times 10^{-6} \text{ (m}^2\text{)} \tag{6.28}$$

Using a ring-shaped E-type core (JIS FEER25.5A standard) of a 7.5 mm diameter center leg and a cross-sectional area of 44.16 mm^2 results in a magnetic flux density of 102 mT, which is within the acceptable range.

4. Consideration of winding specification: From a primary input of 100 V, 100 W, we have the current equal to 0.714 A. If the current density of the winding is considered as 0.714 A, the cross-sectional area of the winding $S_{w'}$ is calculated as follows:

$$S_{w'} = \frac{0.714}{3} = 0.238 \text{ (A/mm}^2\text{)} \tag{6.29}$$

To secure the cross-sectional area with a single copper wire, it is necessary to have a 0.55 mm diameter (0.2376 mm^2) wire. However, considering the skin effect due to the operating frequency, a Litz wire of 0.55 mm diameter (0.2376 mm^2) is used. In that case, a total of 30 twisted wires (0.2356 mm^2) is required.

The cross-sectional area of the primary winding, S_w, is determined as follows when the coil space factor equals 0.5:

$$S_w = \frac{0.2376 \times 70}{0.5} = 33.26 \text{ (mm}^2\text{)} \tag{6.30}$$

To have the same magnetic flux density on both the primary and secondary windings, an absolute same cross-sectional area in both windings is necessary. Having some margin, the window area necessary for the core is 70 mm^2 to accommodate the winding wire. Since in the section of magnetic flux density calculation, we have used a JIS FEER25.5A suitable core, the window area for two of these cores is 76.26 mm^2, which makes the winding wire packable.

Due to the fact that the design has been made on various assumptions, during the actual designing, on the basis of the understanding of the available materials and their characteristics, a better review is absolutely necessary. While designing inductors, permeability is assumed to be constant, irrespective of the flux density, and the gap length is assumed to be 1 mm. However, permeability has a nonlinear character, and ensuring an accurate gap length is difficult. Therefore, it is necessary to keep in mind the gap length has a great influence on results.

Moreover, in the actual calculation, it may not be possible to reach the target value under the conditions assumed above. But in that case, it is necessary to modify the assumptions and recalculate.

6.2 Capacitors

6.2.1 The Position of a Capacitor in Electronic Components

6.2.1.1 The analog AV era

Today, electronic equipment has become absolutely necessary for human life. Looking at the transition of the electronics market, analog equipment has been widely used, from audio, color TV, and video to white goods such as refrigerators and washing machines in ordinary households from the 1960s to the 1980s. With the spread of this equipment, the growing era of electronic components had come. The demand for electronic components rapidly expanded with the development of discrete semiconductors, such as transistors and rectifiers, and analog integrated circuits, both of which made a huge contribution to the spread of analog equipment. Lead terminal–type carbon resistors, aluminum electrolytic capacitors, and tantalum electrolytic capacitors, having excellent general versatility and productivity, had especially grown significantly among passive components.

6.2.1.2 The digital era

Electronic equipment was rapidly digitalized from the late 1980s with the proliferation of personal computers in households. The demand for mobile equipment, such as a digital camera or a cellular phone, increased largely. The existence of large-scale integrated (LSI) circuits (a number of functions integrated in one chip) is the driving force of digitalization. Using components with multiple functions and many input–output terminals, multilayer printed circuit boards were developed, which elevated the mounting configuration to "surface mounting." Surface mounting was also preceded in passive components, such as chip resistors, multilayer capacitors, relatively large aluminum electrolytic capacitors, or magnetic components.

6.2.1.3 The digital network era

The electronic components market, which had grown with the development of electronic equipment, experienced a temporary downturn in demand due to bursting of the information technology (IT) bubble in 2000. The demand for capacitors, as passive components, was also included in this downturn, but it has recovered and keeps expanding, thanks to the driving demand of the IT field afterward, focusing on the Internet (Fig. 6.8).

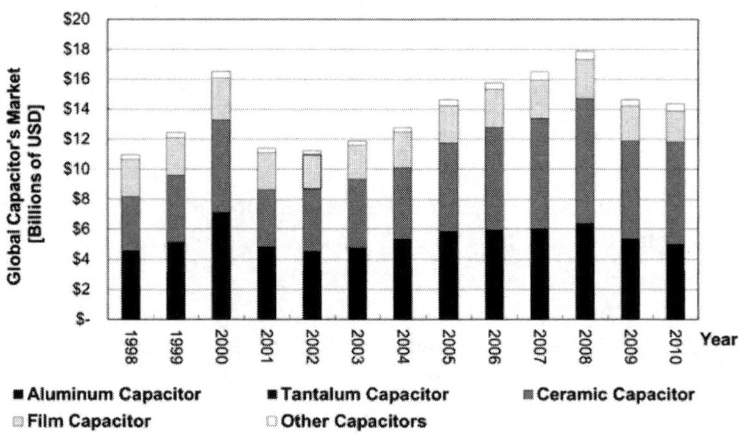

Figure 6.8 Worldwide market scale of capacitors in the past 10 years (amount of money).

It is not too much to say that this firm growth is a contribution of digitalization and goes hand in hand with the growth of logic semiconductors, which had been developed with personal computers. As shown in Fig. 6.9, the growth of capacitors is synchronized with that of logic semiconductors, and it means capacitors, as passive components, are necessary in digital circuits.

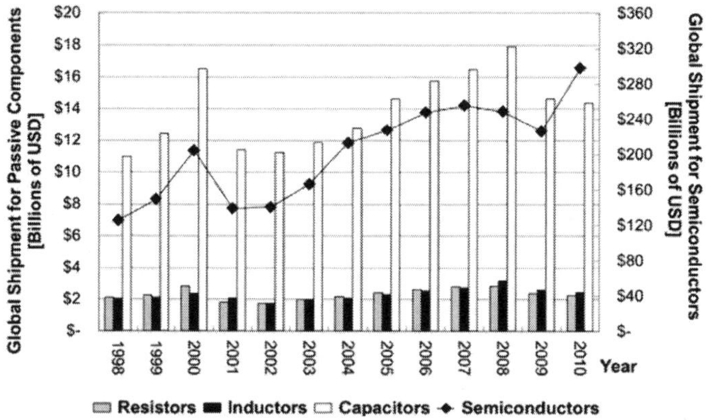

Figure 6.9 Past market scale of passive components and semiconductors (amount of money).

6.2.2 Brief Overview of Various Capacitors

6.2.2.1 Aluminum electrolytic capacitor

An aluminum electrolytic capacitor is the most common capacitor with a wide range of rated voltages and capacitances and high general versatility. Both liquid and solid electrolytes are used for this capacitor, and usage considering life performance is required, especially for the one with a liquid electrolyte. The life performance of a general-purpose product is 1000–2000 hours at 85°C or 105°C, but it has improved with the recent increase in severe-operating-environment applications, such as industrial equipment or vehicles, and even a product with a life performance over 10,000 hours at 105°C has been developed. Regarding heat resistance characteristics, a product for a 150°C ambient temperature, which exceeds the maximum junction temperature of a semiconductor, has been developed. The application field of aluminum electrolytic capacitors is expanding from the point of "reliability" as this: With the recent digitalization of electronic equipment, a capacitor having a conductive polymer for its cathode material is focused, and a low equivalent series resistance (ESR) or equivalent series inductance (ESL; *L* represents inductance) product with this conductive polymer has been developed for the power supply–decoupling application of a digital integrated circuit, which requires transient response performance.

6.2.2.2 Tantalum electrolytic capacitor

A tantalum electrolytic capacitor has large capacitance, taking advantage of the high dielectric constant of tantalum oxide. It was formerly used for analog communication devices as the change in its characteristics by temperature is smaller than that for an aluminum electrolytic capacitor. Currently, a surface-mount product with high capacitance and small size is widely used for mobile equipment, such as cellular phones. In addition, a product using a conductive polymer for its cathode has been developed, like an aluminum electrolytic capacitor, and its demand is increasing, taking advantage of its low-ESR characteristics. On the other hand, a tantalum electrolytic capacitor has disadvantages, such as it requires voltage de-rating and its failure mode is a short circuit. As tantalum, which is the main material of the tantalum electrolytic capacitor, is a rare metal, tantalum metal powder contains the risk of price increase with regard to the amount of its deposit. Therefore, its replacement with a niobium electrolytic capacitor has been partially started. In a niobium electrolytic capacitor voltage de-rating is not required, and it is less flammable than a tantalum electrolytic capacitor.

6.2.2.3 Film capacitor

There are many kinds of film capacitors, depending on the type of dielectric film material. Generally, they have a high rated voltage and a wide temperature range and are used for a wide range of applications, from small signals to high power. A film capacitor has a large market in class X capacitors (across-the-line capacitors) for a normal-mode noise filter and class Y capacitors (line bypass capacitors) for a common-mode noise filter as a countermeasure product against conductive induction noise, which has been actualized as a problem with the spread of electronic equipment. Due to the characteristics of the dielectric, capacitance per unit volume is very small compared to other capacitors. However, its high-power-usage applications, such as in industries or vehicles, progress, thanks to the development of a large-capacitance technology by thin-film technology or the development of a high-heat-resistant material.

6.2.2.4 Ceramic capacitor

A ceramic capacitor has a wide-rated voltage range and is commonly used. Especially, the number of multilayer ceramic capacitors

(MLCCs) used is the largest among all kinds of capacitors. The demand for a ceramic capacitor, using ferroelectric material and excellent high-frequency characteristics, is growing drastically, riding the digitalization wave of electronic equipment. Capacitance of a ceramic capacitor is not as large as that of an aluminum electrolytic capacitor because of the dielectric characteristic, but development of a large-capacitance product progresses, partially competitive to an aluminum electrolytic capacitor, thanks to the thin-film technology of a multilayer sheet. As capacitance of a ceramic capacitor has temperature and voltage bias characteristics, the operating condition needs to be considered to select an appropriate product.

6.2.3 Characteristics and Applications of Various Capacitors

An electronic component can be classified as an "active component," whose function is amplification or rectification, and a "passive component," whose function is consumption, storage, or discharge of power.

A capacitor is part of a passive component, which charges and discharges power, and is used for various applications of electronic circuits, such as voltage smoothing, voltage stabilization, noise removal, and coupling. Therefore, a capacitor is now absolutely necessary for all electronic equipment, including information communication equipment. The following types of capacitors are commonly used for power supply design:

- Aluminum electrolytic capacitor

  - Nonsolid aluminum electrolytic capacitor
  - Winding-type conductive-polymer solid aluminum electrolytic capacitor (radial lead/surface mount)
  - Stacked-type conductive-polymer solid aluminum capacitor

- Tantalum electrolytic capacitor

- Tantalum solid electrolytic capacitor
- Conductive-polymer solid tantalum electrolytic capacitor

• Film capacitor

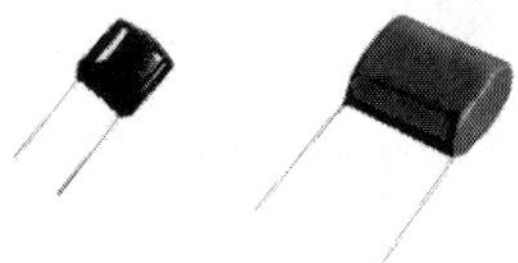

- Foil film capacitor
- Metallized film capacitor

• Ceramic capacitor

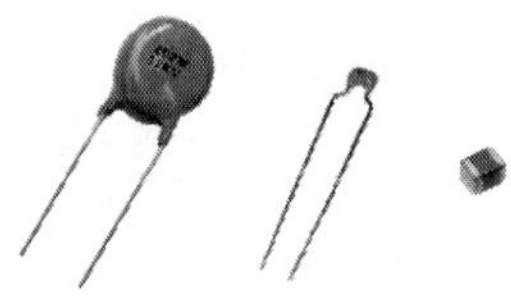

- Lead-type MLCC
- Surface-mount device (SMD)-type MLCC

Generally, various capacitors have a sandwiched construction with two electrical conductors and one dielectric (electric insulator) in between (Fig. 6.10). Each capacitor's name is based on the type of dielectric material. As shown in Eq. 6.31, it is possible to increase the capacitor's capacitance (C) by increasing the surface area of the two electrical conductors (S), decreasing the distance between the terminals (d), and making the dielectric constant of the dielectric (ε_r) high.

$$C = 8.854 \times 10^{-12} \frac{\varepsilon_r S}{d} \tag{6.31}$$

C : Capacitance (F)

ε_r : Dielectric constant

S : Surface area (m^2)

d : Thickness of dielectric (m)

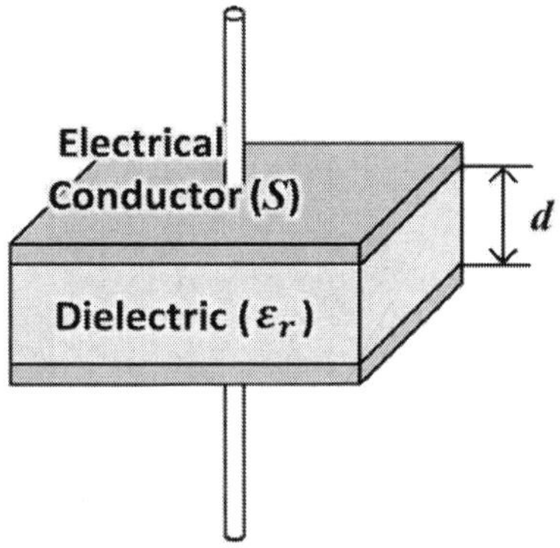

Figure 6.10 Basic model of a capacitor.

As shown in Fig. 6.11, each capacitor has its home ground. However, even in an overlapped area, each capacitor is used with its own characteristics not necessarily competing against each other. Therefore, selecting an appropriate capacitor is required, considering the purpose or usage on the electronic circuit.

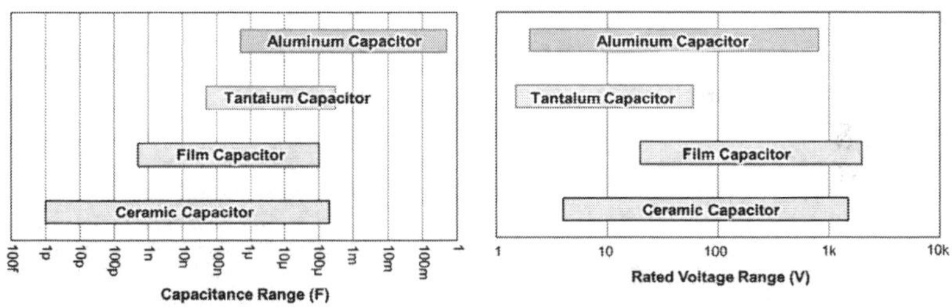

Figure 6.11 Capacitance range/rated voltage range.

The construction and characteristics of various capacitors, which are often used for power supply design, are described next.

6.2.3.1 Aluminum electrolytic capacitor

"Aluminum electrolytic capacitor" is the collective term for capacitors that use aluminum oxide (Al_2O_3) as the dielectric. The dielectric constant (ε_r) of aluminum oxide is about 8 to 10. The anode aluminum surface is treated with surface roughening, called "etching," to increase the surface area (*S*) and to achieve large capacitance (Fig. 6.12).

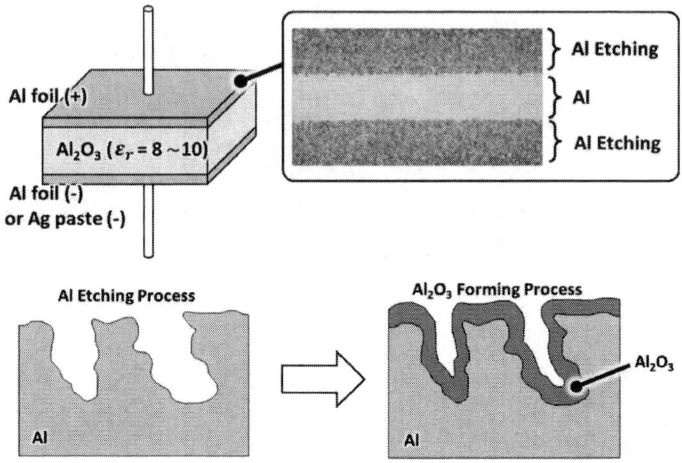

Figure 6.12 Aluminum foil etching and dielectric (Al_2O_3) forming.

A basic model of an aluminum electrolytic capacitor is shown in Fig. 6.13. It is called "electrolytic capacitor" as an electrolyte is used to connect the dielectric and the cathode electrical conductor, and the electrolyte is called a true cathode. The electrolyte material differs between a nonsolid aluminum electrolytic capacitor and a conductive-polymer solid aluminum electrolytic capacitor.

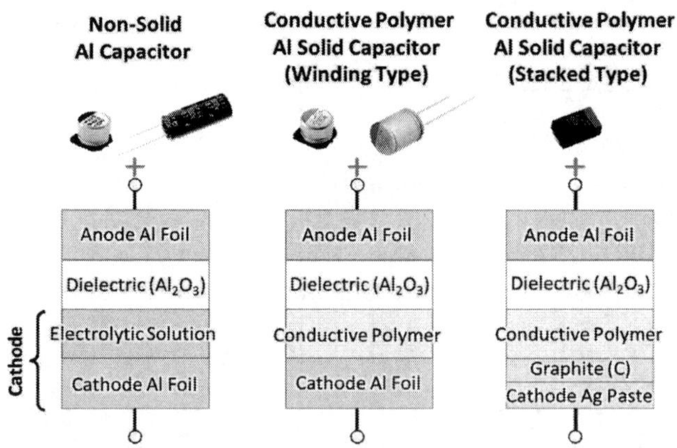

Figure 6.13 Basic model of an aluminum electrolytic capacitor.

A liquid electrolyte solution is used as an electrolyte in a nonsolid aluminum electrolytic capacitor. It is inexpensive and has a wide-rated voltage and capacitance range.

Therefore a nonsolid aluminum electrolytic capacitor is commonly used for a wide variety of electronic equipment.

A solid-state conductive polymer is used as an electrolyte in a conductive-polymer solid aluminum electrolytic capacitor. By using a conductive polymer as an electrolyte, which has extremely higher electrical conductivity than the electrolyte solution, it is possible to make the ESR much lower, which keeps the internal temperature of the capacitor at a very low level. So, a conductive-polymer solid aluminum electrolytic capacitor that is a superlow-ESR capacitor has high ripple current capability (Fig. 6.14). It also has high reliability such as heat resistance or reflow resistance. Figure 6.15 shows the general equivalence circuit of a capacitor, and its factors can be separated into equivalent series capacitance (ESC), ESR, and ESL.

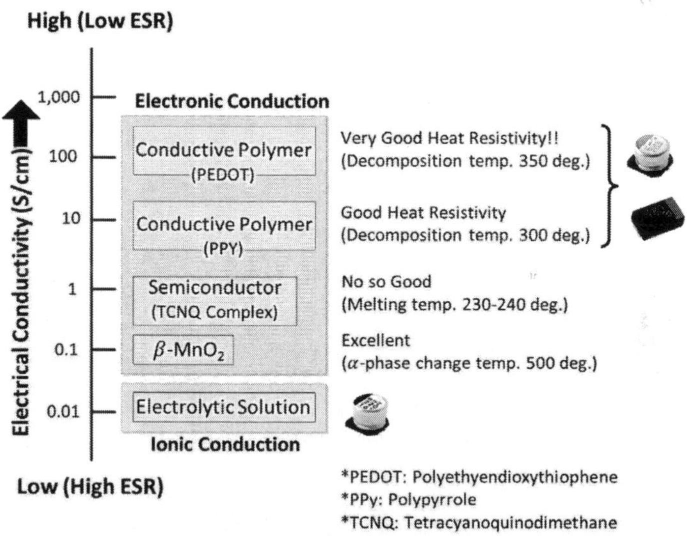

Figure 6.14 Electrical conductivity of an electrolyte (aluminum electrolytic capacitor).

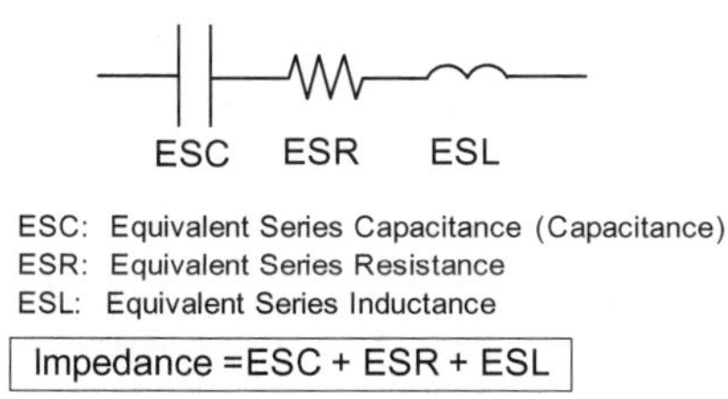

Figure 6.15 Equivalent circuit of a capacitor.

As a conductive-polymer solid aluminum electrolytic capacitor has a relatively low-rated voltage, it cannot be used in a high-voltage field. Therefore it is mainly used at the input–output part of the buck converter for the CPU power supply, whose input voltage is 20 V or less. There are winding-type, radial lead and surface-mount, and stacked-type conductive polymers. An appropriate part should be selected, considering usage area, performance, and cost (Fig. 6.16).

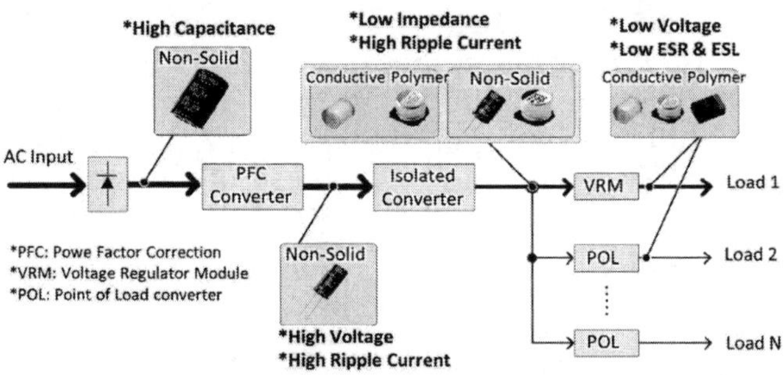

Figure 6.16 Usage of aluminum electrolytic capacitors in a power distribution network.

6.2.3.2 Tantalum electrolytic capacitor

"Tantalum electrolytic capacitor" is the collective term for capacitors that use tantalum oxide (Ta_2O_5) as the dielectric. The dielectric constant (ε_r) of tantalum oxide is about 22 to 27, and this is an appropriate material as a large-capacitance capacitor. The porous material sintered tantalum is used for the anode of a tantalum electrolytic capacitor to increase the surface area (S) (Fig. 6.17).

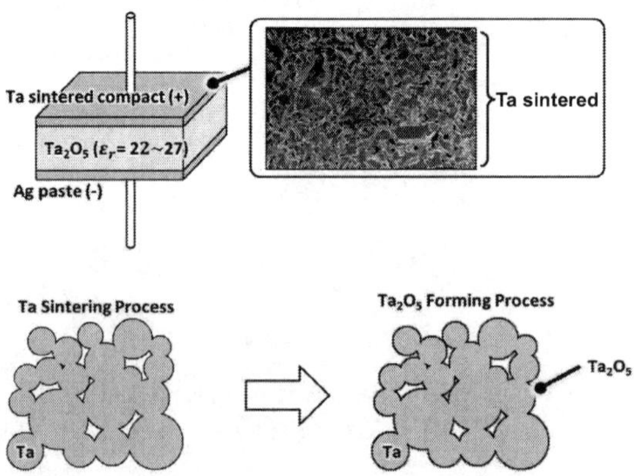

Figure 6.17 Sintered tantalum and dielectric (Ta_2O_5) forming.

A basic model of a tantalum electrolytic capacitor is shown in Fig. 6.18. It is called "electrolytic capacitor" as an electrolyte is used to connect the dielectric and the cathode electrical conductor, and the electrolyte is called a true cathode. The tantalum electrolytic capacitor for power supply mainly uses solid-state material for the electrolyte, such as manganese dioxide or a conductive polymer. The one with manganese dioxide is called a tantalum solid electrolytic capacitor, and the one with a conductive polymer is called a conductive-polymer solid tantalum electrolytic capacitor.

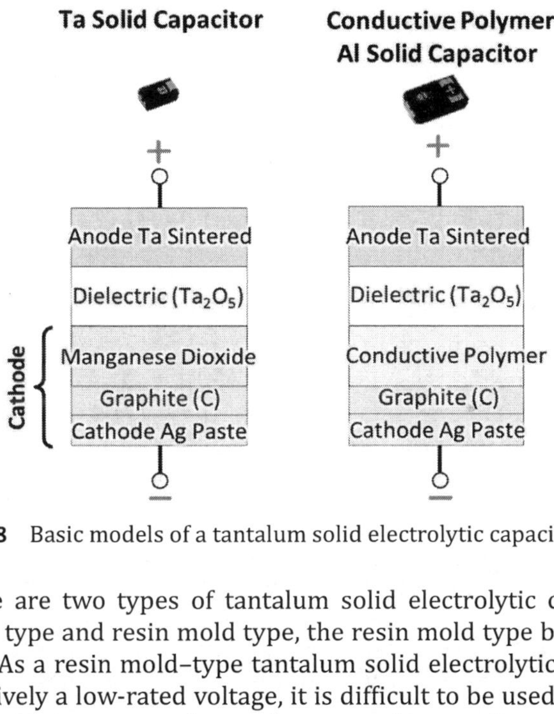

Figure 6.18 Basic models of a tantalum solid electrolytic capacitor.

There are two types of tantalum solid electrolytic capacitors, resin dip type and resin mold type, the resin mold type being more popular. As a resin mold–type tantalum solid electrolytic capacitor has relatively a low-rated voltage, it is difficult to be used in a high-voltage field. However, taking advantage of its small size and high capacitance, it is widely used for mobile equipment, including cellular phones.

At the operation level, voltage de-rating from 30% to 50% of the rated voltage will be required to avoid fire generation from a short circuit because a resin mold–type tantalum solid electrolytic capacitor is very weak for reverse voltage or overvoltage. A conductive-polymer solid tantalum electrolytic capacitor has lower ESR because of using a conductive polymer for the electrolyte, which has higher electrical conductivity (Fig. 6.19).

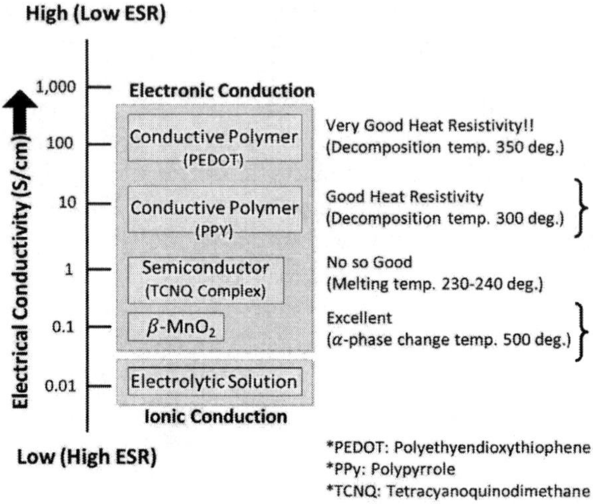

Figure 6.19 Electrical conductivity of an electrolyte (tantalum electrolytic capacitor).

A conductive-polymer solid tantalum electrolyte capacitor is mainly used for the input–output part of the buck converter for small, low-height, high-density-package cellular phones, the input–output part of the buck converter for the CPU power supply, or the input–output part of the series regulator (Fig. 6.20).

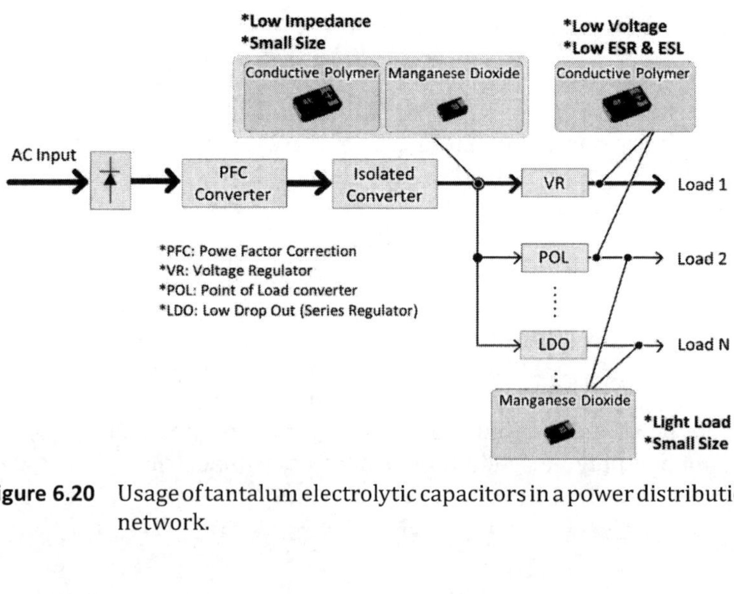

Figure 6.20 Usage of tantalum electrolytic capacitors in a power distribution network.

6.2.3.3 Film capacitor

"Film capacitor" is the collective term for capacitors that use plastic film material as the dielectric. Various substances are used for plastic film material, such as polyethylene terephthalate (PET), polypropylene (PP), polyphenylene sulfide (PPS), polyethylene naphthalate (PEN), polycarbonate (PC), or polystyrene (PS), and each capacitor is called by the name of the film material.

Table 6.2 shows the features of each film capacitor. A PET film capacitor is widely used in various fields due to its wide temperature range and low cost. A PP film capacitor is used for circuits where a large ripple current flows or high accuracy is required because of its low dissipation factor (tan δ). However, it has low heat resistance, which is a disadvantage. A PPS film capacitor is used for circuits that require a severe soldering condition, filter, or oscillating circuit, thanks to its excellent heat resistance and frequency/temperature stability of capacitance (best stability compared to other film capacitors). However, its disadvantage is that it is expensive. A PEN film capacitor is used for surface mounting, which requires a severe soldering condition, like a PPS film capacitor, as it has good heat resistance.

Table 6.2 Features of various kinds of film capacitors

Item	PET	PP	PPS	PEN
Cost	◎	○	Δ	○
Miniaturization	◎	Δ	○	◎
Heat resistance	○	Δ	◎	◎
Moisture resistance	Δ	◎	○	Δ
Solvent resistance	○	○	○	○
Temperature characteristics	Δ	○	◎	Δ
Low loss (tan δ)	Δ	◎	○	Δ

◎: Very good ○: Good Δ: Not so good

A basic construction of a film capacitor is different from that of an aluminum electrolytic capacitor or a tantalum electrolytic capacitor. The electrolyte is not used to connect the dielectric film and the cathode electrode. The dielectric constant (ε_r) of the dielectric film is very low, about 2 to 3, and it is an unfavorable material to actualize large capacitance. However, a film capacitor is widely used, from electric home appliances to electronic vehicle devices, because a film capacitor has no polarity, has higher electric insulation, and has a lower dissipation factor (tan δ) and its capacitance is more stable against frequency and temperature changes than other capacitors (Fig. 6.21).

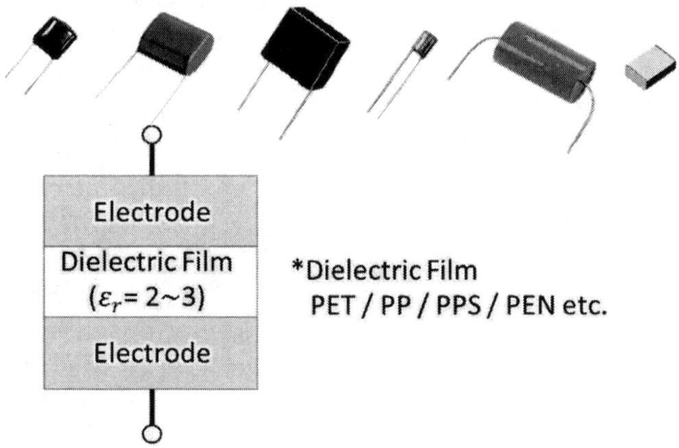

Figure 6.21 Basic model of a film capacitor (nonpolarity).

It is also possible to classify a film capacitor as a "foil film capacitor" and a "metallized film capacitor" according to the element construction (Fig 6.22).

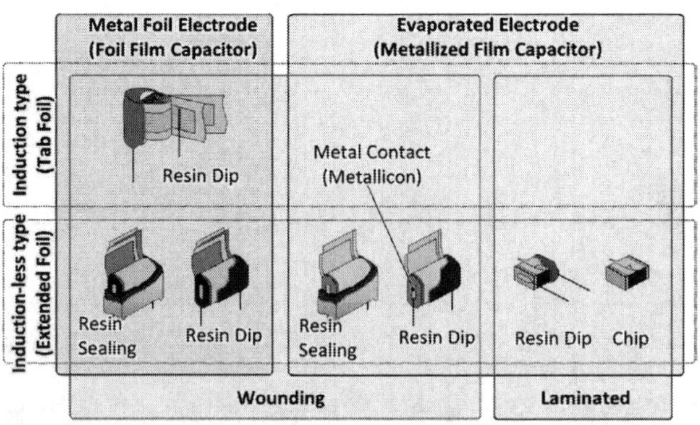

Figure 6.22 Types of electrodes and structures.

A foil film capacitor (metal foil electrode type), which is common among small-capacitance products up to 1 μF, is constructed by winding metal foil, such as aluminum, tin (Sn), or copper (Cu), and a dielectric film together. And the one winding the lead wire together is called "induction type," and the one attaching the lead wire to the sides after winding is called "induction-less type." The dissipation factor (tan δ) and inductance of an induction type tends to be high compared to an induction-less type.

The electrode of a metallized film capacitor (evaporated electrode type) is formed by evaporation of the metal such as aluminum or zinc (Zn) ta o dielectric film. As it is possible to reduce the thickness of the electrode by evaporation, a metallized film capacitor can be smaller and save more weight compared to a foil film capacitor. An evaporated electrode is very thin, about a dozen nanometers, and cannot connect the lead wire to the electrode directly. Therefore, an external electrode is added at the side as a metal contact (metallicon) part, by zinc or solder after winding/stacking. As the contact part is easily removed, care is required so as not to apply stress to the lead wire.

On a power distribution network, a film capacitor is mainly used for normal-mode noise removal of AC input (class X capacitor or across-the-line capacitor), switching-noise removal of a power factor correction (PFC) circuit, or a surge clamp circuit of a transformer (Fig. 6.23).

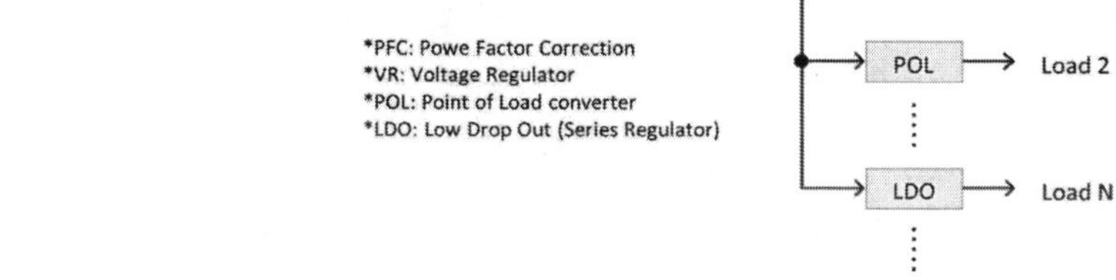

Figure 6.23 Usage of film capacitors in a power distribution network.

6.2.3.4 Ceramic capacitor

"Ceramic capacitor" is the collective term for capacitors that use ceramic material as the dielectric. Ceramic material is classified into type 1, using a low-dielectric-constant material (10 to 200) such as titanium oxide, and type 2, using a high-dielectric-constant material (1000 to 2000) such as barium titanate. The product size, capacitance temperature characteristics, and operating temperature range are defined in detail by electronic industries alliance (EIA) or Japanese industrial standards (JIS), and the appropriate product should be selected, considering the purpose and usage.

A type-1 capacitor, with a low dielectric constant, is mainly used for a feedback circuit or a signal circuit, and a type 2 capacitor, with a high dielectric constant, is mainly used for a decoupling (filtering) or power circuit, which requires large capacitance (Fig. 6.24).

Low Dielectric Constant (Type1)	High Dielectric Constant (Type2)
Dielectric ······ Titanium oxide, etc (ε_r = 10~200)	Dielectric······ Barium Titanium (ε_r = 1000~2000)
Advantage ····· Small capacitance change from temperature.	Advantage ····· Dielectric Constant is high. Capacitance is high.
Disadvantage ··· Dielectric Constant is low. Capacitance is small.	Disadvantage ··· Large capacitance change from temperature.
Usage ····· Temperature compensation (For quartz crystal unit, etc.) High frequency circuits Filter circuits	Usage ····· Smoothing circuits Coupling circuits Decoupling circuits

Figure 6.24 Types of ceramic capacitors by the dielectric.

The operating temperature range should be considered when using a type 2 ceramic capacitor, which is often used for power supply design because its capacitance changes drastically with temperature. In addition, its capacitance decreases as bias voltage increases; therefore the capacitance should be large enough, considering the actual operating voltage.

An MLCC, which occupies more than 95% of the whole ceramic capacitor market, is a typical SMD electronic component that meets the recent SMD needs. The number of capacitors used for a cellular phone is about 200 to 300 and for a laptop computer about 1000, and more than 90% of these are MLCCs. A multilayer capacitor is used not only for mobile-type small electronic devices but also for a wide range of applications, from electric home appliances to electronic vehicle devices, because it has high reliability, no polarity, low ESR, and a wide capacitance range (Fig. 6.25).

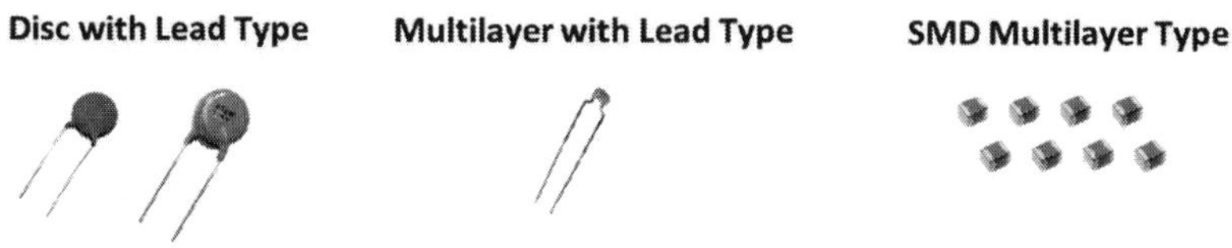

Figure 6.25 Types of ceramic capacitors.

A basic construction of an MLCC is different from that of an electrolytic aluminum capacitor or a tantalum electrolytic capacitor, laminated by a ceramic dielectric sheet with an internal electrode, such as nickel (Ni), and sintered. Therefore, an electrolyte is not used to connect the dielectric ceramic and the cathode (Fig. 6.26).

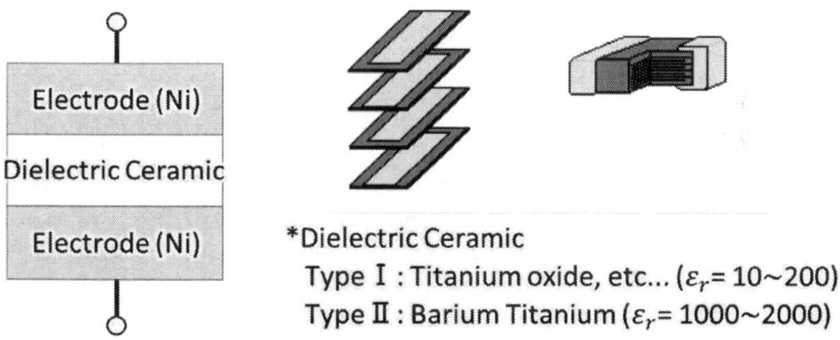

Figure 6.26 Basic model of a ceramic capacitor (nonpolarity).

A high-rated voltage disc lead-type ceramic capacitor, on a power distribution network, is mainly used for common-mode noise removal (class Y capacitor or line bypass capacitor) of AC input, and an SMD-type MLCC is used for the buck converter input–output part of various cellular phones, the buck converter input–output part of the CPU power supply, and the series regulator input–output part (Fig. 6.27). In addition, it is often used for a snubber circuit for noise reduction of switching or the rectifier.

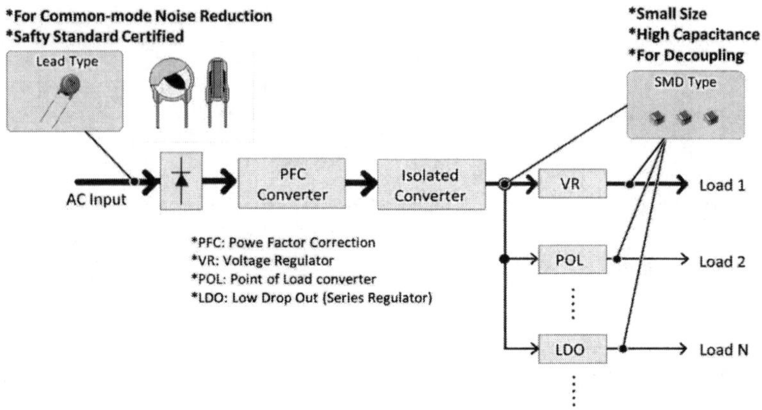

Figure 6.27 Usage of ceramic capacitors in a power distribution network.

On the power supply for a CPU, which requires low voltage and large current, both a large-capacitance capacitor, such as an aluminum electrolytic capacitor, and an MLCC are often used in combination to meet the requirement of a lower-impedance characteristic at a wide frequency range (Fig. 6.28). When a large-capacitance capacitor, such as an aluminum electrolytic capacitor, and an MLCC are used in combination, a phenomenon called antiresonance may occur on their frequency characteristic. Antiresonance is the regional impedance rise by parallel resonance between the inductance factor (L) of a large-capacitance capacitor and the capacitance factor (C) of an MLCC (Fig. 6.29). It makes the power supply transient response (overshoot and undershoot) worse.

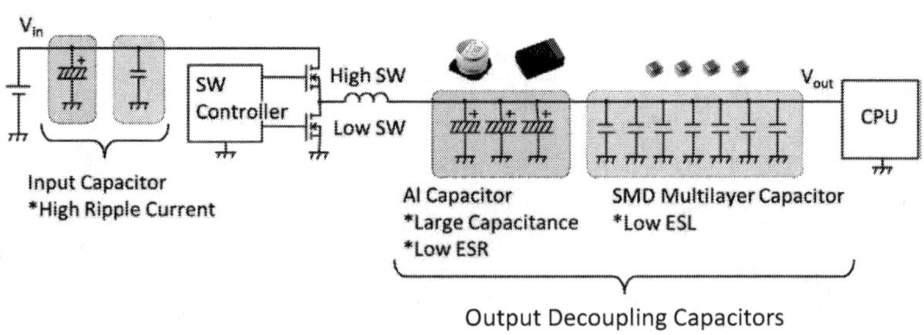

Figure 6.28 Decoupling circuit for a CPU.

A much lower ESR than other capacitors is a feature of an MLCC, but it has a risk of power supply abnormal oscillation, which degrades the transient response performance. This is because of the feedback network phase delay largely due to the excellent performance (low ESR) of an MLCC.

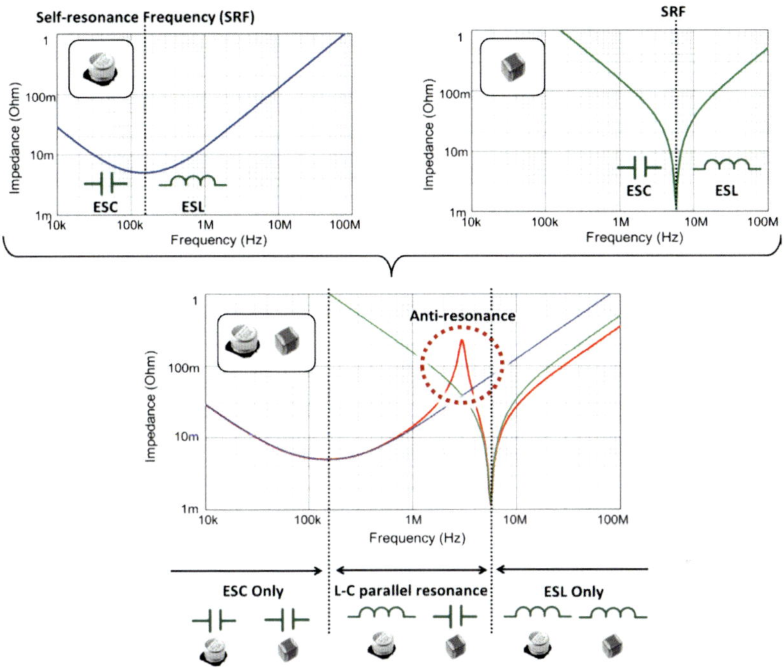

Figure 6.29 Output antiresonance.

A similar phenomenon is expected when using a low-ESR conductive-polymer solid aluminum electrolytic capacitor. Especially for an MLCC, as it is often used for the output of a low dropout (LDO)-type series regulator, it is also required to consider the abnormal oscillation of the LDO. It is generally possible to improve the power supply abnormal oscillation by compensating the phase of the feedback network, but it is not easy to do so because phase compensation requires specialized knowledge, so please refer to the power supply control IC manufacturer or the literature for the phase compensation adjustment method in detail.

6.2.4 Main Roles of a Capacitor in a Power Distribution Network

A power supply circuit in a power distribution network, which is often linked to the circulatory system of the human body, is a circuit that converts the electric power adequate to the load and stably supplies the converted power. This function is crucially important to operate analog or digital equipment. A capacitor is a necessary electronic component to stabilize output voltage in a power distribution circuit, and it is important to lay out the appropriate capacitor to the appropriate mount position. A major power supply system is the "three-stage power distribution network system," in the era of low voltage and large current for CPU technology (Fig. 6.30). The appropriate use of a capacitor for each stage will be described next.

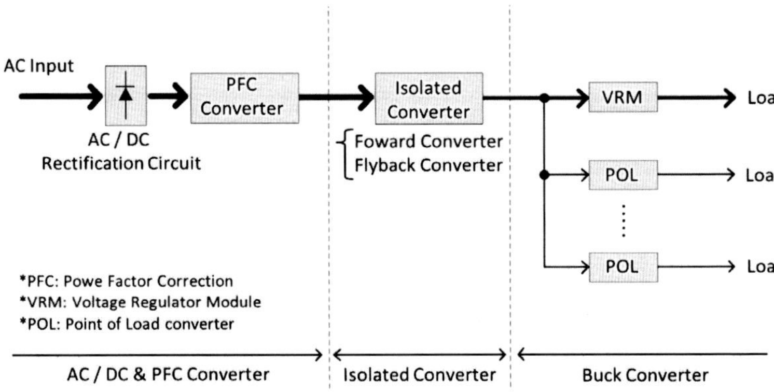

Figure 6.30 Distributed power generation system (three stages).

6.2.4.1 Role of a capacitor in a buck converter (VRM/POL)

VRM and POL converters are buck converters, specialized in low voltage and large current or fast transient response, laid out near each load that operates at low voltage. Figure 6.31 shows the basic circuit and capacitors used for a VRM converter and Fig. 6.32 for a POL converter. A VRM converter is mostly composed of discrete components. As for a POL converter, the power supply module is used as well.

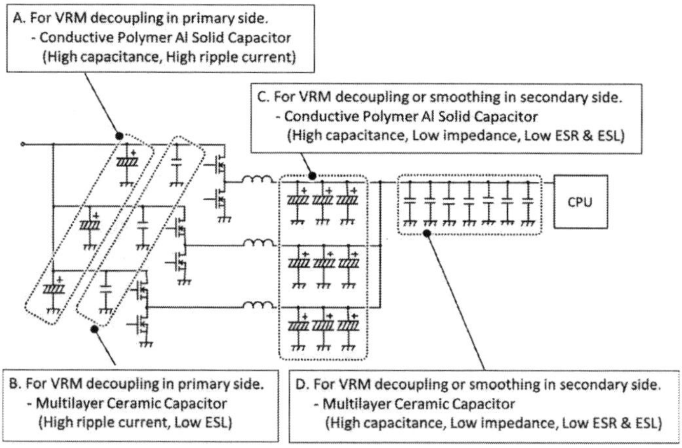

Figure 6.31 VRM converter, ex.3phase VRM.

A. The capacitor to absorb pulsed current noise generated by switching. A conductive-polymer solid aluminum electrolytic capacitor, which has high capacitance and can flow a large ripple current, is appropriate. A simplified equation to calculate the input ripple current is Eq. 6.32. As a large ripple current flows at the input, an appropriate capacitor should be selected, considering the allowable ripple current of the capacitor. Downsizing is possible by selecting a low-impedance capacitor.

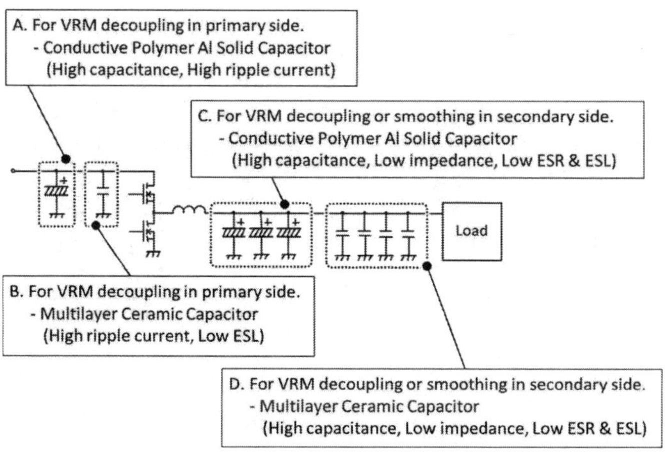

Figure 6.32 POL converter.

$$I_{\text{rip}} = \sqrt{D \cdot (1-D)} \cdot I_{\text{out}} \qquad \text{Here } D = \frac{V_{\text{out}}}{V_{\text{in}}} \qquad \begin{cases} I_{\text{rip}}: \text{Ripple current that flows to input (A)} \\ V_{\text{out}}: \text{Output voltage (V)} \\ V_{\text{in}}: \text{Input voltage (V)} \\ D: \text{Duty ratio} \\ I_{\text{out}}: \text{Output current (A)} \end{cases} \tag{6.32}$$

B. It is possible to absorb switching or radiation noise by laying out a low-ESL MLCC as close as possible to drain the terminal of the high-side metal–oxide–semiconductor field-effect transistor (MOSFET) switch and source terminal of the low-side MOSFET switch. It is desirable to lay out MLCCs as shown in Fig. 6.33.

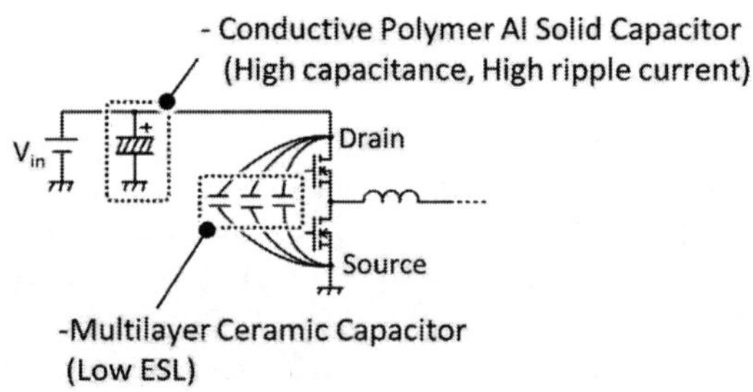

Figure 6.33 Layout position of MLCCs for input.

C. The capacitor to decrease ripple voltage noise to get a good, fast transient response for CPU operation stability. A low-impedance, low-ESR, low-ESL conductive-polymer solid aluminum electrolytic capacitor is appropriate. A conductive-polymer solid tantalum electrolytic capacitor or a multilayer conductive-polymer solid aluminum electrolytic capacitor is also used for a limited layout condition, such as a small laptop computer. It is required to follow the application note of the CPU to decide specifications in detail. Equation 6.33 is the simplified equation to calculate the upper impedance value of the output capacitor. An even lower-impedance capacitor than the calculated result from the equation below for an output capacitor should be selected.

$$Z_{\text{C}} \le \frac{V_{\text{rip}}}{\Delta I} \qquad \begin{cases} Z_{\text{C}}: \text{Impedance of capacitor } (\Omega) \\ V_{\text{rip}}: \text{Output voltage ripple noise (V)} \\ \Delta I: 30\% \text{ of maximum output current (A)} \end{cases} \tag{6.33}$$

D. The decoupling capacitor to meet the requirement of a lower-impedance characteristic at a wide frequency range for stable CPU operation. It is laid out as close as the CPU socket, especially in the case of VRMs, as shown in Fig. 6.34. It is possible to achieve low-impedance performance at a wide frequency range by using several MLCCs that have excellent performance at high frequency. It should be noted that there is a risk of power supply abnormal oscillation, which degrades the transient response performance, caused by the feedback network phase delay due to the excellent performance (low ESR) of an MLCC.

[Top Side] [Bottom Side]

Figure 6.34 Location of a decoupling capacitor (MLCCs) for a VRM.

6.2.4.2 Role of a capacitor in an isolated forward/flyback converter

An isolated converter generates a common intermediate voltage that is required for the input voltage of the VRM or POL converter on a circuit board. The primary and secondary sides of an isolated converter are electrically insulated by a switching transformer to shut conductive induction noise or to prevent an electric-shock accident.

A flyback converter is often used for a cellular phone charger or a laptop computer adapter (10 W to 150 W), a forward converter is for desktop PCs (150 W to 350 W), and a half-bridge converter, or forward converter with two switching devices, is for applications with more than 350 W. Figure 6.35 shows the basic circuit and capacitors used for a forward converter and Fig. 6.36 for a flyback converter.

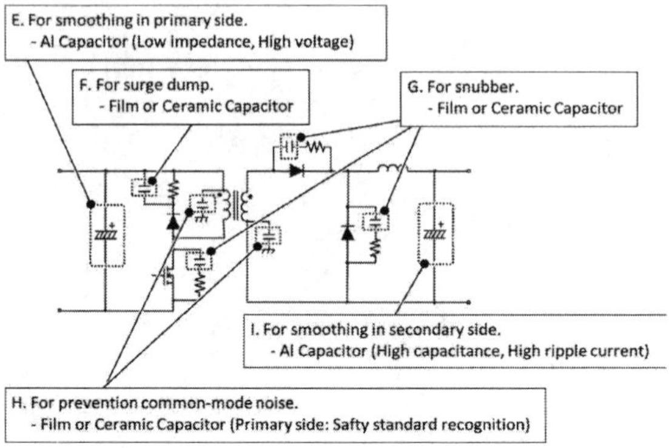

Figure 6.35 Forward converter.

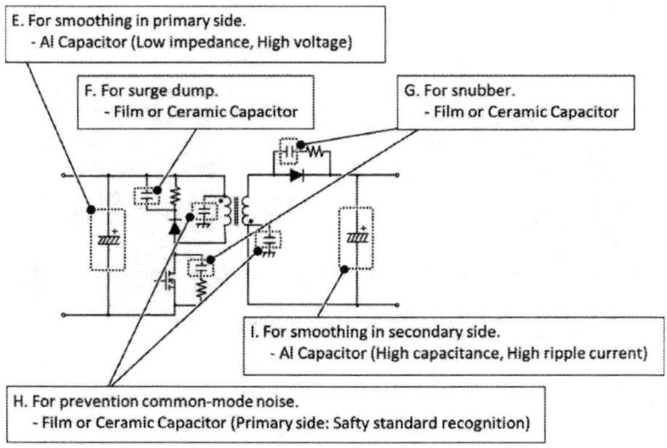

Figure 6.36 Flyback converter.

E. The input capacitor to absorb a high-frequency ripple current that is generated from a PFC circuit-switching operation to prevent the noise from transmitting to the input side. The aluminum electrolytic capacitor with low-impedance characteristics at a high frequency range is used.

F. The capacitor used in a surge cramp circuit to cut the peak of the surge voltage that is generated by leakage inductance of the transformer (Fig. 6.37). A film or ceramic capacitor is generally used.

G. The capacitor used in a snubber circuit to absorb the surge voltage of a switching element or a rectifier with high-speed operation (Fig. 6.38). A film or ceramic capacitor with high heat resistance is used because it is used near the heat-generating component.

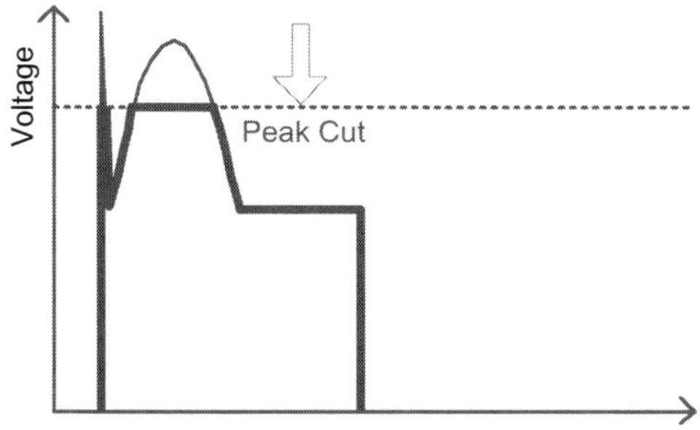

Figure 6.37 Surge clamp circuit waveform.

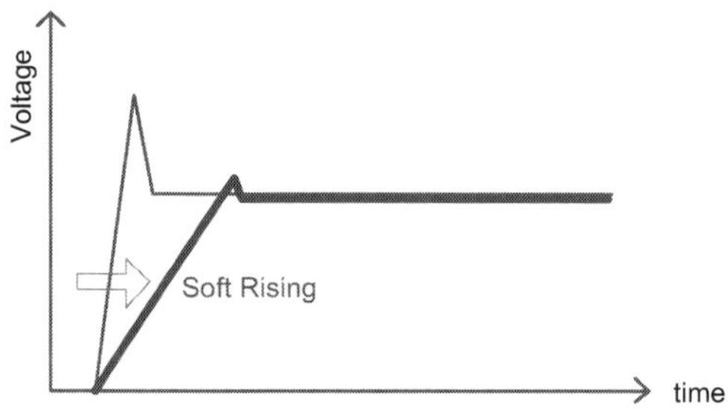

Figure 6.38 Snubber circuit waveform.

H. The capacitor to reduce common-mode noise that is generated from switching transformer stray capacitance (class Y capacitor or line bypass capacitor). It absorbs the conductive induction noise by connecting a capacitor between the primary and secondary non-noisy node sides of the switching transformer (Fig. 6.39). For the primary side, a safety standard–certified film or ceramic capacitor is used.

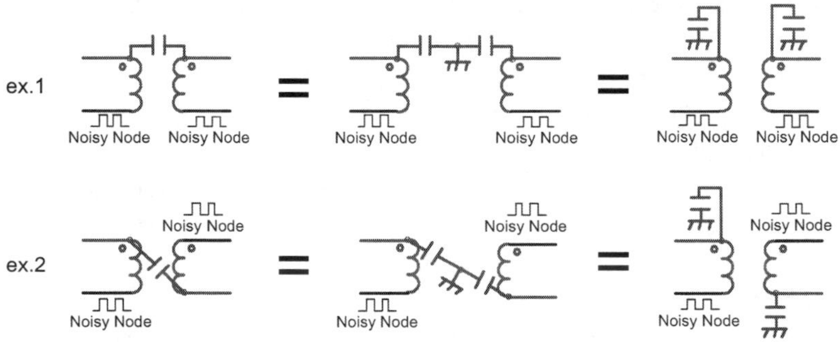

Figure 6.39 Line bypass capacitor for isolated transformer.

I. The output capacitor to stabilize or smooth output voltage to reduce ripple current. An aluminum electrolytic capacitor with large capacitance and low-impedance characteristics at a high frequency range is used.

In the case of a forward converter, it is possible to reduce ripple voltage noise or voltage variation at a rapid load change by laying out the capacitor near the output terminal. In the case of a flyback converter, it is possible to reduce ripple voltage noise or voltage variation at a rapid load change by laying out the capacitor near the switching transformer and also adding an LC filter beside the output terminal (Fig. 6.40).

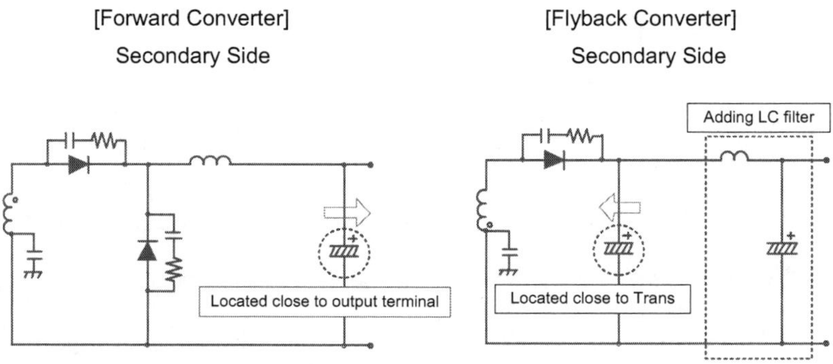

Figure 6.40 Capacitor optimum location for a forward converter and a flyback converter.

6.2.4.3 Role of a capacitor in an AC/DC rectification circuit and a PFC converter

An AC/DC rectification circuit rectifies and converts AC input voltage by a diode bridge into DC voltage. A PFC circuit forms a sine current waveform by correcting the input AC current waveform, including harmonic distortion, to reduce the harmonic distortion current.

Figure 6.41 shows the basic circuit, voltage/current wave, and a capacitor without a PFC circuit. The cause of harmonic distortion of the input AC current waveform is the aluminum electrolytic capacitor for smoothing (capacitor input–type smoothing circuit), which has large capacitance and is laid out beside the output of the diode bridge. A large peak current flows during the charging period of the capacitor, so it makes the input AC current waveform largely distorted. The input AC current waveform, including harmonic distortion, reduces utilization efficiency of the current, and it may cause a large energy loss among the wiring lines of the equipment or generate noise, causing other equipment malfunction.

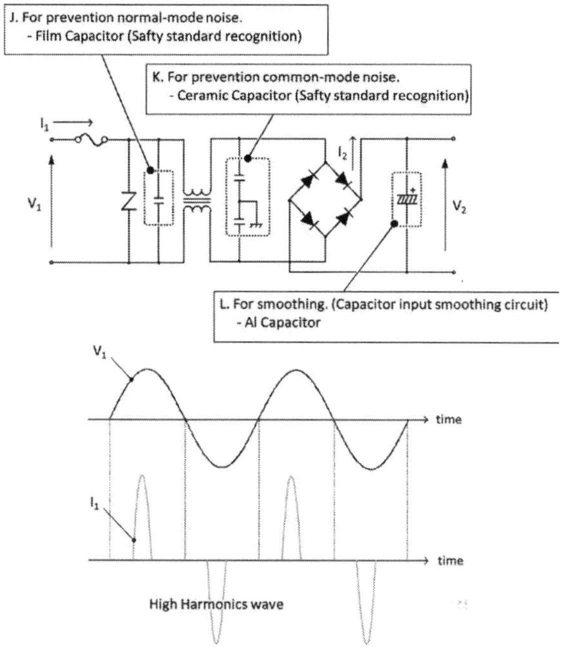

Figure 6.41 Input voltage and current waveform without a PFC circuit.

J. The capacitor to reduce the normal-mode noise among AC inputs (class X capacitor or across-the-line capacitor). A safety standard–certified film capacitor is used.
K. The capacitor to reduce common-mode noise (class Y capacitor or line bypass capacitor). A safety standard–certified ceramic capacitor is used.
L. The capacitor to smooth the voltage that is full-wave-rectified by a diode bridge. It is possible to decrease ripple voltage noise if a large-capacitance aluminum electrolytic capacitor is used, but the input AC current waveform is distorted largely. Figure 6.42 is the output voltage waveform that was smoothed by the output capacitor. It shows that the input AC current waveform is distorted by a large peak current flowing only during the charging period of the capacitor. It is possible to improve the distorted input AC current waveform by adding a PFC circuit.

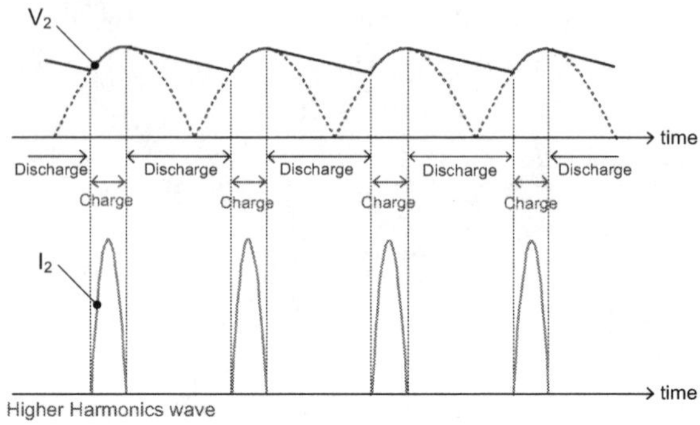

Figure 6.42 Smoothing output voltage (without a PFC circuit).

Figure 6.43 shows the basic circuit, voltage/current wave, and capacitors with a PFC circuit. A PFC circuit corrects the input AC current waveform, including harmonic distortion, into a sine current waveform by pulse width modulation (PWM) switching control. A PFC circuit is basically a boost converter construction, and it is possible to use it as a worldwide specification, accepted by almost all the commercial powers of the world, by setting its output voltage to 380 V to 400 V.

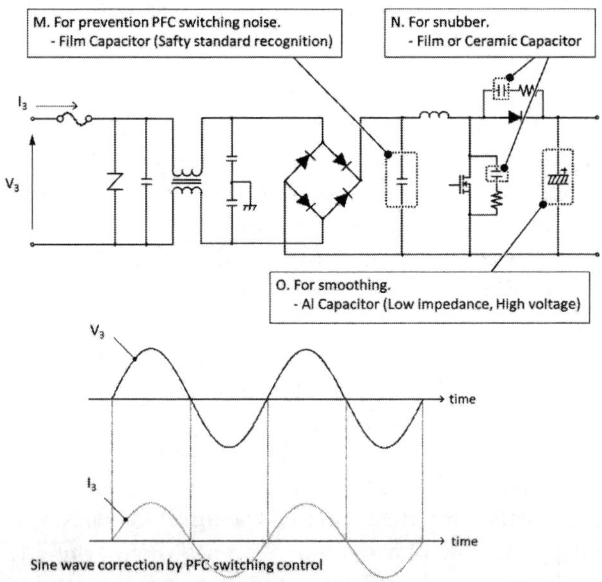

Figure 6.43 Input voltage and current waveform with a PFC circuit.

M. The capacitor to absorb the high-frequency ripple current that is generated from the switching operation of a PFC circuit to prevent noise from transmitting to the input side of the power supply. A film capacitor with low impedance at a high frequency is used.

N. The capacitor used in a snubber circuit to absorb the surge voltage of a switching element or a rectifier with high-speed operation. A film or ceramic capacitor with high heat resistance is used because it is used near the heat-generating component.

O. The output capacitor to stabilize or smooth output voltage to reduce ripple current. An aluminum electrolytic capacitor with large capacitance and low-impedance characteristics at a high frequency range is used.

Chapter 7

On-Chip Voltage Converters

Masashi Horiguchi*
Japan Science and Technology Agency, Tokyo 102-8666, Japan
mhoriguti@sunny.ocn.ne.jp

7.1 Introduction

On-chip voltage converters [1] have been widely used in various large-scale integrations (LSIs) for converting an external supply voltage to internal supply voltages different from the external voltage. They were first introduced in memory LSIs in the late 1980s to the early 1990s [2–12], followed by logic LSIs around 2000 [13–16]. The main object of introducing voltage converters is to bridge a voltage gap between a low voltage for core circuitry and a high voltage for interface (I/O) circuitry. Thus, the interface voltage dictated by systems is maintained at the same level for as long as possible, even if the breakdown voltage of miniaturized devices used in core circuitry is below the external supply voltage.

On-chip voltage converters are designed under the following restrictions:

*Formerly with Renesas Electronics Corp., Kodaira Tokyo 187-8588, Japan

Handbook of Power Management Circuits
Edited by Haruo Kobayashi and Takashi Nabeshima

ISBN 978-981-4613-15-6 (Hardcover), 978-981-4613-16-3 (eBook)
www.panstanford.com

(1) The number of off-chip components required for the converter circuit should be as few as possible, ideally zero. The number of additional terminals for the off-chip components should also be minimized. In particular additional terminals are not allowed for standardized products such as general-purpose dynamic RAMs (DRAMs).
(2) The chip area of the converter circuit should be minimized. Therefore using large passive devices should be avoided.
(3) The converter circuit should be fabricated with the compatible process steps with the LSI in which the circuit is implemented. Additional process steps and/or photomasks will increase the fabrication cost.

In this chapter, the motives of implementing on-chip voltage converters into LSIs are described in Section 7.2. The design issues of on-chip voltage converters are discussed in the following three sections. Section 7.3 presents voltage reference circuits that generate stable reference voltages for voltage converters. Voltage down-converters and voltage up-converters are described in Sections 7.4 and 7.5, respectively.

7.2 On-Chip Voltage Conversion

On-chip voltage converters are implemented into LSIs because of the following reasons:

(1) Voltage gap between external and internal power supply voltages: This is the major motive. The trends in external and internal power supply voltages are shown in Figs. 7.1 and 7.2 for memory LSIs [17] and microcontrollers [15], respectively. After the 5 V standard (transistor–transistor logic [TTL]-compatible) era, the internal (core) voltages have been reduced according to the well-known metal–oxide–semiconductor (MOS) device-scaling rule [18]. On the other hand, LSI users are reluctant to change the supply voltage frequently because of the compatibility with legacy products. The on-chip voltage down-converter is the solution to bridge these conflicting requirements.

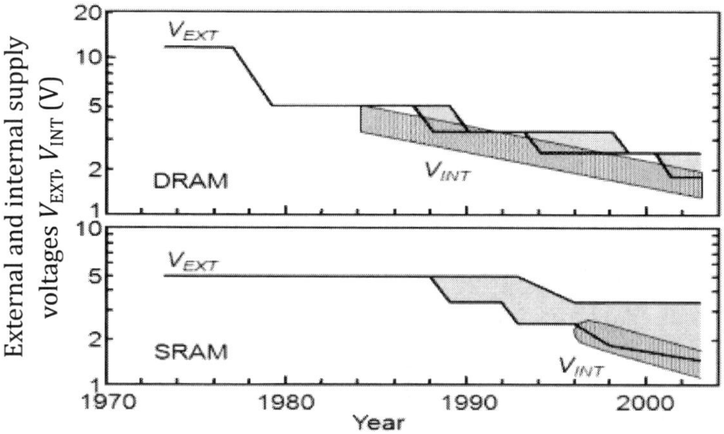

Figure 7.1 Trends in external and internal supply voltages of memory LSIs, dynamic RAMs (DRAMs, top), and static RAMs (SRAMs, bottom) [1].

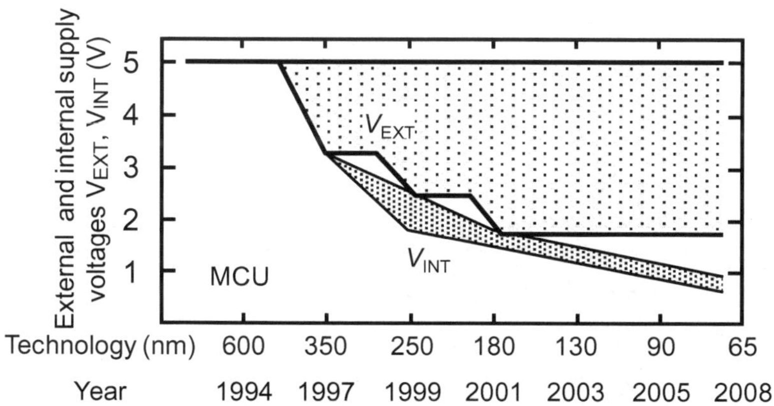

Figure 7.2 Trends in external and internal supply voltages of microcontrollers [15].

(2) Battery operation: With the wide spread of battery-powered electronic equipment, the demand for battery operation of LSIs has increased. However, the output voltage of a battery is not always suitable for circuit operation, because the voltage is chemically determined and cannot be changed. In addition, the output voltage varies with temperature and reduces with discharging. Therefore the conversion from the battery output

voltage into a constant voltage is needed. Figure 7.3 shows an example of a microcomputer with a voltage up-converter (boost converter) to enable single-cell operation [19]. The dry-cell output voltage of 1.5 V is converted to 3 V for internal analog circuits by the on-chip DC–DC boost converter with an off-chip inductor and a capacitor.

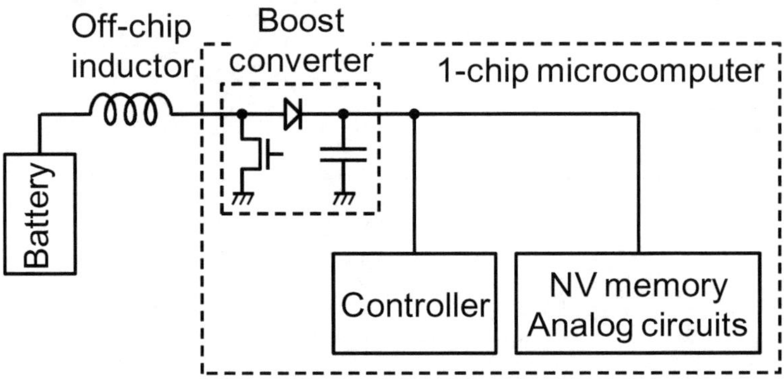

Figure 7.3 Microcomputer with an on-chip DC–DC boost converter [19].

(3) Memory operation: Not only memory LSIs but also logic LSIs with on-chip memories require various voltages for memory operation. Memory cells require various voltages, high/low and positive/negative voltages, as shown in Figs. 7.4 and 7.5 according to operation modes. These voltages are usually generated on-chip voltage up-/down-converters.

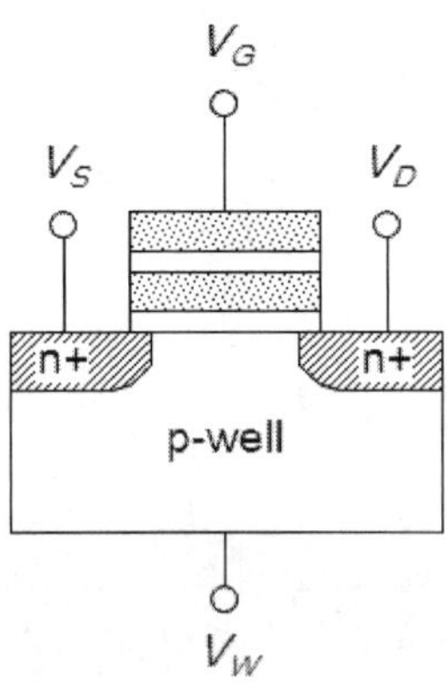

	V_G	V_D	V_S	V_W
Read	3.8	1.0	0.0	0.0
Write	10.0	5.4	0.0	0.0
Erase1	−11.5	open	open	10.0
Erase2	4.7	5.4	0.0	0.0

unit: V

Figure 7.4 Example of voltages used for flash memory.

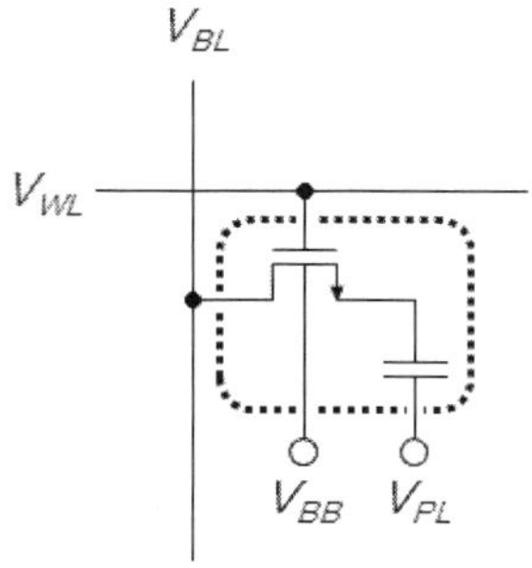

	V_{WL}	V_{BL}	V_{PL}	V_{BB}
Standby	0.0	0.9	0.9	−1.0
Read	3.8	open	0.9	−1.0
Write	3.8	1.8(H) 0.0(L)	0.9	−1.0

unit: V

Figure 7.5 Example of voltages used for DRAM.

(4) Advanced circuit design: The performance of a circuit can be improved by using on-chip voltage converters compared to directly supplying external voltage. Circuits in an LSI are usually designed under a supply voltage variation of ±10% determined by the specifications of the LSI. Using an on-chip voltage converter, however, the variation of internal supply voltage can be less than 10%. Thus, an unnecessary design margin can be eliminated, and the worst-case circuit performance is improved. The worst-case performance can be further improved by a small positive temperature coefficient of the internal supply voltage, because MOS circuit performance generally degrades at high temperature.

(5) Leakage reduction circuits: The leakage current of MOS devices has become one of the critical problems of LSIs with device scaling. The currents include subthreshold current (off current), gate-induced drain leakage (GIDL) current, and gate-oxide tunneling current [1]. Various circuit techniques to reduce these currents are proposed and implemented in LSIs. An example to reduce the leakage currents of an static RAM (SRAM) cell is shown in Ref. [20]. The subthreshold leakage current is reduced by raising the n-channel metal–oxide–semiconductor field-effect transistor (nMOSFET) source voltage, and the tunneling current is reduced due to the electric field relaxation by lowering the bitline voltage. These intermediate voltages are usually generated by on-chip voltage converters.

Figure 7.6 shows the block diagram of a typical on-chip voltage converter, which converts the external supply voltage V_{EXT} into the internal supply voltage V_{INT} lower than V_{EXT}. The converter consists of a reference voltage generator, a voltage-level converter/trimmer, and a voltage down-converter. The reference voltage generator generates a voltage V_{BGR} that is stable against process, voltage, and temperature (PVT) variations. The voltage-level converter generates a voltage V_{REF} based on V_{BGR} because V_{BGR} is not necessarily equal to V_{INT}. At the same time the level of V_{REF} is tuned to cope with the variation of V_{BGR}. The design issues of the reference voltage generator and the voltage-level converter/trimmer are given in Section 7.3. The voltage down-converter generates the internal supply voltage V_{INT} using V_{REF} as a reference. Although the voltage level of V_{INT} is usually equal to that of V_{REF}, the current-driving capability is much larger to provide the internal circuit with a power supply voltage. The details are given in Section 7.4.

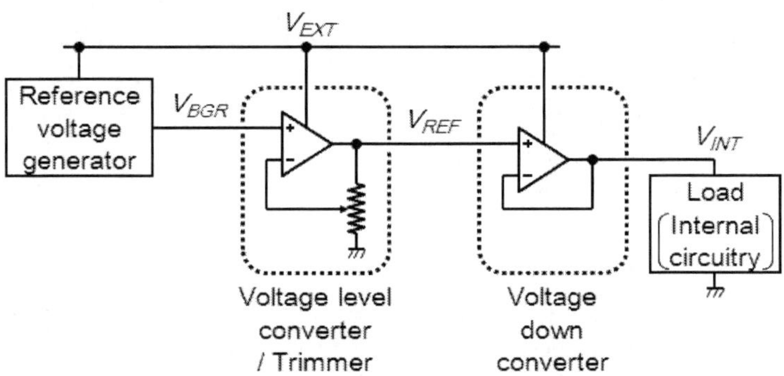

Figure 7.6 Schematic of an on-chip voltage down-converter.

In the case of voltage up-conversion (Fig. 7.7), the voltage down-converter is replaced by a voltage up-converter, which is realized by a charge pump circuit, a switching (boost) converter, etc., as described in Section 7.5. The output voltage V_{PP} is divided and is compared with V_{REF}, and the comparison result is used to control the voltage up-converter. Another implementation of voltage up-conversion is shown in Fig. 7.8. In this case the voltage up-converter is followed by a voltage down-converter (series regulator), which uses the high voltage as a power supply voltage and V_{REF} as a reference. This configuration features a smaller ripple compared to Fig. 7.7 due to the voltage regulation effect of the series regulator. The power efficiency, however, is lower because of the voltage up- and down-conversions.

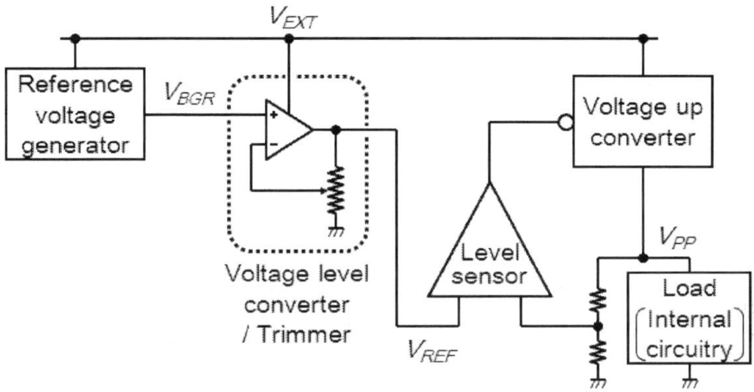

Figure 7.7 Schematic of an on-chip voltage up-converter.

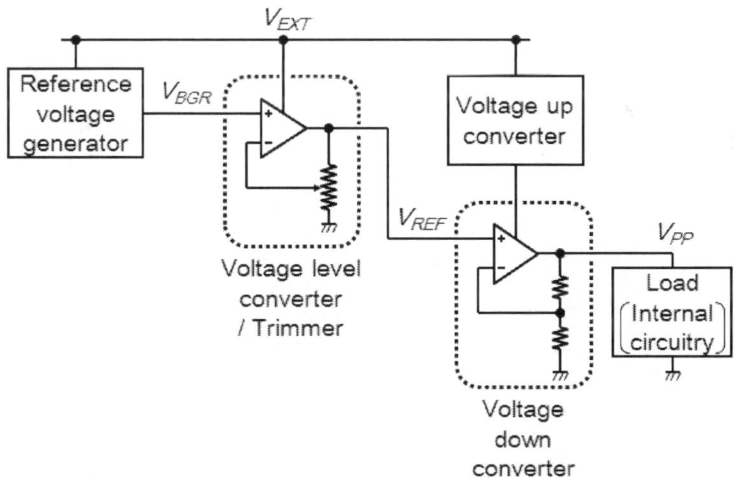

Figure 7.8 Schematic of an on-chip voltage up-converter.

7.3 Voltage Reference Circuits

The design issues for voltage reference circuits are as follows:

(1) The output voltage must be stable against PVT variations.

(2) The circuit area should be minimized. Therefore using large passive devices should be avoided.

(3) If the circuit has a negative feedback loop, the stability should be confirmed. Therefore a sufficient phase margin should be ensured.

(4) Power dissipation should be minimized because the circuit must operate even when an LSI is in standby mode.

Various types of circuits have been proposed for MOS LSIs, and they are roughly classified into three categories, as shown in Table 7.1. A threshold voltage (V_t) reference-type circuit is easy to design and requires no additional process steps. However, its output voltage significantly varies with temperature because V_t has a temperature coefficient of about −2 mV/°C. The voltage also varies due to process fluctuation. A threshold voltage difference (ΔV_t) reference-type circuit [7, 21] has a much smaller temperature variation because the temperature coefficients of two MOSFETs are cancelled by each other. However, the voltage variation due to process fluctuation is still large. A bandgap reference (BGR) circuit using bipolar transistors features a small temperature variation and a relatively small process variation and is mainly used for providing a reference voltage to on-chip voltage converters. A drawback of the BGR circuit is that the output voltage is fixed, around 1.2 V, and the circuit cannot operate on a supply voltage below 1.2 V. However, some solutions to this problem are proposed, as described later.

Table 7.1 Comparison of reference voltage generation circuits

	MOSFET V_t	MOSFET ΔV_t	BGR
Temperature Dependency	Large	Small	Small
Process Dependency	Large	Large	Medium–Small
Additional Process	None	Low/high-V_t MOSFET	None (Triple well)
OutputVoltage	mV_t	$m\Delta V_t$	1.2V
Minimum Operating Voltage	$mV_t + \alpha$	$V_{tn} + \lvert V_{tp}\rvert + \alpha$	1.2V + α

The principle of a BGR circuit is shown in Fig. 7.9. The base emitter voltage (V_{BE}) of a bipolar transistor has a negative temperature coefficient of around −2 mV/°C. On the other hand, the V_{BE} difference (ΔV_{BE}) between two bipolar transistors with different current densities is expressed as

$$\Delta V_{\text{BE}} = V_{\text{BE1}} - V_{\text{BE2}} = \frac{kT \ln N}{q} \tag{7.1}$$

where q is the element charge, k is the Boltzmann constant, T is absolute temperature, and N is the current–density ratio of the bipolar transistors. Thus, ΔV_{BE} has a positive temperature coefficient of 86 μV/°C multiplied by ln$N$. Adding V_{BE} and ΔV_{BE} with appropriate multipliers yields an almost zero-temperature-coefficient voltage expressed as

$$V_{\text{BGR}} = aV_{\text{BE}} + b\Delta V_{\text{BE}} = aV_{\text{BE}} + b \ln N \cdot \frac{kT}{q} \tag{7.2}$$

By setting $a = 1$ and $b\ln N = 20$–23, we can obtain an almost constant voltage V_{BGR} of around 1.2 V.

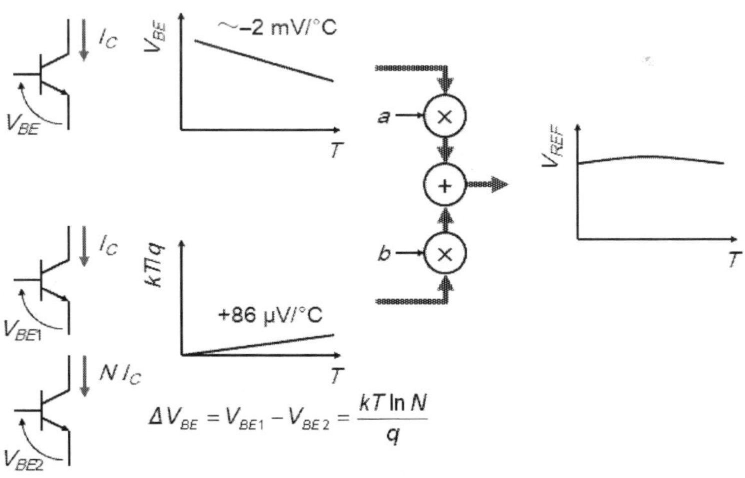

Figure 7.9 Principle of a bandgap reference (BGR).

Various kinds of circuits that realize the above principle have been proposed and implemented in LSIs. Some of the circuits are described in the following text.

Figure 7.10 shows a BGR circuit utilizing parasitic pnp transistors Q_1 and Q_2 [22]. Although a BGR circuit requires bipolar transistors, we can use parasitic bipolar transistors even in MOS LSIs. The cross section of the pnp transistor is shown in Fig. 7.11. The transistor has an emitter of a p+ diffusion area, a base of an n-type well, and a collector of a p-type substrate and is compatible with the MOS process. Since the substrate is usually connected to ground, we must design a circuit with the collector grounded. The circuit in Fig. 7.10 satisfies this condition. The operational amplifier A with a sufficiently large gain equalizes the potentials of nodes N_1 and N_2 (virtual short circuit). If we design $R_1 = R_2$, the currents I_1 and I_2 are equalized. Since the emitter junction size of Q_2 is N times that of Q_1, the current density of Q_2 is $1/N$ that of Q_1. This current–density ratio generates a base-emitter-voltage difference, which is expressed by Eq. 7.1. Since this voltage difference appears at the both ends of R_3, the current is expressed as

$$I_1 = I_2 = \frac{kT \ln N}{qR_3} \tag{7.3}$$

Therefore the output voltage V_{BGR} is given by

$$V_{BGR} = V_{BE1} + I_1 R_1 = V_{BE1} + \frac{kTR_1 \ln N}{qR_3} \tag{7.4}$$

Compared to Eqs. 7.2 and 7.4, we can find that $a = 1$ and $b = R_1/R_3$. The temperature coefficient is cancelled out by choosing appropriate values of R_1, R_3, and N.

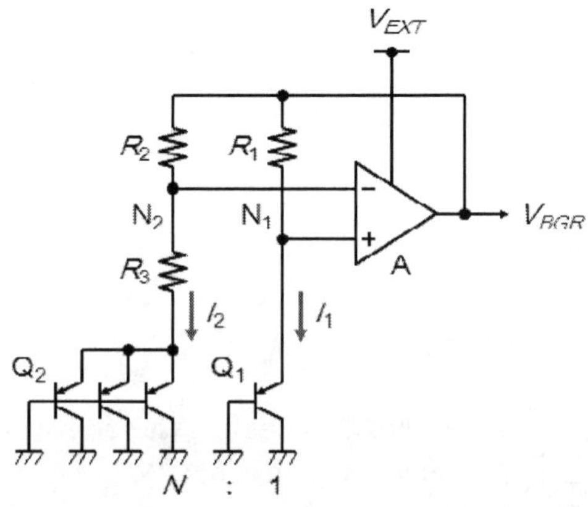

Figure 7.10 Bandgap reference circuit using pnp bipolar transistors [22].

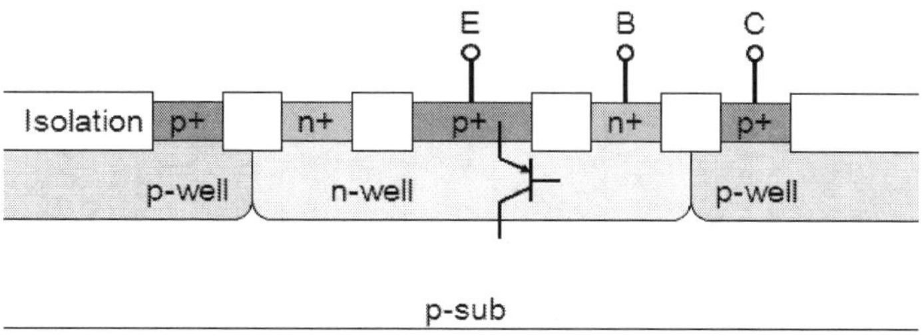

Figure 7.11 Cross section of a parasitic pnp bipolar transistor.

Another BGR circuit using parasitic npn transistors Q_1 and Q_2 [23] is shown in Fig. 7.12. Figure 7.13 shows the cross section of an npn transistor that can be fabricated by a triple-well complementary metal–oxide–semiconductor (CMOS) process. The device has an emitter of an n+ diffusion area, a base of a p-type well, and a collector of an n-type deep well. Since all three terminals can be at any potential, we can design a circuit without the collector-grounded restriction. The potentials of nodes N_1 and N_2 are equalized by the operational amplifier A. If we design $R_1 = R_2$, and Q_1 and Q_2 have identical current gain, the currents I_1 and I_2 are equalized. The current–density ratio between Q_1 and Q_2 generates a base-emitter-voltage difference, which is expressed by Eq. 7.1. Since this voltage difference appears at the both ends of R_3, the current is expressed as

$$I_1 = I_2 = \frac{kT \ln N}{\alpha q R_3} \tag{7.5}$$

where α is the common-base current gain of the npn transistor. Since $I_1 + I_2 = 2I_1$ flows through R4, the output voltage V_{BGR} is given by

$$V_{\mathrm{BGR}} = V_{\mathrm{BE1}} + 2I_1 R_4 = V_{\mathrm{BE1}} + \frac{kTR_4 \ln N}{\alpha q R_3} \tag{7.6}$$

Therefore the temperature coefficient is cancelled out by choosing appropriate values of R_3, R_4, and N. This circuit features better voltage accuracy than the circuit in Fig. 7.10 because of less sensitivity on the offset voltage of amplifier A [1]. However, the minimum operating voltage is higher due to the voltage drop across the collector resistors.

The output voltage of the above BGR circuits is around 1.2 V, and they cannot operate under a supply voltage below 1.2 V. Some BGR circuits for low-voltage operation have been proposed. By adopting a smaller value of $a < 1$ in Eq. 7.2, and maintaining the ratio a/b, we can obtain a lower output voltage than 1.2 V.

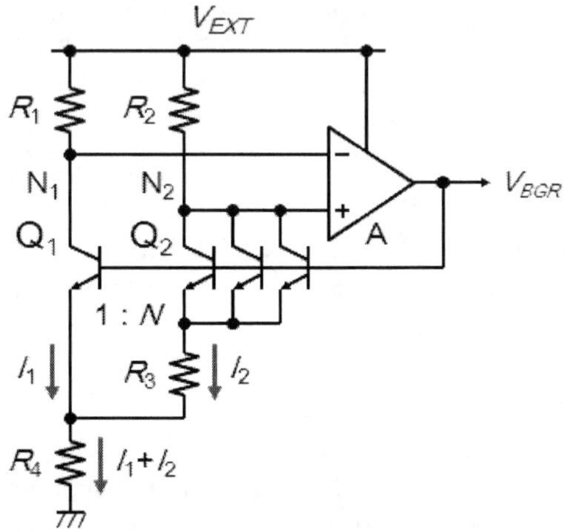

Figure 7.12 Bandgap reference circuit using npn bipolar transistors [23].

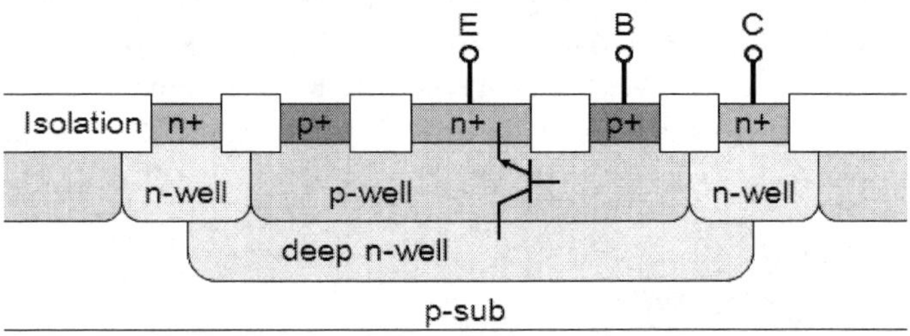

Figure 7.13 Cross section of a parasitic npn bipolar transistor.

The circuit shown in Fig. 7.14 is proposed in Ref. [24]. The node voltages of N_1 and N_2 are equalized by the operational amplifier A. The MOSFETs M_1, M_2, and M_3 have the same size and form a current mirror. The V_{BE} difference appears across R_2 because the currents of two pnp transistors Q_1 and Q_2 have the same emitter current expressed as

$$I_1 = I_2 = \frac{kT \ln N}{qR_2} \tag{7.7}$$

The same current flows through M_3. The output voltage V_{BGR} is expressed as

$$V_{BGR} = V_{BE3} + (I_3 - I_R)\, R_1 = I_R R_3 \tag{7.8}$$

From Eqs. 7.7 and 7.8 and $I_1 = I_2 = I_3$, we obtain

$$V_{BGR} = \frac{R_3}{R_1 + R_3}\left(V_{BE3} + \frac{R_1}{R_2} \cdot \frac{kT \ln N}{q}\right) \tag{7.9}$$

Comparing Eq. 7.9 with Eq. 7.2, we can find that $a = R_3/(R_1 + R_3)$, which is less than unity. If R_3 does not exist, a will be unity. However, by connecting R_3 in parallel with R_1 and Q_3, a lower value of a and a lower V_{BGR} are thus obtained.

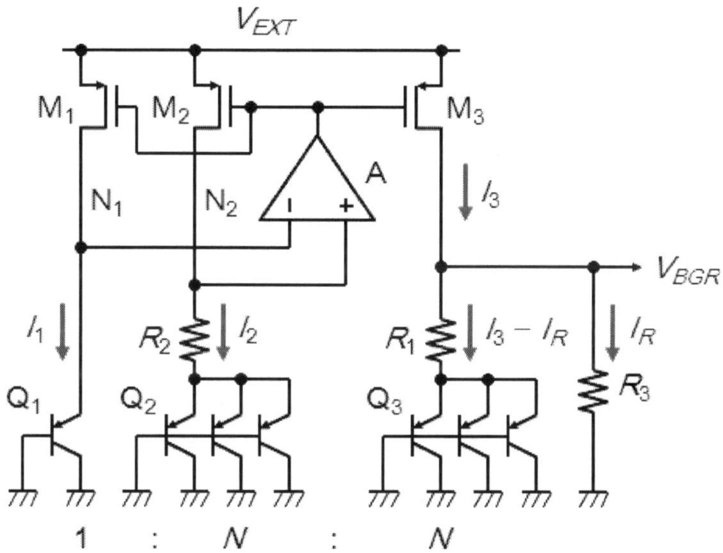

Figure 7.14 Low-voltage bandgap reference circuit [24].

Figure 7.15 shows another low-voltage BGR circuit [25]. The node voltages of N_1 and N_2 are equalized by the operational amplifier A. This circuit features resistors R_1 and R_2 connected in parallel with the bipolar transistors. The currents flowing in these resistors are proportional to V_{BE} given by

$$I_{1A} = I_{2A} = \frac{V_{BE1}}{R} \tag{7.10}$$

where $R_1 = R_2 = R$. On the other hand, the currents flowing in the bipolar transistors Q_1 and Q_2 are proportional to absolute temperature given by

$$I_{1B} = I_{2B} = \frac{kT \ln N}{qR_3} \tag{7.11}$$

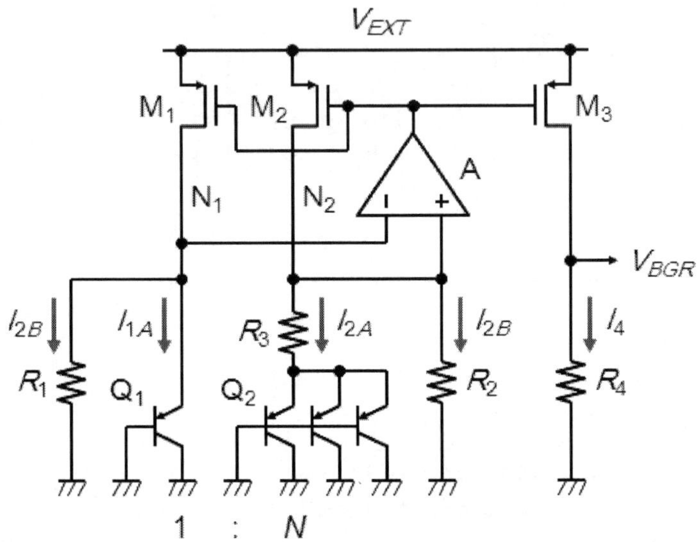

Figure 7.15 Low-voltage bandgap reference circuit [25].

These currents are summed in MOSFETs M_1 and M_2. Since the same current flows in R_4, the output voltage is expressed as

$$V_{BGR} = I_4 R_4 = (I_{1A} + I_{1B})R_4 = \frac{R_4}{R} V_{BE1} + \frac{R_4}{R_3} \cdot \frac{kT \ln N}{q} \tag{7.12}$$

If we design $R_4 < R$, a lower output voltage V_{BGR} is obtained. This circuit is called as a current-mode BGR circuit because currents with positive and negative temperature coefficients are summed.

Another low-voltage BGR circuit shown in Fig. 7.16 [26] is very similar to that in Fig. 7.14, except for M_4. Similar analysis yields

$$V_{BGR} = \frac{R_3}{R_1 + R_3}\left(V_{BE3} + \frac{R_1}{R_2} \cdot \frac{kT \ln N}{q}\right) \tag{7.13}$$

which is identical to Eq. 7.9. This circuit can output a voltage lower than V_{BE3}, while the circuit in Fig. 7.14 cannot. This is because the current I_3–I_R may be positive or negative according to temperature, while I_3 must be always larger than I_R in Fig. 7.14. Therefore this circuit is more suitable for low-voltage operation.

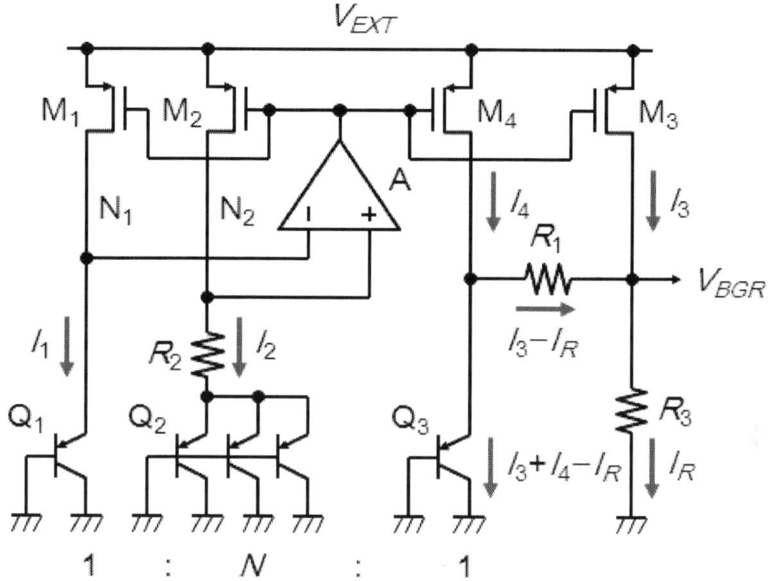

Figure 7.16 Low-voltage bandgap reference circuit [26].

The reference voltages, even those from BGR circuits, fluctuate due to fabrication process variations. Therefore a trimming circuit is usually inserted after a reference voltage generator. Figure 7.17 shows a trimming circuit, which converts V_{BGR} to V_{REF}. The conversion ratio is expressed as

$$\frac{V_{\mathrm{REF}}}{V_{\mathrm{BGR}}} = \frac{R_1 + R_2}{R_1} \tag{7.14}$$

where R_1 and R_2 are the upper and lower resistances of the tap and the resistor string, respectively. The ratio can be adjusted by changing the tap position. The voltage V_{REF} (or the internal supply voltage) of each LSI is tested, and the data to select an appropriate tap position is stored in the programmable devices before shipping. The trimming data read from the devices is decoded by the decoder and used for selecting the tap. The programmable devices must be nonvolatile to retain the trimming data, even during power-down. Fuses are usually used for memory LSIs [7], while flash memory cells are used for logic LSIs [15].

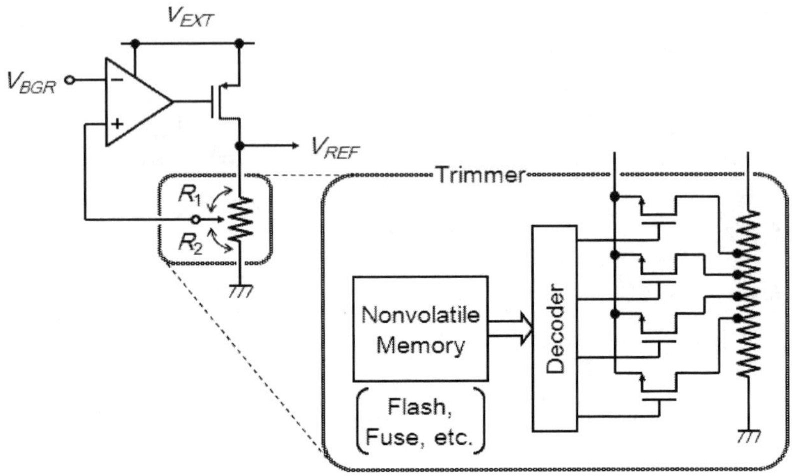

Figure 7.17 Voltage converter/trimmer circuit.

7.4 Voltage Down-Converters

On-chip voltage down-converters, which generate an internal supply voltage V_{INT} lower than the external supply voltage V_{EXT}, are indispensable for low-voltage LSIs. The design issues for the voltage down-converters are as follows:

(1) The circuit must supply a sufficient current necessary for the load (internal core circuit) with minimal voltage fluctuation.
(2) The number of off-chip components should be minimized, ideally zero, for reducing system cost as well as the number of LSI terminals.
(3) If the circuit has a negative feedback loop, the stability should be confirmed. Therefore a sufficient phase margin should be ensured.
(4) The power conversion efficiency should be maximized.

Voltage down-converters are categorized as series (linear) regulators, switching regulators (buck converters), and switched-capacitor regulators, as shown in Fig. 7.18. The characteristics of the regulators are summarized in Table 7.2, and a comparison of the power conversion efficiencies is shown in Fig. 7.19 [1]. Since the horizontal and vertical axes are current and voltage, respectively, the area of each rectangle corresponds to power. The efficiency of a series regulator is poor because the input current I_{EXT} is almost equal to the output current I_{INT}, except for a small idling current. Therefore the voltage drop due to the on resistance of the output transistor (M in Fig. 7.18a), $V_{EXT} - V_{INT}$, corresponds to the power loss, and the power efficiency is lower than the voltage conversion ratio, V_{INT}/V_{EXT}, as shown in Fig 7.19a. The switching and switched-capacitor regulators achieve much higher power conversion efficiency, as shown in Fig. 7.19b,c because they utilize reactance elements (inductor or capacitor) that can temporarily store electric energy. The switching regulator, however, requires off-chip components (an inductor, a capacitor, etc.), increasing the board area and system cost. The switching regulator has another problem, electromagnetic interference (EMI), because of switching of the inductor current. The switched-capacitor regulator also requires off-chip capacitors. If the required supply current is small, the capacitors can be implemented on a chip [27]. The most important problem of the switched-capacitor regulator is its fixed voltage conversion ratio, while it can be changed by changing the reference voltage V_{REF} (series regulator) or by changing the duty ratio (switching regulator). The voltage conversion ratio is 1/2 in the case of Fig. 7.18c irrespective of the capacitor ratio. If another conversion ratio is required, we must change the circuit topology.

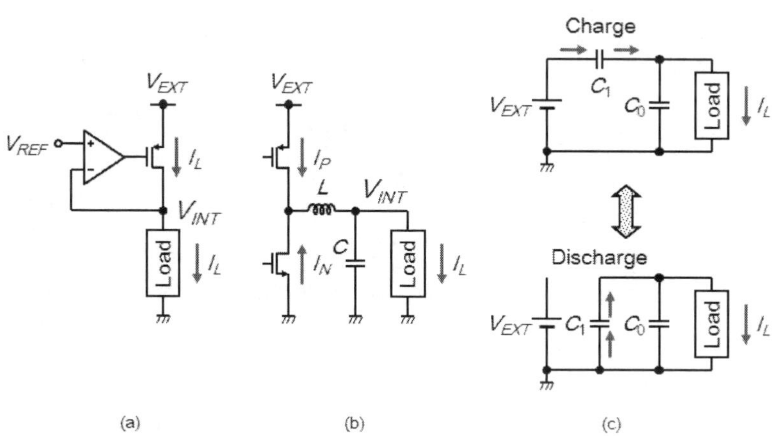

Figure 7.18 Voltage down-converters.

Table 7.2 Comparison of voltage down-converters

	Series	Switching	Switched C
Votage-conversion ratio	Any ratio [except for $V_{EXT} \sim V_{INT}$]	Any ratio	Fixed ratio
Power efficiency	$< \frac{V_{INT}}{V_{EXT}}$	>90%	>80%
Off-chip components	None (1 C)	1 L, 1 C, 1 Diode, (2 MOSTs)	n C (None)
Additional terminals	0-1	≥2	$2n$–1
Transient response	Amplifier UGF	Clock frequency	Clock frequency
EMI	Low	High	Medium

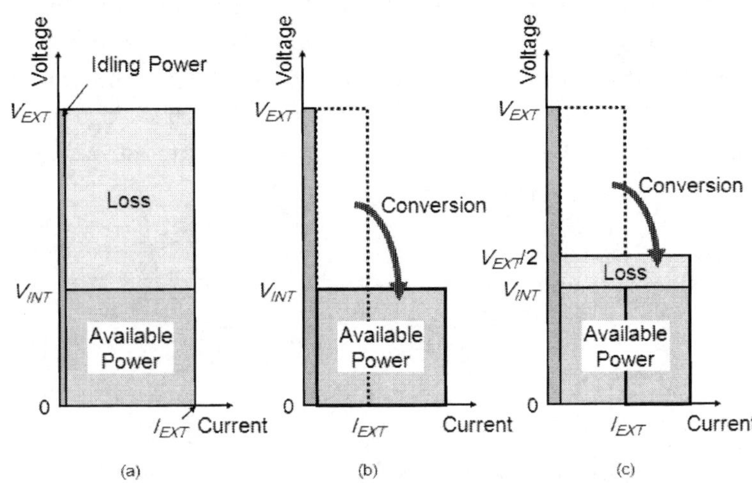

Figure 7.19 Comparison of power conversion efficiencies of voltage down-converters.

Series regulators are mainly used as on-chip regulators of memory LSIs as well as logic LSIs despite the low power conversion efficiency because few off-chip components are required. In particular off-chip components are not allowed in the case of standardized products, such as general-purpose DRAMs. A typical series regulator is illustrated in Fig. 7.20. It consists of a differential amplifier (error amplifier) and an output stage with a compensation circuit. The error amplifier compares the internal supply voltage V_{INT} with the reference voltage V_{REF} and controls the gate of the output p-channel MOSFET (pMOSFET) M_0 in the output stage so that V_{INT} is equal to V_{REF} irrespective of the load current I_L. Since this circuit contains a negative feedback loop, the circuit stability is an important design issue. Various compensation techniques to ensure sufficient stability are utilized [28–32], though the details are not described here.

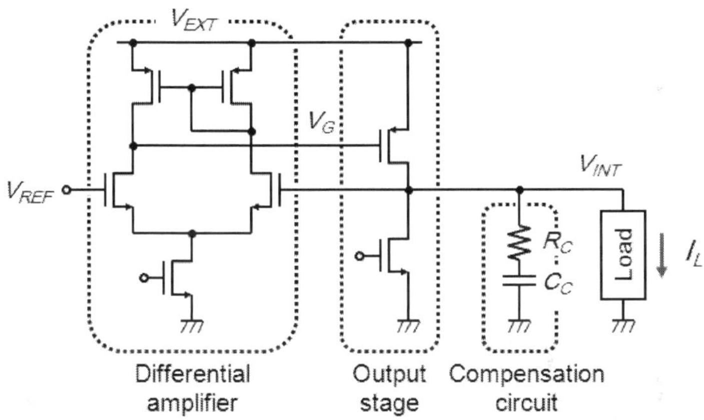

Figure 7.20 Typical series regulator.

A simplified circuit diagram of the series regulator and its large-signal equivalent circuit are shown in Fig. 7.21. The DC characteristics of the series regulator are calculated using this circuit. The load current is supplied from M_0, as expressed by

$$I_L = \frac{\beta_p}{2}\left(V_{EXT} - V_G - |V_{thp}|\right)^2 \tag{7.15}$$

where β_p and V_{thp} are the gain coefficient and threshold voltage of M_0, respectively. The gate voltage of M_0 is expressed as

$$V_G = A\,(V_{INT} - V_{REF}) \tag{7.16}$$

From these equations we obtain

$$V_{INT} = V_{REF} - \frac{1}{A}\sqrt{\frac{2I_L}{\beta_p}} \tag{7.17}$$

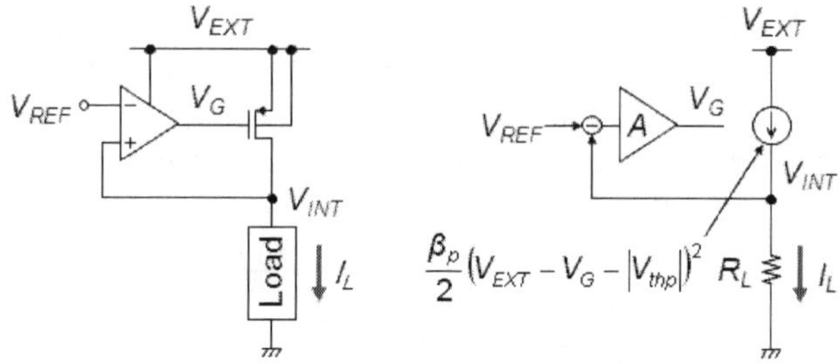

Figure 7.21 A pMOSFET output series regulator (a) and its large-signal equivalent circuit (b).

Thus, V_{INT} is equal to V_{REF} if $I_L = 0$ and decreases as I_L increases. If the output transistor is an nMOSFET, as shown in Fig. 7.22, Eqs. 7.15–7.17 are replaced by

$$I_L = \frac{\beta_n}{2}(V_G - V_{INT} - V_{thn})^2 \tag{7.18}$$

$$V_G = A\,(V_{REF} - V_{INT}) \tag{7.19}$$

$$V_{INT} = V_{REF} - \frac{1}{A+1}\sqrt{\frac{2I_L}{\beta_n}} \tag{7.20}$$

where β_n and V_{thn} are the gain coefficient and threshold voltage of M_0, respectively. The output voltage lowering of this circuit is almost the same as that of Fig. 7.21 if the gain coefficient of the output transistor is the same. This circuit features a smaller loop gain due to the source-follower output stage and therefore better stability. However, a low dropout voltage (V_{EXT} – V_{INT}) is difficult due to the V_{thn} drop of M_0, unless $V_{thn} \leq 0$ [11] or the error amplifier is supplied with a higher voltage than V_{EXT} [33].

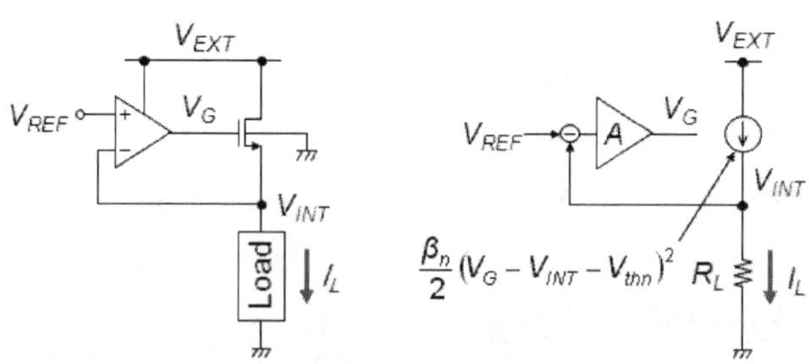

Figure 7.22 An nMOSFET output series regulator (a) and its large-signal equivalent circuit (b).

The above discussion is the DC characteristics when load current I_L changes. In practical LSI applications, however, the transient behavior of V_{INT} when I_L changes is more important. If the level of V_{INT} temporarily drops at a sudden change of I_L, a malfunction of the internal circuit might occur. The drop voltage is estimated by Tanaka [30] using the equivalent circuit of Fig. 7.23. When the load current steps from 0 up to I_L, the maximum drop voltage ΔV_{INT} is expressed as

$$\Delta V_{INT} = \frac{4\sqrt{2} I_L^{1.5} C_{OX} WL}{9 C_L I_S \sqrt{\beta_p}} = \frac{4\sqrt{2} I_L^{1.5} C_{OX} L^{1.5} \sqrt{W}}{9 C_L I_S \sqrt{\beta_{p0}}} \tag{7.21}$$

where C_{OX} is the gate-oxide capacitance per unit area, L and W are the channel length and width of M_0, I_S is the tail current of the error amplifier, and β_{p0} is the gain coefficient of a pMOSFET with unit W/L. We can determine W, L, and I_S using this equation so that ΔV_{INT} does not exceed a target value.

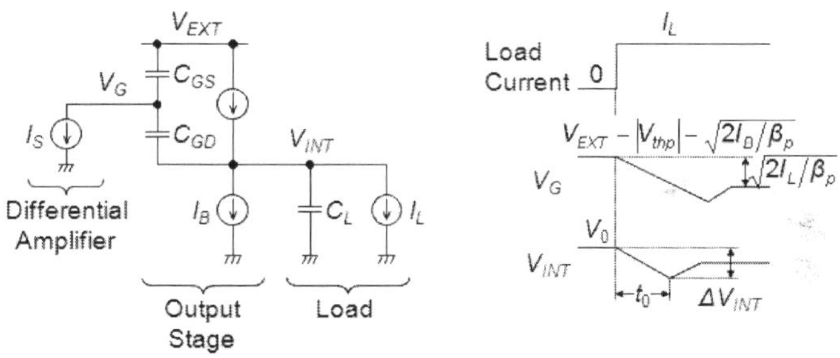

Figure 7.23 Large-signal equivalent circuit of a series regulator (a) and step-response waveforms (b).

7.5 Voltage Up-Converters

On-chip voltage up-converters and voltage inverters have been used in memory LSIs [34–42] as well as in logic LSIs with on-chip memories. The voltage up-converter generates an internal supply voltage V_{PP} higher than the supply voltage V_{EXT}, while the voltage inverter generates a negative voltage V_{BB}. The voltages include write/erase voltages for flash memories and wordline/substrate voltages for DRAMs. The design issues for the voltage up-converters are as follows:

(1) The circuit must supply a sufficient current necessary for the load (internal circuit) with minimal voltage fluctuation.

(2) The number of off-chip components should be minimized, ideally zero, for reducing system cost as well as the number of LSI terminals.

(3) The voltage applied to each device used in the circuit must not exceed a maximum limit to prevent device breakdown or reliability problems. This is an inherent design issue of voltage up-converters.

(4) The power conversion efficiency should be maximized.

Voltage up-converters are categorized as charge pumps, switching (boost) converters, and switched-capacitor converters, as shown in Fig. 7.24. The characteristics of the regulators are summarized in Table 7.3, and a comparison of the power conversion efficiencies is shown in Fig. 7.25. The charge pump and the switched-capacitor converter generate a high voltage using capacitors, which are off chip or are implemented in a chip if the required current is not large. On the other hand the switching converter utilizes an inductor and a capacitor. The charge pump generates a high voltage step by step, first generating double voltage, next generating triple voltage, and so on. On the other hand, the switched-capacitor and switching converters generate a high voltage at a stroke. The difference between the power conversion efficiencies of the circuits is not so large.

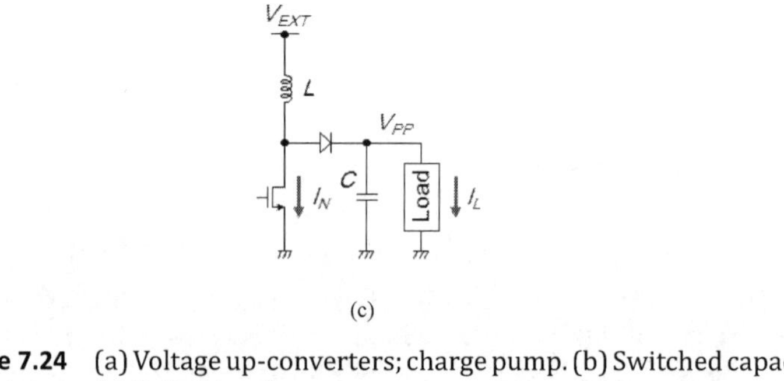

Figure 7.24 (a) Voltage up-converters; charge pump. (b) Switched capacitor. (c) Switching (boost) converter.

Table 7.3 Comparison of voltage up-converters

	Charge pump	Switched C	Switching
Conversion method	$V_{EXT} \rightarrow 2V_{EXT}$ ... $\rightarrow nV_{EXT}$	$V_{EXT} \rightarrow nV_{EXT}$	$V_{EXT} \rightarrow nV_{EXT}$
Conversion ratio	Fixed ratio	Fixed ratio	Any ratio
Power efficiency	>80%	>80%	>80%
Off-chip components	None (n C)	None (n C)	1 L, 1 C, 1 Diode, (2 MOSFETs)
Max. voltage of MOSFET	$2V_{EXT}$	$(n-1)V_{EXT}$	nV_{EXT}
Max. voltage of C	$(n-1)V_{EXT}$	V_{EXT} (except of C_0)	nV_{EXT}

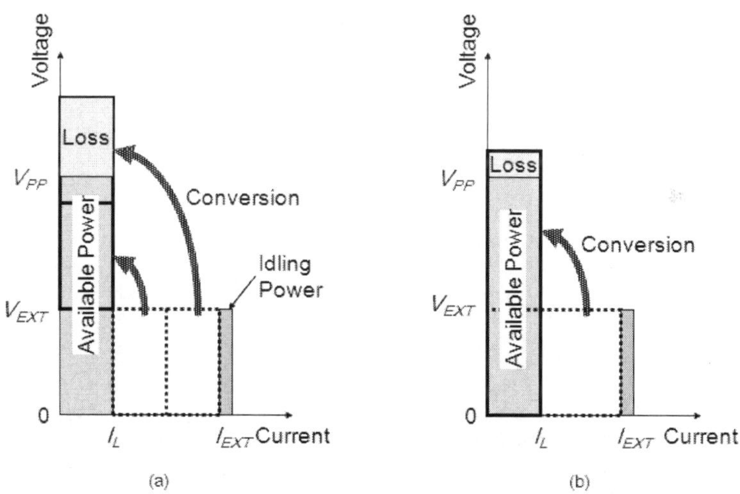

Figure 7.25 Comparison of power conversion efficiencies of voltage up-converters; (a) charge pump and switched capacitor, (b) switching converter.

The charge pump circuit and the switched-capacitor circuit are sometimes confused. Both utilize capacitors to generate a high voltage. However, they should be distinguished because the voltage conversion methods and voltages applied to devices are different, as shown in Table 7.3. For voltage doubling, however, the charge pump is almost the same as the switched capacitor.

The charge pumps are mainly used as on-chip voltage up-converters, especially those for memories. The principle of voltage doubling is shown in Fig. 7.26 [24]. This circuit generates a high voltage V_{PP} using a capacitor C_1 and four switches SW_1–SW_4. During period T_1, SW_1 and SW_4 are on and SW_2 and SW_3 are off. Capacitor C_1 is charged by the external supply voltage V_{EXT}, as shown in Fig. 7.26c. The load current I_L is supplied by capacitor C_0 during this state. During period T_2, SW_1 and SW_4 are off and SW_2 and SW_3 are on. C_1 is connected in series with V_{EXT}, as shown in Fig. 7.26d, generating a high voltage V_{PP}, and the charge in C_1 is supplied to the load. The analysis of this circuit yields the average output voltage expressed as [24]

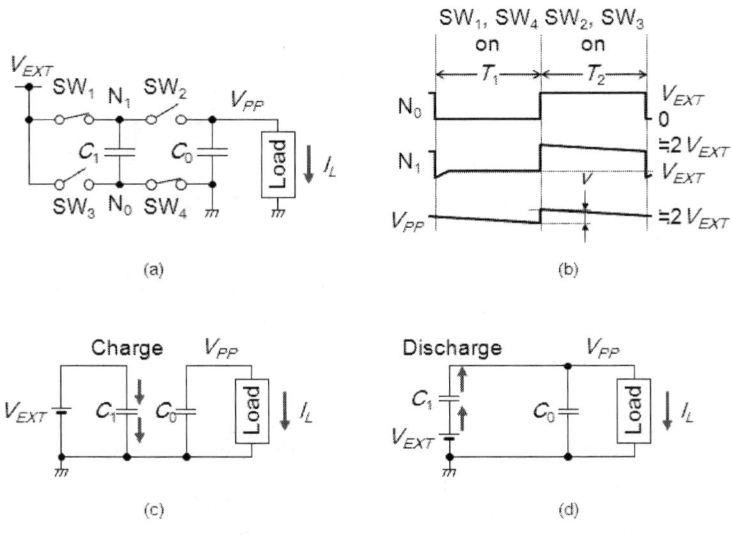

Figure 7.26 Principle of voltage doubling; circuit diagram (a), operating waveforms (b), equivalent circuit during T_1 (c), and equivalent circuit during T_2 (d).

$$\overline{V_{PP}} = 2V_{EXT} - I_L\left(T_1 + \frac{T_2 C_0}{C_0 + C_1}\right)\left(\frac{1}{2C_0} + \frac{1}{C_1}\right)$$

$$\cong 2V_{EXT} - \frac{I_L T}{C_1} \quad \left(C_0 >> C_1, T_1 = T_2 = \frac{T}{2}\right) \tag{7.22}$$

Thus, V_{PP} is equal to twice the supply voltage without load current I_L and decreases with I_L at a rate of T/C_1. Therefore this circuit is equivalent to a voltage source of $2V_{EXT}$ with an output resistance of

T/C_1. To enhance the current-driving capability, a smaller output resistance is required by reducing T and using a large capacitor C_1. The ripple v and power conversion efficiency η are expressed as

$$v = \frac{I_L}{C_0}\left(T_1 + \frac{T_2 C_0}{C_0 + C_1}\right) \cong \frac{I_L T}{C_0} \tag{7.23}$$

$$\eta = \frac{V_{PP}}{2V_{EXT}} \tag{7.24}$$

Thus, a shorter period T or a larger capacitor C_0 is desired for a smaller ripple.

A practical implementation of the principle is shown in Fig. 7.27. The switches SW_1 and SW_2 are replaced by rectifiers (nMOSFETs with a gate–drain connection), thus eliminating the control signals for the switches. On the other hand SW_3 and SW_4 are replaced by a CMOS inverter. This circuit is called a charge pump from the analogy with a water pump. A thrice or higher voltage can be generated by a multiple-stage charge pump circuit shown in Fig. 7.24a. The output voltage of the charge pump circuit, however, is lower than Eq. 7.22 by twice the threshold voltage (V_{th}) because two nMOSFETs are inserted along the path from V_{EXT} to V_{PP}:

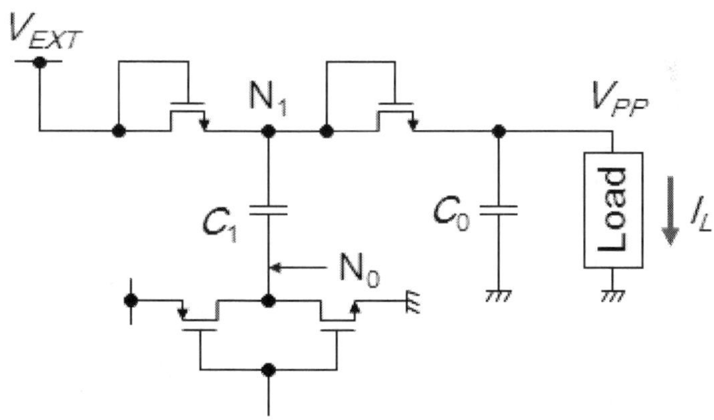

Figure 7.27 Charge pump voltage doubler.

$$\bar{V}_{PP} \cong 2V_{EXT} - 2V_{th} - \frac{I_L T}{C_1} \tag{7.25}$$

In the case of a multiple-stage charge pump, the output voltage is expressed as

$$\bar{V}_{PP} \cong nV_{\text{EXT}} - nV_{\text{th}} - \frac{(n-1)I_{\text{L}}T}{C_1} \tag{7.26}$$

where n is the number of nMOSFETs [34]. The output voltage lowering due to V_{th} is especially serious for low-voltage operation. In addition, the nMOSFETs of latter stages have higher V_{th}s because of body effect. To overcome this problem, boosting the gate voltages of nMOSFETs [38, 41, 43] and using pMOSFETs [44, 45] are proposed.

A voltage inverter, which generates a negative internal voltage V_{BB} from a positive external supply voltage V_{EXT}, can be constructed by applying the design principle of a voltage up-converter. The principle of the voltage inversion is shown in Fig. 7.28. When we need a negative voltage $V_{\text{BB}} = -V_{\text{EXT}}$, as shown in Fig. 7.28a, we assume V_{EXT} as ground and the ground as a negative supply voltage $-V_{\text{EXT}}$, as shown in Fig. 7.28b. We can obtain V_{BB} by doubling $-V_{\text{EXT}}$. Thus, voltage inversion is equivalent to negative voltage doubling. Similarly we can obtain a negative voltage $V_{\text{BB}} = -nV_{\text{EXT}}$, by multiplying the negative voltage $-V_{\text{EXT}}$ by $(n + 1)$, as shown in Fig. 7.28c,d. Thus, voltage inversion followed by multiplication by n is equivalent to negative voltage multiplication by $(n + 1)$.

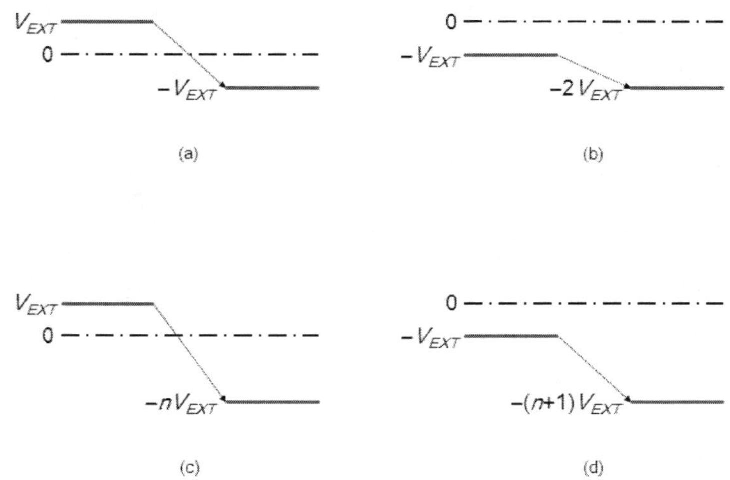

Figure 7.28 Principle of voltage inversion.

References

1. K. Itoh, M. Horiguchi, and H. Tanaka, *Ultra-Low Voltage Nano-Scale Memories*, Springer, New York, 2007.

2. T. Mano, J. Yamada, J. Inoue, and S. Nakajima, Circuit techniques for a VLSI memory, *IEEE J. Solid-State Circuits*, **SC-18**, 463–470, 1983.

3. H. K. Itoh, J. Etoh, S. Asai, N. Hashimoto, K. Yagi, and H. Sunami, An high S/N design, *IEEE J. Solid-State Circuits*, **SC-19**, 634–640, 1984.

4. M. Takada, T. Takeshima, M. Sakamoto, T. Shimizu, H. Abiko, T. Katoh, M. Kikuchi, S. Takahashi, Y. Sato, and Y. Inoue, A 4-Mbit DRAM with half-internal-voltage bit-line precharge, *IEEE J. Solid-State Circuits*, **SC-21**, 612–617, 1986.

5. S. Fujii, M. Ogihara, M. Shimizu, M. Yoshida, K. Numata, T. Hara, S. Watanabe, S. Sawada, T. Mizuno, J. Kumagai, S. Yoshikawa, S. Kaki, Y. Saito, H. Aochi, T. Hamamoto, and K. Toita, A 45-ns 16-Mbit DRAM with triple-well structure, *IEEE J. Solid-State Circuits*, **24**, 1170–1175, 1989.

6. D. Chin, C. Kim, Y. Choi, D.-S. Min, H. S. Hwang, H. Choi, S. Cho, T. Y. Chung, C. J. Park, Y. Shin, K. Suh, and Y. E. Park, An experimental 16-Mbit DRAM with reduced peak-current noise, *IEEE J. Solid-State Circuits*, **24**, 1191–1197, 1989.

7. M. Horiguchi, M. Aoki, J. Etoh, H. Tanaka, S. Ikenaga, K. Itoh, K. Kajigaya, H. Kotani, K. Ohshima, and T. Matsumoto, A tunable CMOS-DRAM voltage limiter with stabilized feedback amplifier, *IEEE J. Solid-State Circuits*, **25**, 1129–1135, 1990.

8. H. Hidaka, K. Arimoto, K. Hirayama, M. Hayashikoshi, M. Asakura, M. Tsukude, T. Oishi, S. Kawai, K. Suma, Y. Konishi, K. Tanaka, W. Wakamiya, Y. Ohno, and K. Fujishima, A 34-ns 16-Mb DRAM with controllable voltage down-converter, *IEEE J. Solid-State Circuits*, **27**, 1020–1027, 1992.

9. A. L. Roberts, J. H. Dreibelbis, G. M. Braceras, J. A. Gabric, L. E. Gilbert, R. B. Goodwin, E. L. Hedberg, T. M. Maffitt, L. G. Meunier, D. S. Moran, P. K. Nguyen, D. E. Reed, D. R. Reismiller, and R. A. Sasaki, A 256K SRAM with on-chip power supply conversion, *ISSCC Dig. Tech. Papers*, 252–253, February 1987.

10. S. Hayakawa, K. Sato, A. Aono, T. Yoshida, T. Ohtani, and K. Ochii, A process-insensitivity voltage down converter suitable for half-micron SRAM's, *Symp. VLSI Circuits Dig. Tech. Papers*, 53–54, August 1988.

11. K. Ishibashi, K. Sasaki, and H. Toyoshima, A voltage down converter with submicroampere standby current for low-power static RAM's, *IEEE J. Solid-State Circuits*, **27**, 920–926, 1992.

12. K. Itoh, *VLSI Memory Chip Design*, Springer-Verlag, New York, 2001.

13. G. W. den Besten and B. Nauta, Embedded 5 V-to-3.3 V voltage regulator for supplying digital IC's in 3.3 V CMOS technology, *IEEE J. Solid-State Circuits*, **33**, 956–962, 1998.
14. R. Nair and B. Tang, Low cost on-die CMOS distributed voltage regulation for microcontrollers, *Proc. ESSCIRC*, 188–191, September 1998.
15. M. Hiraki, K. Fukui, and T. Ito, A low-power microcontroller having a 0.5-μA standby current on-chip regulator with dual-reference scheme, *IEEE J. Solid-State Circuits*, **39**, 661–666, 2004.
16. P. Hazucha, T. Karnik, B. A. Bloechel, C. Parsons, D. Finan, and S. Borkar, Area-efficient linear regulator with ultra-fast load regulation, *IEEE J. Solid-State Circuits*, **40**, 933–940, 2005.
17. Y. Nakagome, M. Horiguchi, T. Kawahara, and K. Itoh, Review and future prospects of low-voltage RAM circuits, *IBM J. Res. Dev.*, **47**, 525–552, 2003.
18. R. H. Dennard, F. H. Gaensslen, H.-N. Yu, V. L. Rideout, E. Bassous, and A. R. LeBlanc, Design of ion-implanted MOSFET's with very small physical dimensions, *IEEE J. Solid-State Circuits*, **SC-9**, 256–268, 1974.
19. Y. Nakase, S. Hirose, T. Goda, K. Hu, H. Onoda, Y. Ido, H. Kono, W. Kong, W. Zhang, T. Oishi, S. Mori, and T. Shimizu, 0.8V start-up 92% efficiency on-chip boost DC-DC converters for battery operation micro-computers, *Proc. A-SSCC*, 21–24, November 2011.
20. K. Osada, Y. Saito, E. Ibe, and K. Ishibashi, 16.7-fA/cell tunnel-leakage-suppressed 16-Mb SRAM for handling cosmic-ray-induced multierrors, *IEEE J. Solid-State Circuits*, **38**, 1952–1957, 2003.
21. H. Tanaka, Y. Nakagome, J. Etoh, E. Yamasaki, M. Aoki, and K. Miyazawa, Sub-1-μA dynamic reference voltage generator for battery-operated DRAM's, *IEEE J. Solid-State Circuits*, **29**, 448–453, 1994.
22. K. E. Kuijk, A precision reference voltage source, *IEEE J. Solid-State Circuits*, **SC-8**, 222–226, 1973.
23. A. P. Brokaw, A simple tree-terminal IC bandgap reference, *IEEE J. Solid-State Circuits*, **SC-9**, 388–393, 1974.
24. H. Neuteboom, B. M. J. Kup, and M. Janssens, A DSP-based hearing instrument IC, *IEEE J. Solid-State Circuits*, **32**, 1790–1806, 1997.
25. H. Banba, H. Shiga, A. Umezawa, T. Miyaba, T. Tanzawa, S. Atsumi, and K. Sakui, A CMOS bandgap reference circuit with sub-1-V operation, *IEEE J. Solid-State Circuits*, **34**, 670–674, 1999.
26. M. Sugawara, Regulated low-voltage generation circuit, US Patent No. 6,384,586, May 2002.

27. J. Kwong, Y. Ramadass, N. Verma, M. Koesler, K. Huber, H. Moormann, and A. Chandrakasan, A 65nm sub-Vt microcontroller with integrated SRAM and switched-capacitor DC-DC converter, *ISSCC Dig. Tech. Papers*, 318–319, February 2008.

28. P. R. Gray, P. J. Hurst, S. H. Lewis, and R. G. Meyer, Chapter 9, in *Analysis and Design of Analog Integrated Circuits*, Wiley, New York, 2001.

29. R. Razavi, Chapter 10, in *Design of Analog CMOS Integrated Circuits*, McGraw-Hill, 2001.

30. H. Tanaka, M. Aoki, J. Etoh, M. Horiguchi, K. Itoh, K. Kajigaya, and T. Matsumoto, Stabilization of voltage limiter circuit for high-density DRAM's using pole-zero compensation, *IEICE Trans. Electron.*, **E75-C**, 1333–1343, 1992.

31. H. Tanaka, M. Aoki, J. Etoh, M. Horiguchi, K. Itoh, K. Kajigaya, and T. Matsumoto, Stabilization of voltage limiter circuit of high-density DRAM's using Miller compensation, *IEICE Trans.*, **J75-C-II**, 425–433, 1992 (in Japanese).

32. B. K. Ahuja, An improved frequency compensation technique for CMOS operational amplifiers, *IEEE J. Solid-State Circuits*, **SC-18**, 629–633, 1983.

33. H. J. Shin, S. K. Reynolds, K. R. Wrenner, T. Rajeevakumar, S. Gowda, and D. J. Pearson, Low-dropout on-chip voltage regulator for low-power circuits, *Symp. Low Power Electron. Dig. Tech. Papers*, 76–77, 1994.

34. J. F. Dickson, On-chip high-voltage generation in MNOS integrated circuits using an improved voltage multiplier technique, *IEEE J. Solid-State Circuits*, **SC-11**, 374–378, 1976.

35. K. C. Hardee and R. Sud, A fault-tolerant 30 ns/375 mW 16k x 1 NMOS static RAM, *IEEE J. Solid-State Circuits*, **SC-16**, 435–443, 1981.

36. E. M. Lucero, N. Challa, and J. Fields Jr., A 16 kbit smart 5 V-only EEPROM with redundancy, *IEEE J. Solid-State Circuits*, **SC-18**, 539–544, 1983.

37. R.-A. Cernea, G. Samachisa, C.-S. Su, H.-F. Tsai, Y.-S. Kao, C.-Y. M. Wang, Y.-S. Chen, A. Renninger, T. Wong, J. Brennan Jr., and J. Haines, A 1Mb flash EEPROM, *ISSCC Dig. Tech. Papers*, 138–139, February 1989.

38. Y. Nakagome, H. Tanaka, K. Takeuchi, E. Kume, Y. Watanabe, T. Kaga, Y. Kawamoto, F. Murai, R. Izawa, D. Hisamoto, T. Kisu, T. Nishida, E. Takeda, and K. Itoh, An experimental 1.5-V 64-Mb DRAM, *IEEE J. Solid-State Circuits*, **26**, 465–472, 1991.

39. K. Sato, K. Kenmizaki, S. Kubono, T. Mochizuki, H. Aoyagi, M. Kanamitsu, S. Kunito, H. Uchida, Y. Yasu, A. Ogishima, S. Sano, and H. Kawamoto, A

4Mb pseudo SRAM operating at 2.6 +- 1 V with 3-uA data retention current, *IEEE J. Solid-State Circuits*, **26**, 1556–1562, 1991.

40. R. C. Foss, G. Allan, P. Gillingham, F. Larochelle, V. Lines, and G. Shimokura, Application of a high-voltage pumped supply for low-power DRAM, *Symp. VLSI Circuits Dig. Tech. Papers*, 106–107, June 1992.
41. A. Umezawa, S. Atsumi, M. Kuriyama, H. Banba, K. Imamiya, K. Naruke, S. Yamada, E. Obi, M. Oshikiri, T. Suzuki, and S. Tanaka, A 5-V-only operation 0.6-um flash EEPROM with row decoder scheme in triple-well structure, *IEEE J. Solid-State Circuits*, **27**, 1540–1546, 1992.
42. T. Tanzawa and T. Tanaka, A dynamic analysis of the Dickson charge pump circuit, *IEEE J. Solid-State Circuits*, **32**, 1231–1240, 1997.
43. J.-T. Wu and K.-L. Chang, MOS charge pumps for low-voltage operation, *IEEE J. Solid-State Circuits*, **33**, 592–597, 1998.
44. P. Favrat, P. Deval, and M. J. Declercq, A high-efficiency CMOS voltage doubler, *IEEE J. Solid-State Circuits*, **33**, 410–416, 1998.
45. T. Myono, A. Uemoto, S. Kawai, E. Nishibe, S. Kikuchi, T. Iijima, and H. Kobayashi, High-efficiency charge-pump circuits with large current output for mobile equipment applications, *IEICE Trans. Electron.*, **E84-C**, 1602–1611, 2001.

Chapter 8

Applications of DC–DC/AC–DC Switching Converters

Yasunori Kobori
Department of Innovative Electrical and Electronic Engineering,
National Institute of Technology, Oyama College, Oyama Tochigi 323-0806, Japan
kobori@oyama-ct.ac.jp

8.1 Noninverted Buck–Boost DC–DC Converter with Dual Delta-Sigma Modulators

8.1.1 Introduction

Secondary batteries used in mobile devices continue to undergo improvements in their output voltage range and output capacity. Accordingly, development of a buck–boost power source that can be controlled continuously to output a stable voltage despite changes in input current over a broad range is urgently needed.

For instance, in a cell phone, the output of a conventional lithium-ion battery is 4.2 V to 3.0 V; to drive a 2.5 V load, a step-down power source is sufficient. However, in practical batteries, the battery

Handbook of Power Management Circuits
Edited by Haruo Kobayashi and Takashi Nabeshima

ISBN 978-981-4613-15-6 (Hardcover), 978-981-4613-16-3 (eBook)
www.panstanford.com

output is extended toward the lower voltage range of 4.2 V to 2.2 V, and the need arises to switch automatically and continuously from step-down to step-up operation.

8.1.2 Full-Bridge Configuration Buck–Boost Power Source

8.1.2.1 Mixed-control method

In a switching power supply, the output voltage is normally controlled by a pulse width modulation (PWM) signal. The duty cycle ratio of this PWM signal usually has a limited range, for instance, 10% to 90%. Therefore, when the input voltage drops slowly and the power source control switches from buck to boost, there is an input voltage range (voltage gap) in which control is not possible around an ideal switching point.

We evaluated a control method to successively alter the mixing ratio by mixing the buck and boost control methods in this voltage gap. Figure 8.1 shows the concept of operation in this method. The voltage difference (gap voltage) in the gap is measured on the basis of the difference between the input and output voltages, and the buck and boost control ratios are switched using the PWM cycle. In other words, the mix ratio (boost:buck) is switched successively in accordance with the gap voltage. Figure 8.2 shows the configuration using the proposed method. The Up/Down (U/D) controller in the figure is the equivalent of the voltage gap detector and mix ratio counter.

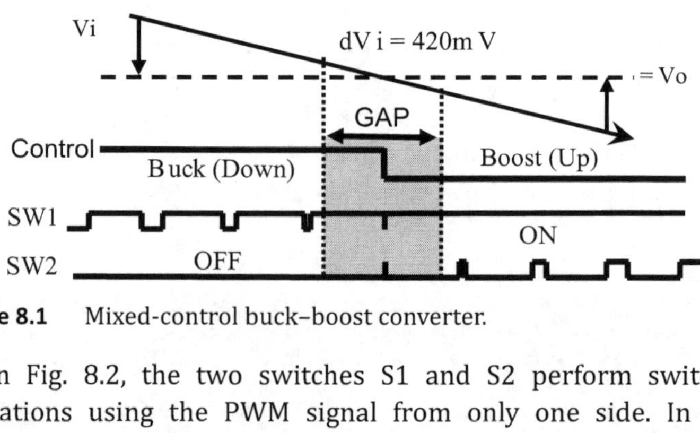

Figure 8.1 Mixed-control buck–boost converter.

In Fig. 8.2, the two switches S1 and S2 perform switching operations using the PWM signal from only one side. In buck switching, S2 is set to off at all times and S1 performs switching. In boost switching, S1 is set to on at all times and S2 performs switching. Switching is controlled in the U/D SW for every PWM cycle.

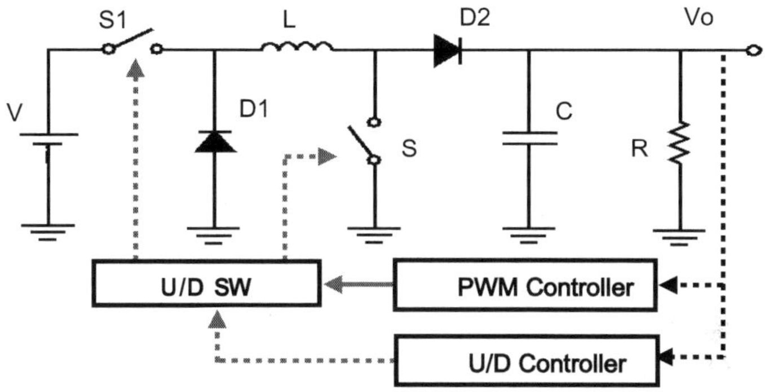

Figure 8.2 Buck–boost converter with U/D controller.

8.1.2.2 Voltage conversion equation in a buck and boost power source

The voltage conversion equation $M = V_o/V_i$ in a buck power source or a boost power source is ideally given by the following equations. Here, M and V are variables and the subscripts o and i are labels.

$$\text{Buck: } M_D = D \quad [<1] \tag{8.1}$$

$$\text{Boost: } M_U = 1/(1 - D) = 1/D' \quad [>1] \tag{8.2}$$

Here, D is a variable that is the duty ratio and the subscripts D and U are labels.

In a real power source, the switching elements, coil internal resistance, and load resistance R influence the circuit operation, and the voltage conversion ratio can be expressed by the following, more complex, equations:

$$\text{Buck: } M_D'' = D/(1 + Z_o/R) \quad [<M_D] \tag{8.3}$$

$$\text{Boost: } M_U' = 1/D'\cdot(1 + Z_o'/R) \quad [<M_U] \tag{8.4}$$

Here, Z_o is the various internal resistance. When r_L is taken to be the coil internal resistance, r_D is the diode connection resistance and r_S is the switch connection resistance, and then Z_o can be represented using the equations below:

$$Z_o = r_L + D\cdot r_S + D'\cdot r_D \tag{8.5}$$

$$Z_o' = (r_L + D \cdot r_S + D' \cdot r_D)/D'^2 \tag{8.6}$$

Here, r is a variable and the subscripts L, S, and D are labels.

Therefore, when outputting a fixed voltage, the actual input voltage range must have a further margin and the voltage gap expands more.

8.1.2.3 Voltage conversion equation in the mixed-control method

The voltage conversion ratio in a buck and boost power source using the mixed-control method can be considered as follows using the state space averaging method. In a mixed state, both the buck voltage and the boost voltage are regulated to their maximum values in the duty ratio in the voltage gap. Therefore, if the respective voltage conversion ratios are designated as M_{UM} and M_{DM}, and the mixing ratio (buck:boost) is M:N, then the voltage conversion equation using buck–boost operation is given as follows:

$$M_{UD} = M \cdot M_{UM} + N \cdot M_{DM}/(M + N) \tag{8.7}$$

Here, M and N are variables and the subscripts UM and DM are labels.

8.1.3 ΔΣ Modulated Mixed-Control Method

In the mixed-control method, detection of the voltage gap and control of the mixed ratio in the gap are essential. In this case, the mixing ratio must be controlled in sequence, and the circuit configuration and control procedure are complicated. Thus, we evaluated a ΔΣ modulated mixed-control method that uses a ΔΣ modulator circuit in the U/D controller (the configuration is the same as that in Fig 8.2). Utilizing this method, it is not necessary to detect the voltage gap, and the mixing ratio is automatically switched across the entire input voltage range.

8.1.4 Dual ΔΣ Modulated Control Method

8.1.4.1 Configuration of the dual ΔΣ modulated method

In the ΔΣ modulated mixed-control method in Fig. 8.2, both switches S1 and S2 are switched and controlled separately using the PWM signal. With respect to the operation of each switch, buck–boost operation becomes possible by primarily operating S1 for buck switching and S2 for boost switching, even when controlling the switches individually in the PWM cycle. Thus, we evaluated a method of controlling the switches by setting up two independent ΔΣ modulation circuits, as shown in Fig. 8.3. Note that the response speed is improved by using the clock in both modulation circuits as the inverse phase.

In the configuration shown in Fig. 8.3, the circuit parameters are L = 1.6 μH and C = 200 μF, and the ΔΣ modulation clock set to f_{ck} = 2 MHz. First, the operation of buck–boost control is checked using an open loop. Figure 8.4 shows the output pulse (metal–oxide–semiconductor field-effect transistor [MOSFET] gate control pulse) in each modulation circuit and then the output voltage when the same sine wave is added to the input edge of each ΔΣ modulation circuit. Here, the MOSFET is on when the pulse is at the H level. On the basis of the figure, the on state for both MOSFETs is long when the output voltage is in boost mode, and the off state for both MOSFETs is sustained for a long time in buck mode. The minimum pulse width and clock cycle are both the same, 0.5 μs.

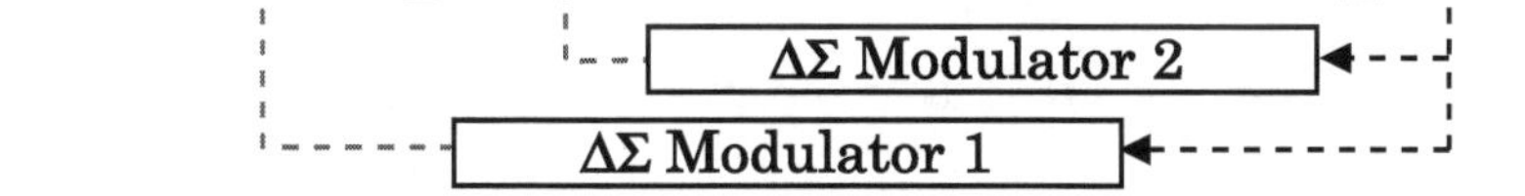

Figure 8.3 Buck–boost converter with dual ΔΣ modulators.

8.1.4.2 Characteristics of the ΔΣ modulation control method

In feedback control of the power supply under the PWM control method, the pulse width (i.e., the duty ratio) of the PWM signal is controlled at high resolution and the output voltage is stabilized using the amplified error voltage. On the other hand, in the dual ΔΣ modulation method, a binary pulse that is controlled to on or off in the clock cycle is output on the basis of the amplified error voltage. Therefore, control precision is poor for a signal pulse, but after successive feedback of the control error component, the later pulse precision rises. In this instance, the output of the ΔΣ modulation circuit determines the on/off output for each clock cycle on the basis of the sum of the power source output error component and the modulation circuit error component. The minimum pulse width is fixed at the clock cycle T_0, and the cycle T is determined by the number of on/off continuous pulses. In this case, the duty ratio is varied by adjusting the switching period using pulse frequency modulation (PFM) control. Additionally, the average voltage conversion ratio is determined by the duty ratio, similar to PWM control, and is represented in Eqs. 8.3 and 8.4 in accordance with the buck operation and boost operation at each instant.

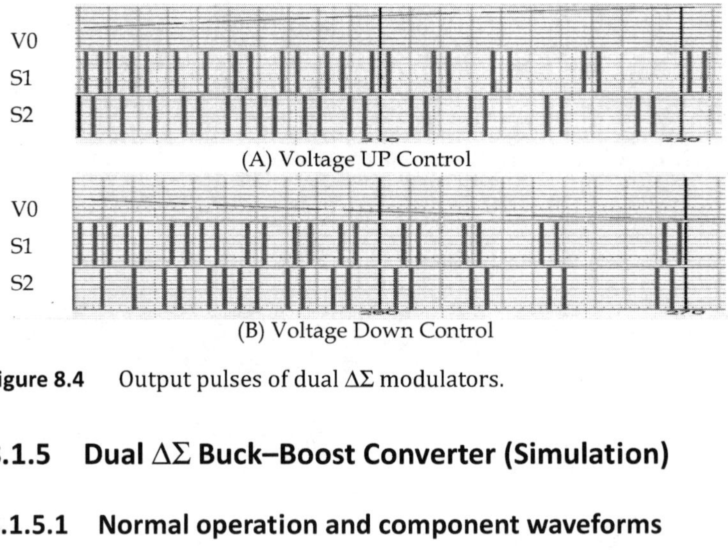

Figure 8.4 Output pulses of dual ΔΣ modulators.

8.1.5 Dual ΔΣ Buck–Boost Converter (Simulation)

8.1.5.1 Normal operation and component waveforms

We evaluated the operation and characteristics of the dual ΔΣ modulated buck–boost converter shown in Fig. 8.3 through simulations. Figure 8.5 shows the gate pulses in both MOSFETs and the input current i_{d2} and the output voltage V_o across diode D2. The buck/boost modes in the output voltage and the pulse waveforms for each MOSFET are operating appropriately. Here the steady-state ripple at an input voltage of V_i = 2.5 V, an output voltage of V_o = 2.5 V, and a load current of I_o = 0.5 A is extremely low at V_{rip} = 3 mVpp.

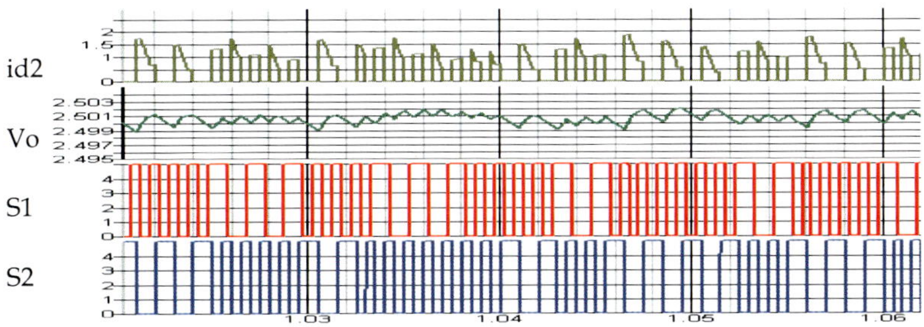

Figure 8.5 Waveform of a dual ΔΣ converter.

Figure 8.6 shows the output ripple waveform when the amount of fluctuation in the current is ΔI_o = 0.5, 1.0, and 1.5 A with respect to a current I_o = 0.5 A at low load.

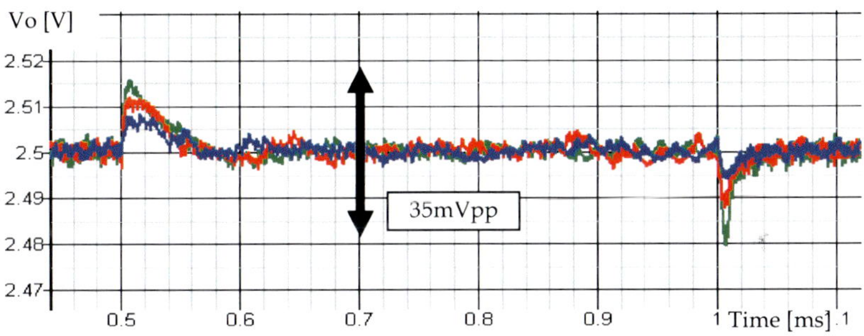

Figure 8.6 Improvement of output voltage ripple.

8.1.5.2 Load fluctuation response and ripple

We confirmed the response characteristics (output ripple voltage) with respect to load fluctuations in the same configuration. Initially, the ripple was large, and a large positive and negative imbalance occurred. The ripple was smoothed to 3.5 mVpp with good high and low balance by combining the characteristics of the two ΔΣ circuits (gain and phase compensation characteristics) for each ripple. Figure 8.7 shows a comparison of the two proposed methods in regard to improvement of the ripple with respect to the load current step ΔI_o at this point. It is clear that the ripple was reduced even when ΔI_o was high under the dual ΔΣ modulation method.

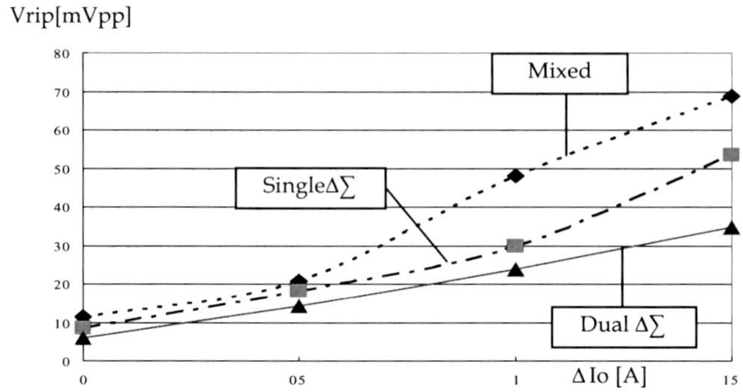

Figure 8.7 Output ripple versus load current step.

8.1.5.3 Evaluation of efficiency

In addition to the ripple in the power source performance, the efficiency η is also important. Figure 8.8 shows the results of a simulation of the efficiency η with respect to the output current I_o for each input voltage V_i when the output voltage is V_o = 2.5 V, the clock frequency f = 500 kHz, L = 1.5 μH, c = 300 μF, and the connection resistance for each SW·Di is R_{on} = 50 mΩ. In the figure, a typical example of the input voltage is shown in sequence starting from the characteristics at the top. In a real circuit, a buck–boost power source using an asynchronous rectification method would be used because control under synchronous rectification method using the MOSFET switches in parallel would be difficult. The synchronous rectification method is used for the purpose of improving efficiency, and as a result, the efficiency drops by several percent under the asynchronous rectification method.

In the characteristics shown in Fig. 8.8, the efficiency is good overall when the input voltage is near 3.0 V, and the maximum efficiency reaches 83% at I_o = 0.5 A. In the operation of the buck power source in this state, buck control switches to the main current from the buck–boost control state. When the input voltage is below 3 V, the proportion of boost-type operation increases, the equivalent internal resistance increases, the connection loss rises, and the efficiency drops. Furthermore, if the input voltage rises above the buck-state limit voltage of 3 V, then the voltage conversion efficiency and the duty ratio steadily decrease, and the efficiency tends, in theory, to fall slightly. When the efficiency is considered with respect to the output current, if the output current drops below 0.3 A, then the proportion of switching loss unrelated to the output current rises, and as a result, the efficiency steadily drops. On the other hand, if the output current is greater than 0.5 A, then the connection loss in the switch and diode increases proportionally and the efficiency tends to drop slightly. Under the simulation conditions here, the maximum efficiency is obtained at about I_o = 0.5 A. Note that to increase the output current I_o to obtain the maximum efficiency, the internal resistance must be reduced.

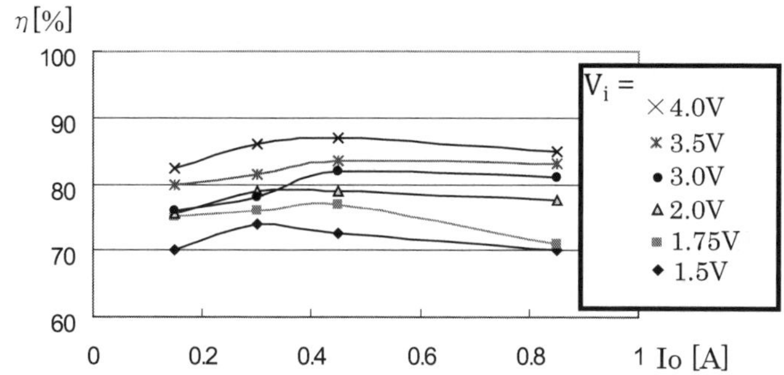

Figure 8.8 Efficiency versus input voltage.

8.1.6 Confirmation Experiments (Duty Ratio ΔΣ Control Method)

8.1.6.1 Experimental circuit

Figure 8.9 shows the prototype circuit for the ΔΣ modulator, an A/D converter and a D/A converter are required, and 1 bit is sufficient for switching control. This can be achieved using one D latch. Moreover, the adder handles operational amplifier–reversed addition, and the added resistance is assumed to be the same.

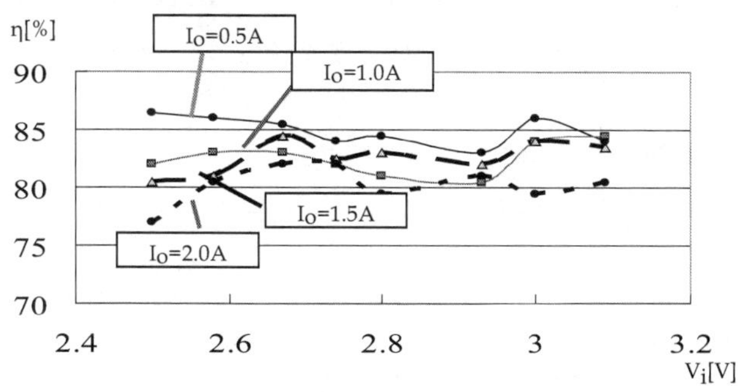

Figure 8.9 Efficiency versus input voltage.

8.1.6.2 Efficiency improvement in the experimental circuit

We confirmed the operation and measured the performance by experiments on the dual ΔΣ control method. We initially used a small MOSFET and diode in the experiments, and as a result, the loss due to the on resistance was high. Moreover, the prototype coil diameter was small, and the internal resistance was high. Thus, the internal diameter of the MOSFET was set to 10 mΩ, and the forward voltage V_F of the diode was set to 0.3 V. The coil had its internal resistance halved to r = 50 mΩ. In this way we revised each element and attempted to improve the efficiency, but on average the efficiency was only about 75%.

Thus, we evaluated the factors other than circuit elements that reduced efficiency. First, we evaluated the increase in switching loss caused by an increase in the diode capacity due to the larger switch S1. If the clock frequency is reduced, then the efficiency should rise. However, in reality there was no significant change in efficiency. Next, we evaluated the various output pulse states in dual ΔΣ modulation and realized that models not found in the buck or boost operation were present in the two switching states. That is, because each modulator operates independently, there are four modes, SW1:SW2 = (on:on), (on:off), (off:on), and (off:off), resulting from the switching states.

Here, the (off:on) mode is the state that maintains the coil current or the coil energy and is an idle state that hardly contributes to the control of the output voltage. Also, it is clear that the coil current flows in the form D1 ⇒ L ⇒ S2 ⇒ D1, and a large loss occurs due to the internal resistances of each element. Logically, efficiency can be improved by more than 5% on average by forcibly allocating this mode to the other modes.

8.1.6.3 Measurements of efficiency

We removed the idle state shown above. Figure 8.10 shows the measured efficiency with respect to the load current when V_{in} = 2.5 V. In the measurements reported here, we created a prototype using a discrete power MOSFET, and used a handmade coil. In this case, efficiency was reduced by increasing the switching loss slightly as a result of adding wiring and IC pin capacity to the gate wiring and also by free inductance.

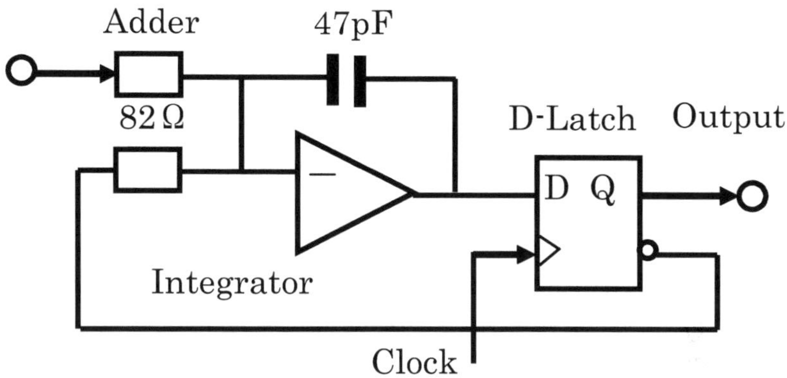

Figure 8.10 Circuit of a ΔΣ modulator.

In the measurements reported here, the efficiency dropped significantly when $I_o < 0.2$ A, but the power source was similar to one without the conventional measures to improve efficiency under low load. On the other hand, at $I_o = 0.9$ A, $\eta = 82\%$ or more, and the efficiency tends to rise slightly with respect to a further increase in I_o, with performance similar to that of an asynchronous controlled power source.

The circles in the upper part where $I_o = 0.5$ A in the figure represent the efficiency for a buck unit ($V_{in} = 2.5$ V) with a conventional configuration in the same circuit, and the circles below represent the efficiency for a boost unit ($V_{in} = 2.5$ V). Measurements were performed with unnecessary switches and diodes removed. The efficiency of a conventional boost and buck series-connected power source is the product of the two efficiencies, with $\eta = 66\%$ expected. In comparison to this, it is clear that the dual ΔΣ control method has an efficiency adequate for use.

Figure 8.11 shows the measured values for the efficiency η with respect to the input voltage V_{in} when the load current is constant ($I_o = 0.46$ A). A maximum efficiency of $\eta = 81\%$ is reached near where the mixing ratio is highest due to ΔΣ modulation. The average efficiency is about 80%, and further improvements in efficiency due to synchronous rectification of the two diodes can be expected.

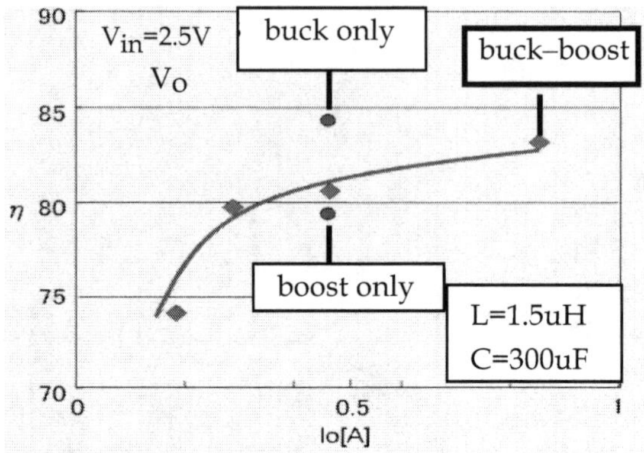

Figure 8.11 Efficiency versus I_o (experiment).

Figure 8.12 shows the relationship between the clock frequency for each input voltage when the load current is constant (I_o = 0.47 A). When the frequency rises, the switching loss increases and the efficiency drops slightly, as is the case in a normal switching power supply. Furthermore, in the lower region, the operation of the ΔΣ modulator is slow, the loss due to increased ripple in the coil current rises, and efficiency drops.

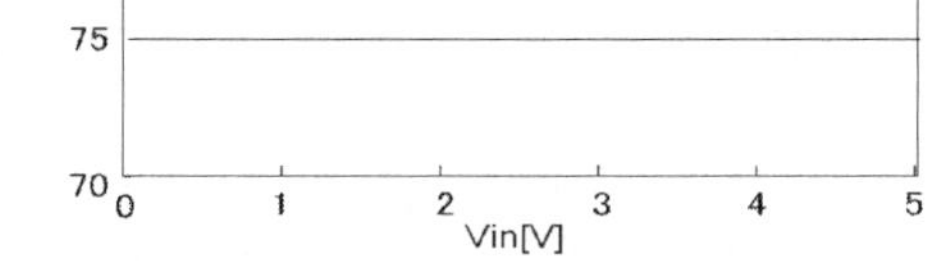

Figure 8.12 Efficiency versus V_{in} (experiment).

8.1.6.4 Voltage ripple versus load current fluctuations

We switched the load resistance in the prototype circuit periodically and measured the output voltage ripple (see Fig. 8.13). The voltage variation is ΔV_o = 2 mV with respect to a load current fluctuation of ΔI_o = 0.25 A, which represents sufficient load regulation characteristics. Furthermore, no overshoot is seen, although harmonic oscillation during switching generates approximately $\Delta V_o = 15$ mV. However, this is large because the circuit has been created with discrete components. It can probably be resolved by using IC and modularization, as found in ordinary power supplies.

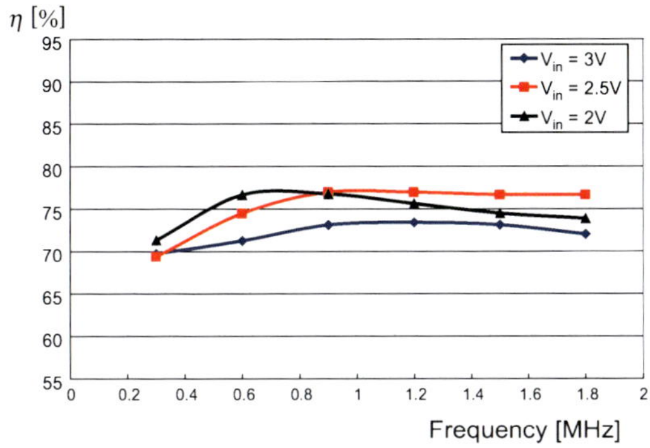

Figure 8.13 Efficiency versus clock frequency.

For the loop characteristics of the buck–boost power source using the dual ΔΣ modulation in the prototype described here, the cutoff frequency is 3.0 kHz, as see in Figs. 8.14 and 8.15. If the inductance and the output capacitance are reduced, then the response frequency characteristics can be increased, but the current and voltage rippled also rise. A determination based on the level of variation in the load current and the ripple specifications should be made.

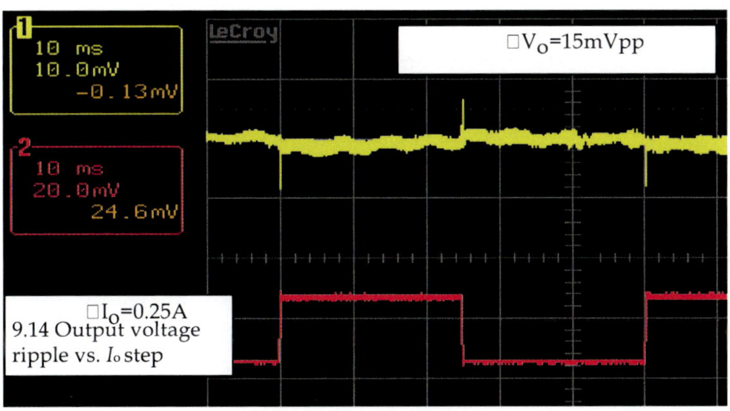

Figure 8.14 Output voltage ripple versus I_o step.

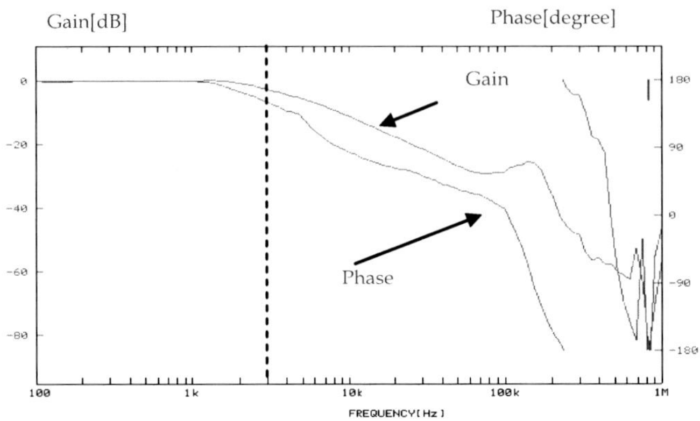

Figure 8.15 Closed-loop characteristics of an experimental circuit.

8.2 Nonisolated AC–DC Direct Converters

8.2.1 Introduction

AC–DC converters are indispensable for virtually all electronic devices, from cell phones to large manufacturing machinery. AC–DC converters produce steady direct current (DC) from alternating current (AC). In a typical converter, the AC input is rectified and drives a high-voltage high-frequency switching circuit connected to a transformer, which outputs the desired DC voltage. However, this type of converter is bulky and has low efficiency because it contains a switching DC–DC converter, a transformer, and a rectifier.

To realize a new AC–DC converter without a transformer, a direct nonisolated AC–DC conversion is implemented using a noninverted buck–boost converter with an H bridge circuit comprising four switches. These switches are operated according to the input voltage polarity to make current flow in the inductor in one direction.

Usually in an AC–DC converter, a power factor correction (PFC) circuit is used. There are two kinds of PFC circuits, boundary conduction mode (BCM) and continuous conduction mode (CCM). The PFC circuit usually needs one multiplexer and two operational amplifiers. A new PFC circuit with a new multiplexer is provided for a novel direct nonisolated AC–DC converter.

A nonisolated buck–boost AC–DC converter with an H bridge circuit is introduced for a high-voltage and a middle-voltage DC output. Next, a direct buck AC–DC converter with a diode bridge and a single switch is described for a low-voltage DC output. Their operating principles and simulation results are introduced to verify their basic operation and performance. Also the voltage conversion ratio is calculated for the direct buck–boost converter.

8.2.2 Direct Buck–Boost AC–DC Converter with an H Bridge

8.2.2.1 Basic circuit and principle operation

The direct buck–boost AC–DC converter is shown in Fig. 8.16 and Fig. 8.17, where the red solid line shows current flow when the inductor is charged and the blue dashed line shows the current flow when the inductor is discharged. Five switches operate at a frequency of 200 kHz, and the operation for these switches varies with the case of charging or discharging mode of the inductor and the input voltage polarities.

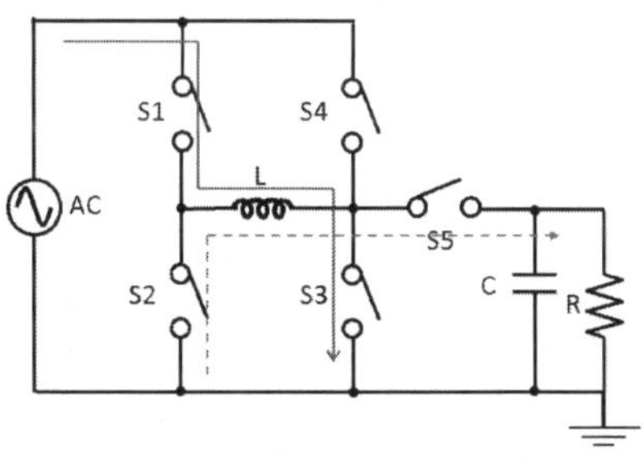

Figure 8.16 H bridge AC–DC converter (current when $V_{in} < 0$).

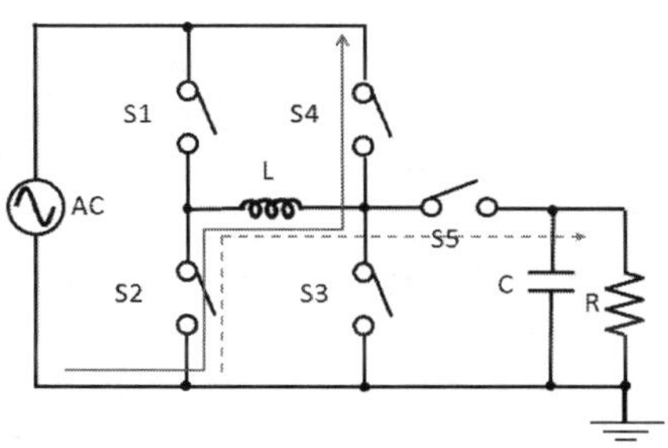

Figure 8.17 H bridge AC–DC converter (current when $V_{in} > 0$).

The operations of these switches are decided with a polarity of input voltage and the duty mode of the control signal. In case the polarity of the input voltage is positive, as shown in Fig. 8.16 and Fig. 8.18b, S4 is off. First, S1 and S3 are on and S2 and S5 are off for a period of $D \cdot T$ (D is the duty ratio, and T represents the switching period) and the current in the inductor is charged. Next, S1 and S3 are turned off and S2 and S5 are turned on so that the energy of the inductor is discharged into the capacitor and the load resistor. For the period when the polarity of input voltage is positive, S1/S3 and S2/S5 are alternatively turned on and off, as shown in Fig. 8.18b. In this case, when the control signal for the switch is at a high level, the controlled switch is on.

In case the polarity of input voltage is negative, as shown in Fig. 8.16 and Fig. 8.18c, S1 and S3 are always off and S2 is on. First, S4 is on and S5 is off for a period of $D \cdot T$ and the current in the inductor is charged. Next, S4 is turned off and S5 is turned on so that the energy of the inductor is discharged into the capacitor and the load resistor. In this case, S4 and S5 are alternatively turned on and off, as shown in Fig. 8.18c.

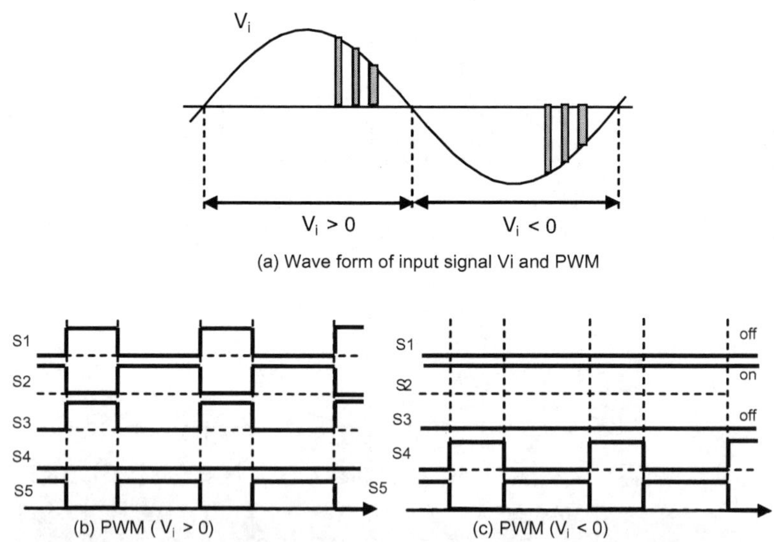

Figure 8.18 Timing chart of control signals for switches. (a) Waveform of input signal V_i and PWM. (b) PWM ($V_i > 0$). (c) PWM ($V_i < 0$).

The duty ratio D is controlled to make the output DC voltage constant. For a short time, the input voltage is almost a constant voltage V_i and the output DC voltage is steady at V_o. In this case, the duty ratio D should be controlled to be nearly equal to Eq. 8.8. The operation of these switches is just like the normal buck–boost DC–DC converter, and a steady DC output voltage is obtained.

$$M = V_o / V_i = D/(1 - D) \tag{8.8}$$

Here, D and M are variables and M represents a voltage conversion ratio.

8.2.2.2 Simulation results

The circuit schematic for simulation is illustrated in Fig. 8.19. The input voltage is 100 Vrms with a frequency of 50 Hz, and we use PWM operating at 100 kHz. The other parameters are shown in Table 8.1. We set the output voltage to 50 V and the output resistor current to I_o = 0.5 A.

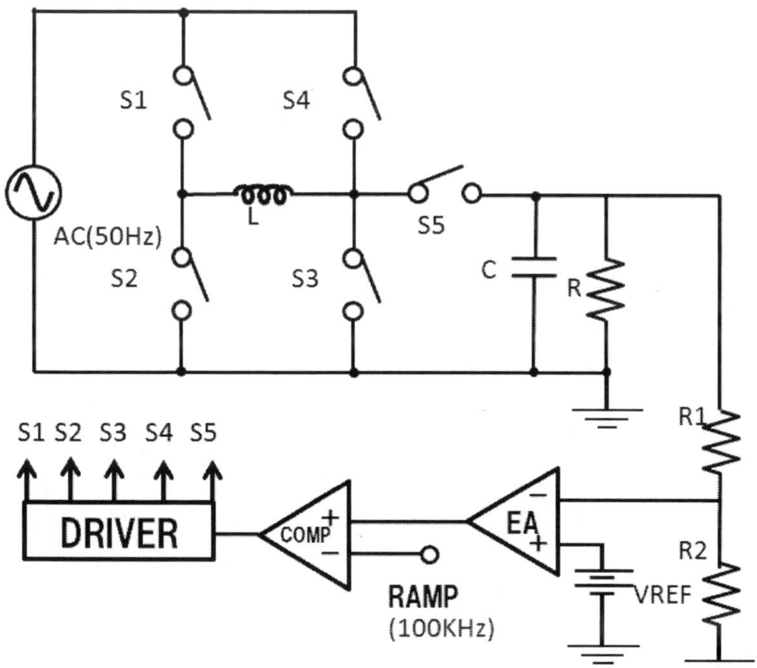

Figure 8.19 Simulation circuit.

Table 8.1 Simulation parameters of Fig. 8.18

C	220 uF
R_1	9 kΩ
R_2	1 kΩ
L	220 uH
V_{REF}	5.0 V

The waveforms of input voltage V_i and output voltage, output voltage ripple, the inductor current waveform, and load transient response are shown in Fig. 8.20, Fig. 8.21, Fig. 8.22, and Fig. 8.23, respectively. These figures show the transient responses when the input voltage is near the peak. The output voltage ripple is 6 mVpp, which is very small, and the inductor current ripple is under 1.7 App.

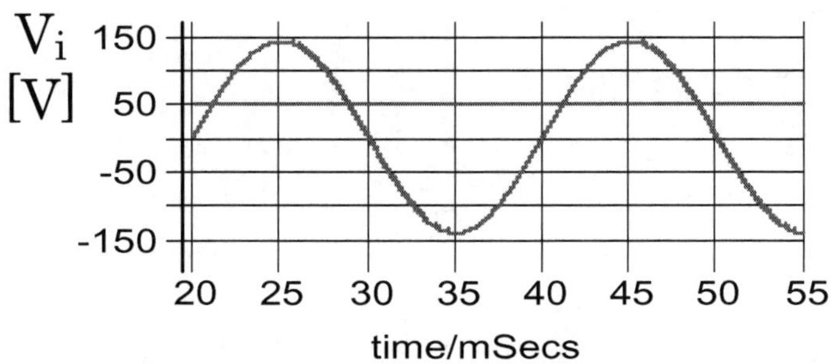

Figure 8.20 Waveform of input voltage and output voltage.

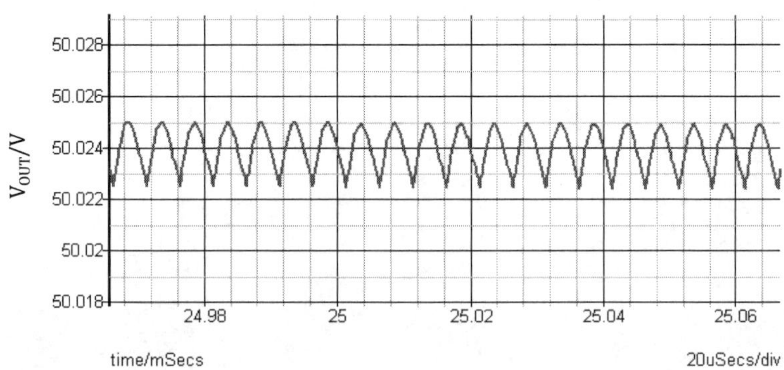

Figure 8.21 Output voltage ripple.

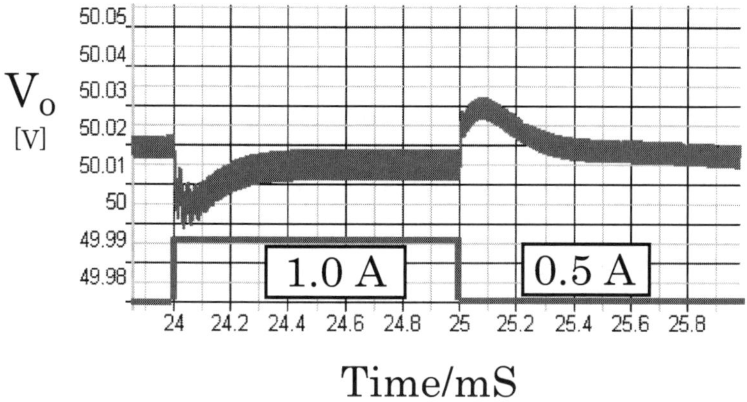

Figure 8.22 Load transient response.

For the transient response, we set the current change as $\Delta I = 1.0 \times 0.5$ A. The voltage ripple is 15 mVpp/0.5 A, as shown in Fig. 8.22, which is very small compared to the output voltage. We see in Fig. 8.23 that when the inductor current is large (1.0 A), the circuit operates in continuous mode. When the inductor current is small (0.5 A), the circuit operates in intermittent mode.

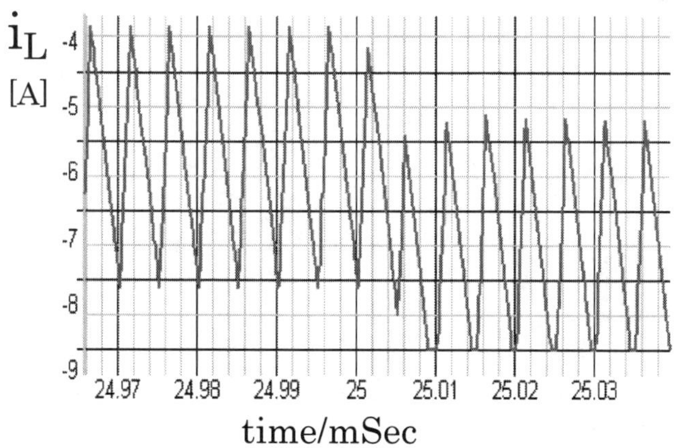

Figure 8.23 Waveform of inductor current.

8.2.2.3 Voltage conversion ratio

Compared to the PWM clock frequency, the frequency of the input sine wave is very low; hence the instantaneous input voltage can be considered to be almost constant. Then the output voltage V_o can be calculated as follows:

$$V_o = \frac{D}{1-D} \cdot V_i = \sqrt{2} \cdot \frac{D}{1-D} V_{rms} \sin(\theta) \quad (8.9)$$

$$D(\theta) = \frac{1}{1 + \left[\sqrt{2}/M\right] \sin(\theta)} \quad (8.10)$$

Here D is the duty ratio, and M is given by

$$M = V_o / V_{rms} \quad (8.11)$$

Thus the average duty ratio D^* in a half period is obtained as follows:

$$D^* = \frac{1}{\pi} \int D(\theta) d\theta = \frac{1}{\pi} \int_0^{\pi} \frac{d\theta}{1 + \left[\sqrt{2}/M\right] \sin(\theta)} \quad (8.12)$$

Since we cannot solve the above equation analytically, we solved it approximately by using interval integration. In Fig. 8.24 we compare the result with that of a commonly used noninverting buck–boost converter, where the x axis indicates the average duty ratio and the y axis shows output voltage. We see that, compared to the common buck–boost converter, the output voltage is a little bit smaller for the same duty ratio. In other words, a larger duty ratio is used for a given low output voltage, which makes it possible for our circuit to convert to a low output voltage directly. This feature is an advantage over the commonly used PWM-controlled buck–boost converter.

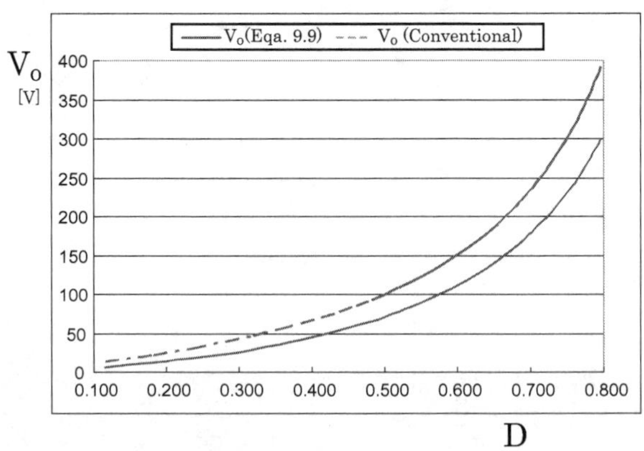

Figure 8.24 Average D versus V_o (V_{rms} = 100 V).

8.2.3 Inverted Direct AC–DC Converter

8.2.3.1 Circuit and operation

We can obtain negative outputs with the same circuit topology by reversing the direction of the inductor current. The circuit schematic is shown in Fig. 8.25. We can reverse the inductor current just by changing the operation of the switch groups.

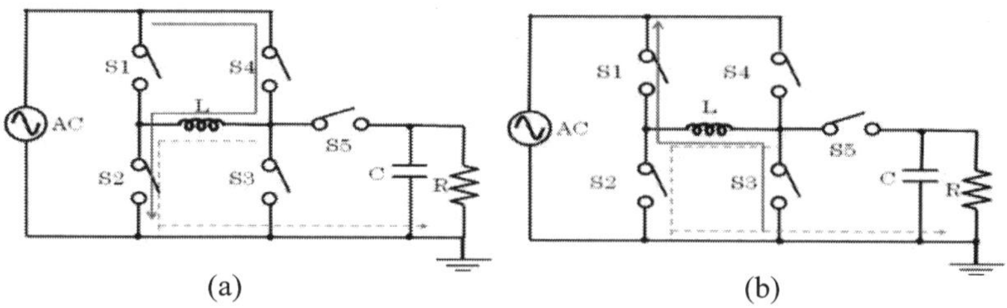

Figure 8.25 Inverting H bridge AC–DC converter. (a) Current flow when $V_i > 0$. (b) Current flow when $V_i < 0$.

The operations of the switch groups are as follows: (1) When $V_i > 0$, first S4–S2 are on and then S2–S5 are on; (2) for $V_i < 0$, first S3–S1 are on and then S2–S5 are on.

8.2.3.1 Simulation results

We have performed circuit simulations to check the operation and performance of the proposed inverting H bridge AC–DC converter (Fig. 8.25) with the same simulation conditions as the noninverting one. The waveforms of input and output voltages and output voltage ripple for a load current of 220 mA are shown in Fig. 8.26 and Fig. 8.27. We see that the output ripple is under 3 mV, which is very small, and the characteristics are similar to those of the noninverting H bridge AC–DC converter.

The direct AC–DC converters described above have no problems when the input voltage is high enough. However, when the input is lower than 20 V, the output ripple becomes a bit larger than 3 mV. We will find out the reason and solve this problem in the near future.

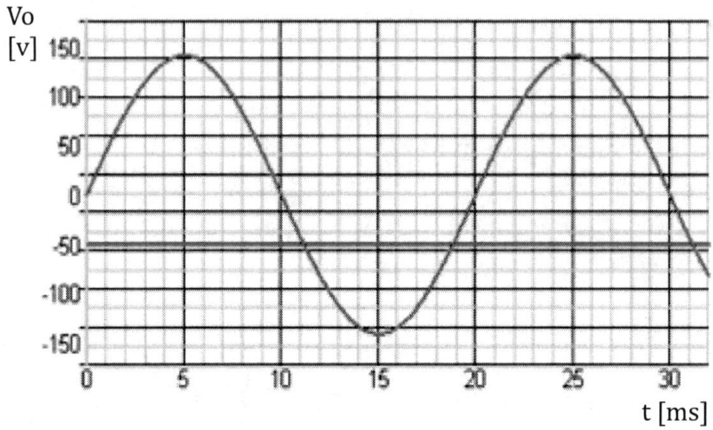

Figure 8.26 Output voltage of the converter.

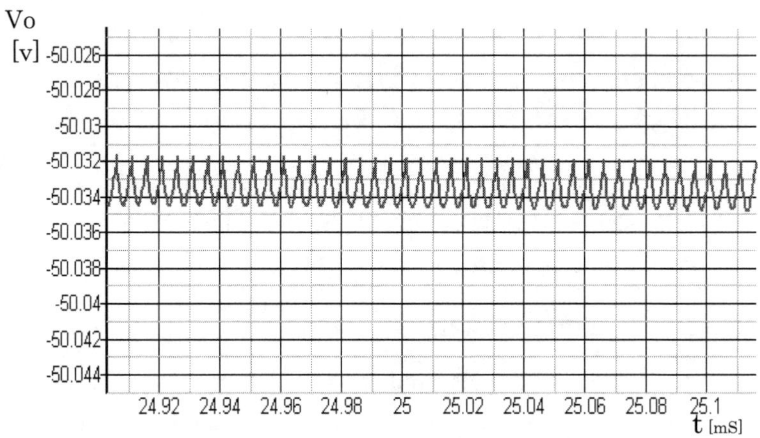

Figure 8.27 Output ripple of the converter.

8.3 Power Factor Correction Circuit for a Direct AC–DC Converter

For AC–DC converters, distortion of the input current and spurious current at clock frequencies should be reduced below the level permitted by electromagnetic interference (EMI) regulations, because AC–DC converters are connected directly to power lines. We have designed PFC circuits to meet this requirement.

8.3.1 New PFC Circuit in Boundary Conduction Mode

8.3.1.1 Conventional BCM PFC circuit with a diode bridge

In conventional AC–DC converters, a boost-type PFC circuit with an active filter is frequently used, as shown in Fig. 8.28. This circuit consists of an analog multiplier, an operational amplifier, two comparators, and D, L, and C components. In this circuit, the on time of the PWM signal should be constant to keep the waveform of the input current similar to that of the input voltage sine wave. The waveform of the inductor current is shown in Fig. 8.29, which is a series of triangle waveforms in BCM. The current is zero at switching timing from off to on. The solid line represents the charge current to the inductor, and the dashed line shows the discharge. During each triangle waveform, the input charges $Q_{\text{in}}(t)$ and the voltage source are shown below:

$$Q_{\text{in}}(t) = T\cdot(T_{\text{on}}\cdot V_{\text{i}}\cdot \sin\omega t)/2L \tag{8.13}$$

The on time T_{on} of the PWM signal is designed to be constant, but the off time is variable, and thus the clock frequency varies in phase. In this case, the PWM period T is given below:

$$\begin{aligned} T &= T_{\text{on}} + T_{\text{off}} \\ &= T_{\text{on}} + \text{L}\cdot T_{\text{on}}\cdot V_{\text{i}}\cdot \sin\theta/(V_{\text{o}} - T_{\text{on}}\cdot V_{\text{i}}\cdot \sin\theta) \end{aligned} \tag{8.14}$$

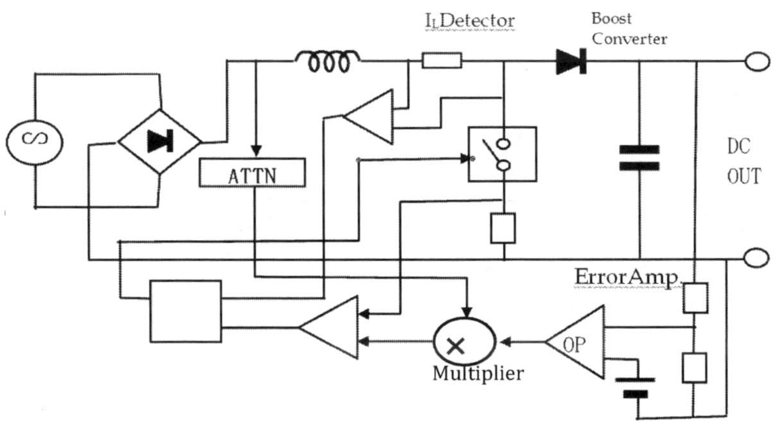

Figure 8.28 Conventional PFC circuit in BCM.

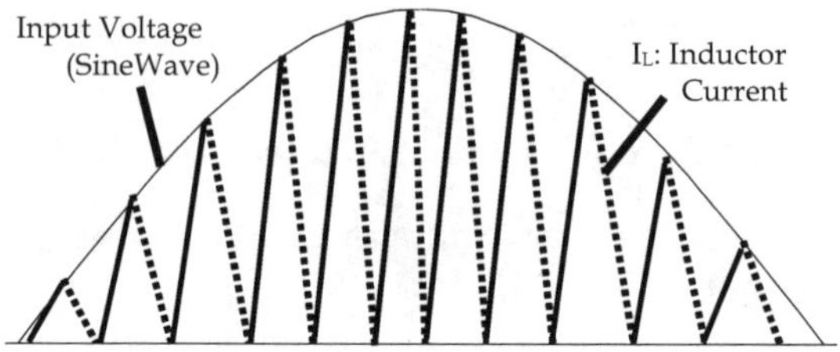

Figure 8.29 Waveform of the inductor current in BCM.

8.3.1.2 New BCM PFC in a buck–boost converter with an H bridge

Since our proposed circuit is a buck–boost converter different from the above boost converter, it needs a new PFC circuit. In our proposed circuit, the input current is not equal to the inductor current, because the on-time current is the input current and the off-time current is the load current. Thus, the on time is constant and the off time is given by

$$I_{\text{off}}(t) = I_p - t \cdot V_o/L = T_{\text{on}} \cdot V_i \cdot \sin(\theta i)/L - t \cdot V_o/L$$

$$\therefore \quad T_{\text{off}} = (V_i/V_o) \cdot T_{\text{on}} \cdot \sin(\theta i) \tag{8.15}$$

Here, I_p represents the peak current of I_L.

Equation 8.15 tells us that T_{off} is proportional to the input $\sin(\theta i)$ wave. Thus the input current is shaped nearly the same as the input voltage because the average of V_i/V_o is much larger than 1 in Eq. 8.15. This means that a multiplier is not needed in the new PFC system shown in Fig. 8.30. We note that conventional AC–DC PFC correction requires large capacitors to hold the input AC power and to output the DC power, and our proposed converter also requires a large capacitor of 47 mF.

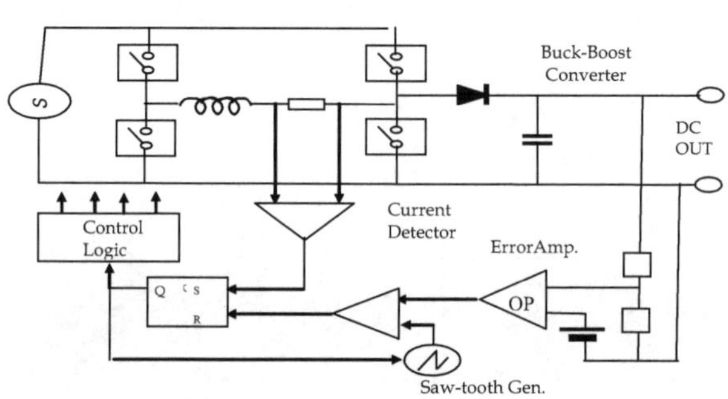

Figure 8.30 New BCM PFC.

In general, AC–DC converters have many output voltages, and the most popular level is 24 V output. Figure 8.31 shows the input voltage and the output voltage as well as the input current. The input current is a sawtooth shape with a clock frequency of about 100 kHz. In Fig. 8.31, the input current represents the waveform of the source current through a low-pass filter.

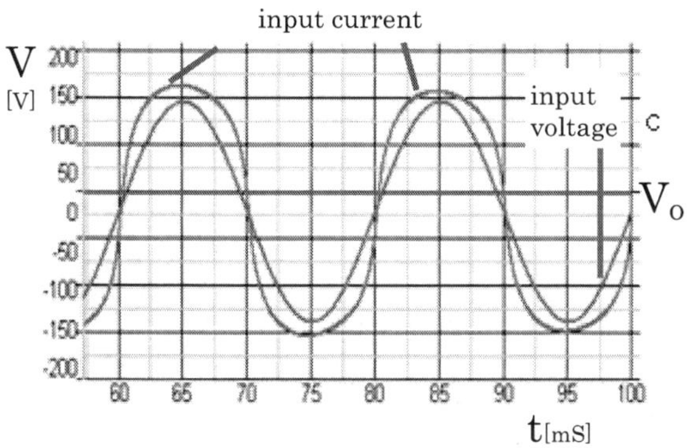

Figure 8.31 Waveforms of V_i, V_o, and I_i.

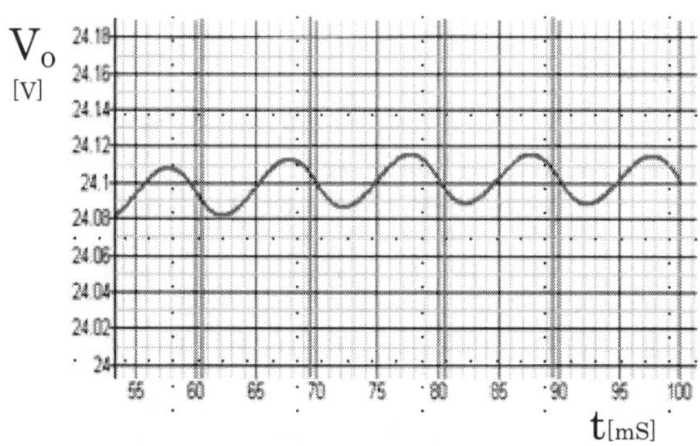

Figure 8.32 Output voltage ripple (100 Hz).

In this waveform, the power factor calculated from simulation is about 0.97. The output voltage ripple caused by clock signals is small enough, and the ripple caused by input signals is 25 mVpp at I_o = 0.25 A and 60 mVpp at I_o = 1.0 A. The ripple frequency is 100 Hz. Figure 8.33 shows the waveform of the input voltage and the inductor current, while Fig. 8.34 shows the wide-scope waveform of the inductor current.

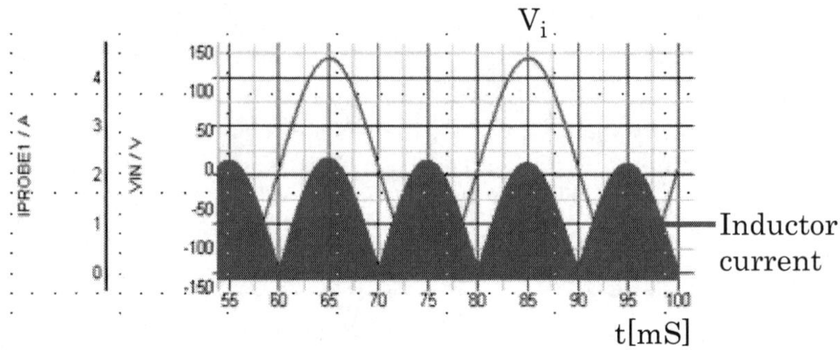

Figure 8.33 V_i and inductor current.

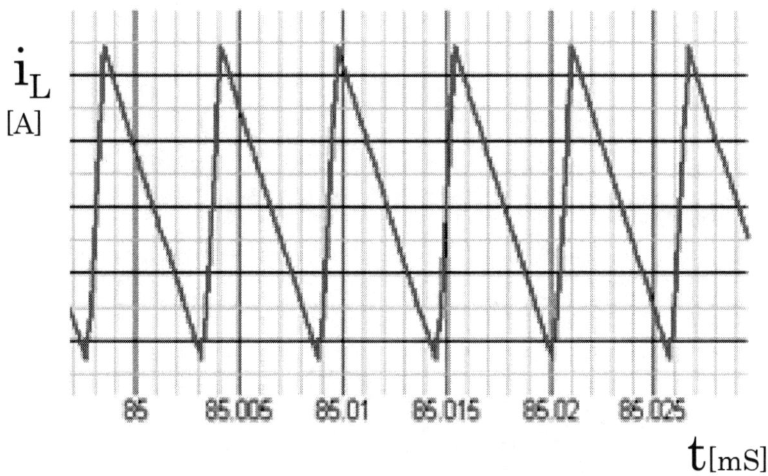

Figure 8.34 Inductor current in BCM.

8.3.2 New PFC Circuit in Continuous Conduction Mode

8.3.2.1 Conventional CCM PFC in a boost converter with a diode bridge

Figure 8.35 shows a conventional AC–DC converter with a CCM PFC. This circuit consists of an analog multiplier, two operational amplifiers, a comparator, and several components like in a normal boost converter. In this circuit, the inductor current and the input current are the same and they flow as shown in Fig. 8.36. Usually the frequency of the PWM signal is kept constant, and the off-time T_{off} is controlled to be proportional to the input voltage wave.

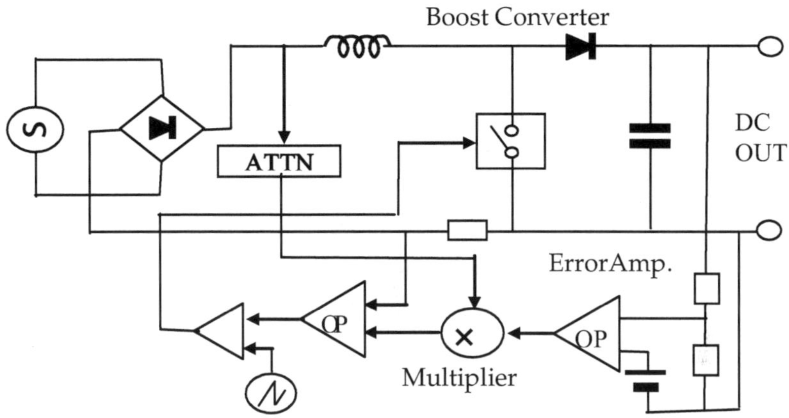

Figure 8.35 Conventional CCM PFC circuit.

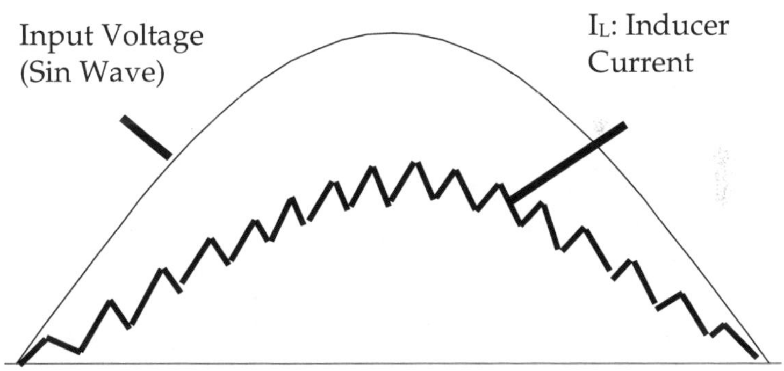

Figure 8.36 Waveform of the inductor current in CCM.

8.3.2.2 New CCM PFC in a buck–boost converter with an H bridge

We have developed a new PFC circuit for our AC–DC direct converter shown in Fig. 8.37. This PFC circuit consists of a new multiplier, two operational amplifier–reversed additions, and a comparator to generate the PWM signal. The multiplier circled by the red dashed line consists of a pulse generator controlled by the output of the error amplifier, a time-to-voltage converter.

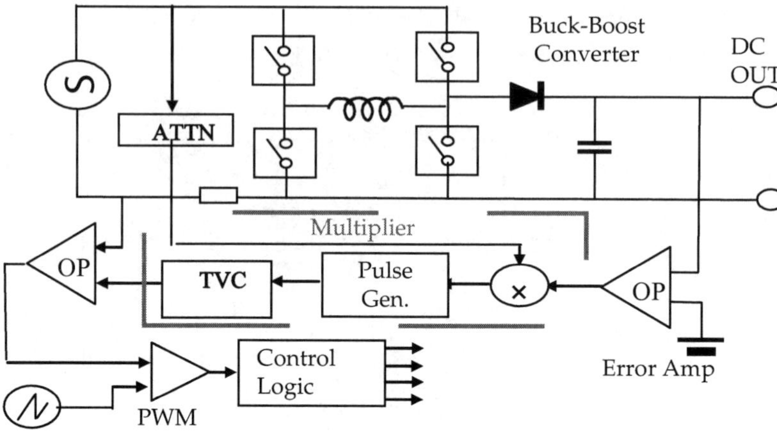

Figure 8.37 New CCM PFC with an H bridge.

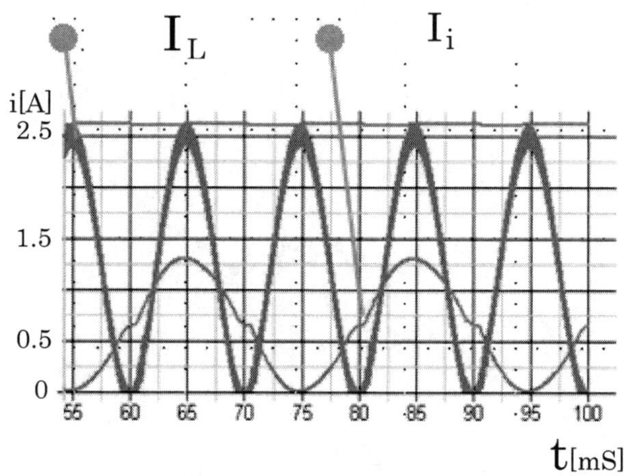

Figure 8.38 I_i and inductor current.

Figure 8.38 shows the output voltage, the input current through a low pass filter (LPF), and the inductor current. The input current is almost similar to the input sine wave shape except for zero-crossing points. The output voltage ripple caused by the input voltage is 60 mVpp at $I_o = 1.0$ A shown in Fig. 8.39. The waveform of the inductor current is shown in Fig. 8.40 at the timing of the inductor current peak at $t = 95$ ms. The ripple of the inductor current is about 0.2 App. In this case, the power factor in our simulation is 0.99.

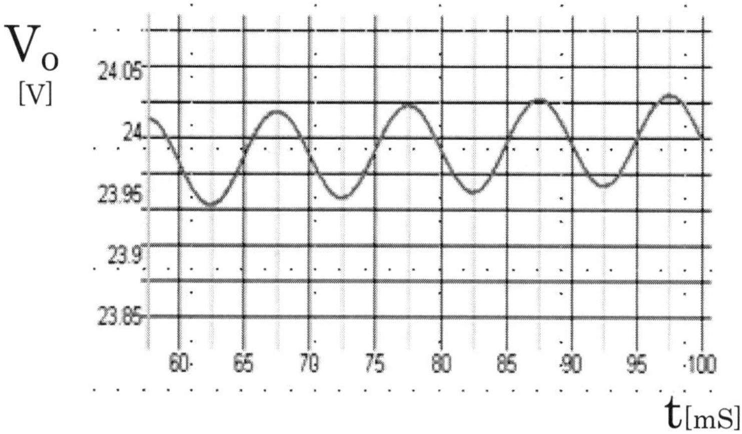

Figure 8.39 Output voltage ripple.

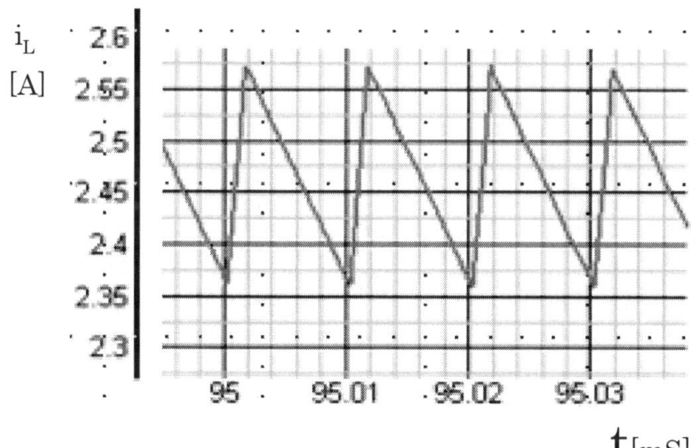

Figure 8.40 Inductor current in CCM.

8.4 Conclusions

A noninverted buck–boost DC–DC converter using dual ΔΣ modulation was described. We established its operation through simulations. We also confirmed stable operation in a real circuit and measured the ripple and efficiency. The voltage variation was 2 mV and the overshoot was 8 mV for a load current fluctuation of 0.25 A. The efficiency was about 80% on average when not using the synchronous rectification method, demonstrating the possibility of practical use.

Additionally, a direct AC–DC converter with H bridge topology and PFC circuits was described. We investigated and proposed two PFC circuits in BCM and CCM. We explained their principles of operation and verified their basic operation by simulations. Simulation results show that the output voltage ripple is 60 mVpp for BCM and 50 mVpp for CCM at I_o = 1.0 A. Furthermore, we developed a new PFC circuit in a BCM converter and a new multiplier with a time-to-voltage converter in CCM. Our simulations show that the power factor is 0.96 in BCM and 0.99 in CCM at V_i =100 Vrms, V_o = 24 V, and I_o = 1.0 A.

References

1. T. Furuya, Y. Kobori, M. Tsugane, and H. Kobayashi, Evaluation of a switching control method in a DC-DC converter for portable phones, *IEEJ Electron. Circuit Res. Rep.*, **ECT-05-53**, 61–66, 2005.
2. Y. Kobori, T. Furuya, M. Kono, T. Shimizu, and H. Kobayashi, A new control method for switched buck–boost DC-DC converters with delta-sigma modulation for mobile equipment, *2006 Int. Symp. Intelligent Signal Proc. Commun. Syst.*, 127–130, 2006.
3. Y. Kobori, T. Furuya, M. Kono, T. Shimizu, and H. Kobayashi, A new control method for buck–boost DC-DC converters using dual delta-sigma modulations for mobile equipment applications, *IEEJ VLSI Workshop*, CD-ROM, 2006.
4. Y. Kobori, T. Furuya, H. Kobayashi, M. Kono, and T. Shimizu, A seamless control buck–boost converter with dual delta-sigma modulation over a wide input range, *IEEJ Natl. Conf.*, **4**, 82–83, 2007.
5. Y. Kobori, M. Kono, H. Kobayashi, and T. Shimizu, Advanced seamless control for buck–boost converters with dual delta-sigma modulations, *IEEJ Int. Analog VLSI Workshop*, CD-ROM, 2007.
6. L. Xing, G. Hong, Y. Kobori, N. Okamoto, M. Ohshima, K. Wakabayashi, T. Okada, M. Onozawa, H. Kobayashi, N. Takai, and K. Niitsu, Novel DC-DC converter design, *IEEJ Papers Tech. Meeting Electron. Circuits*, **ECT-11-047**, 59–64, 2011.
7. L. Xing, G. Hong, Y. Kobori, K. Murakami, M. Onozawa, H. Kobayashi, N. Takai, and K. Niitsu, Direct AC–DC converter with small number of

components, *IEEJ Papers Tech. Meeting Electron. Circuits*, **ECT-12-005**, 25–30, 2012.

8. Y. Kobori, L. Xing, G. Hong, T. Shishime, M. Ohshima, H. Kobayashi, N. Takai, and K. Niitsu, Novel AC–DC converter design with PF correction, *Int. Conf. Power Electron. Power Eng.*, CD-ROM, 2011.
9. K. Harada, T. Ninomiya, and B. Ko, *Fundamentals of Switched-Mode Converters*, Corona, 2004.
10. R. W. Erickson and D. Maksimovic, *Fundamentals of Power Electronics (Second Edition)*, Kluwer Academic, 2001.

Chapter 9

Single-Inductor Multi-Output DC–DC Converter

Nobukazu Takai
Division of Electronics and Informatics, Gunma University, 1-5-1 Tenjin-cho, Kiryu, Gunma 376-8515, Japan
takai@el.gunma-u.ac.jp

9.1 Introduction

9.1.1 Background and Motivation

Portable devices such as cellular phones, personal desktop assistants (PDAs), game appliances, etc., have become a large and lucrative market for switching-power ICs. A switching regulator is suitable for the power supply circuit of mobile equipment because of its high efficiency, small size, and low power consumption. Low-cost, high-efficiency, and extremely small system solutions are critical to success, but the demands are quite conflicting.

In many types of electric equipment, there are a lot of circuit blocks for operation. The power supply voltage of a digital circuit block is unique and now less than 1 V, while analog circuit blocks,

Handbook of Power Management Circuits
Edited by Haruo Kobayashi and Takashi Nabeshima

ISBN 978-981-4613-15-6 (Hardcover), 978-981-4613-16-3 (eBook)
www.panstanford.com

such as signal process blocks, power amplifiers, display panels, etc., require multiple power supplies with different regulated voltages. For example, the active matrix organic electroluminescence (AMOEL) display requires sophisticated positive and negative voltage supplies for each color (red, green, or blue) to optimize display efficiency and display quality with brightness, contrast, and vividness. On the other hand, the power source of these types of equipment is unique. For instance, the output voltage of a lithium-ion battery is approximately equal to 3.6 V. Therefore, buck and/or boost switching converters that can supply multiple outputs for application are important. Usually, to supply these multiple outputs, many DC–DC converters are embedded. This causes the printed circuit board (PCB) areas to be occupied and increases the implementation cost because a DC–DC converter requires one energy storage component, usually an inductor, for each output.

Single-inductor multiple-output (SIMO) switching converters can support more than one output, while requiring only one off-chip inductor, which yields many appealing advantages for mass production and applications. The SIMO boost switching converter is reported in Refs. [1–10]. The SIMO converter works in pseudo-continuous conduction mode (PCCM) with a freewheel period, trying to handle large load currents and eliminate cross regulation [11–13]. The PCCM technique is suitable for a SIMO converter because of its advantage of cross regulation.

9.1.2 Organization

This section describes a SIMO DC–DC converter. Section 9.2 introduces basic topologies of a DC–DC switching converter and discusses the basic operation mode for a DC–DC converter. Section 9.3 introduces topologies of the SIMO converter. Section 9.4 describes control methods and circuitry for the SIMO converter realization. Finally, Section 9.5 provides conclusions.

9.2 Basics of a DC–DC Converter

Figure 9.1 shows the basic schematic of a switching regulator. A switching regulator is composed of a DC–DC switching converter and controller as well as function circuits such as a start-up circuit, an overcurrent/overvoltage protection circuit, etc. The DC–DC switching converter converts the DC input voltage to an arbitrary DC output voltage. The output voltage of the converter is applied to the controller to stabilize the output voltage. The controller controls the DC–DC switching converter to minimize the difference between the output voltage and the reference voltage. The DC–DC switching converter handles power and is the most important part of the switching regulator. The DC–DC switching converter has various circuit topologies. In the next subsection, basic circuit topologies of the DC–DC switching converter are described.

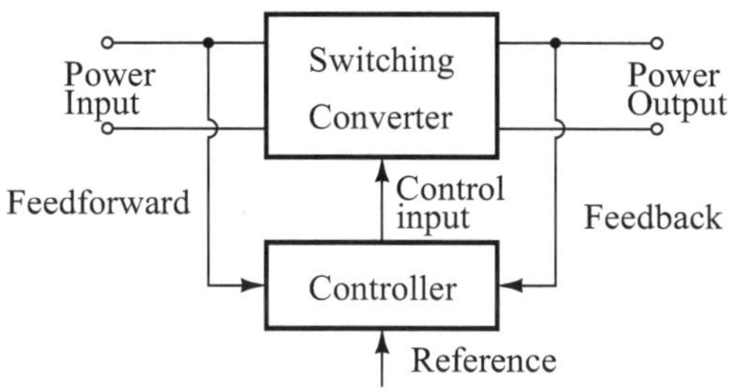

Figure 9.1 Basic schematic of a switching regulator.

9.2.1 Basic Topologies

9.2.1.1 Buck converter

Figure 9.2 illustrates the basic schematic of a buck converter as the DC–DC switching converter. The switch output voltage $v_s\,(t)$ is equal to the converter input voltage V_{in} when the switch is in position 1 and is equal to zero when the switch is in position 2. The switch position is varied periodically, as shown in Fig. 9.3, such that $v_s\,(t)$ is a rectangular waveform having frequency f_s and period $T_s = 1/f_s$. The duty cycle D is defined as the fraction of time in which the switch occupies position 1. The switch changes the DC component of the voltage. Recall from Fourier analysis that the DC component of a periodic waveform is equal to its average value. Hence, the DC component of $v_s\,(t)$ is given as

$$V_s = \frac{1}{T_s}\int_0^{T_s} v_s(t)dt = DV_{in} \tag{9.1}$$

Because $0 \leq D \leq 1$, V_s is less than and equal to the input voltage of the converter, which means the switching converter shown in Fig. 9.2 is a buck converter. Components L and C are the low-pass filter to remove undesirable harmonics of the switching frequency.

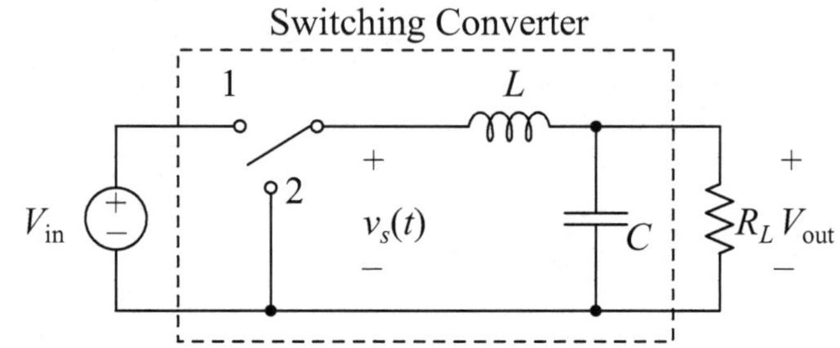

Figure 9.2 Schematic of a buck converter.

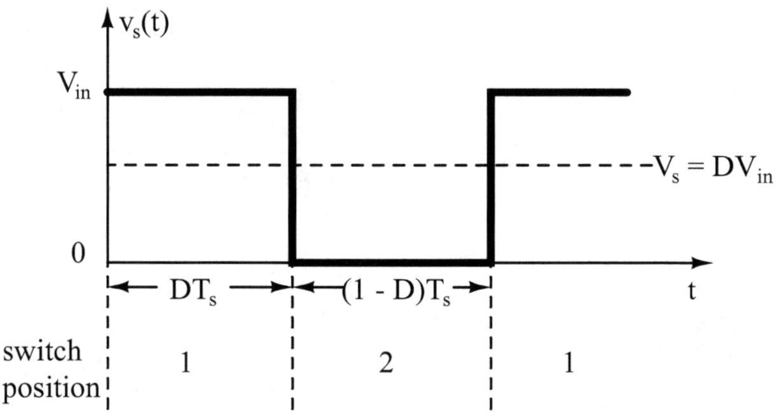

Figure 9.3 Switch output voltage waveform.

From the energy point of view, the inductor L stores energy from the voltage source V_{in} as a current when the switch is in position 1. Again, the inductor supplies stored energy to load resistance, which causes the output voltage V_{out} when the switch is in position 2.

9.2.1.2 Boost converter

Figure 9.4 illustrates a circuit known as the boost converter, in which the positions of the inductor and the switch are interchanged. The routine calculation gives the output voltage of Fig. 9.4 as

$V_{\text{out}} = \dfrac{1}{1-D} V_{\text{in}}$. This switching converter is capable of producing output voltages that are greater in magnitude than the input voltage of the converter, because $0 \le D \le 1$.

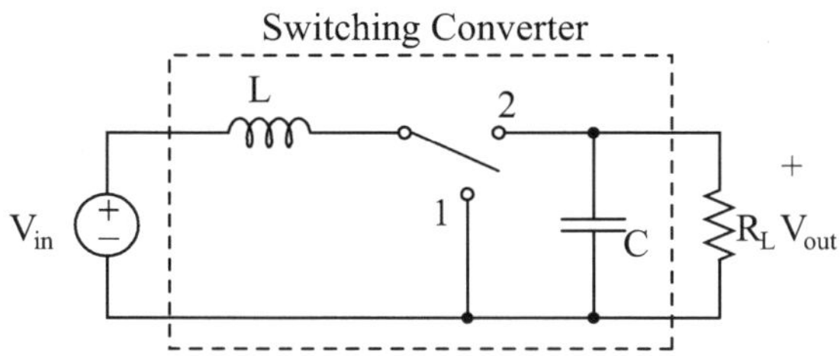

Figure 9.4 Schematic of a boost converter.

9.2.1.3 Buck–boost converter

Figure 9.5 shows a basic schematic of a buck-boost converter. The output voltage of Fig. 9.5 is found as $V_{\text{out}} = \dfrac{-D}{1-D} V_{\text{in}}$. The buck–boost converter can either increase or decrease the magnitude of the input voltage. With this topology, the polarity of the output voltage is inverted. Thus the buck–boost converter of this topology can produce a negative output voltage of any magnitude.

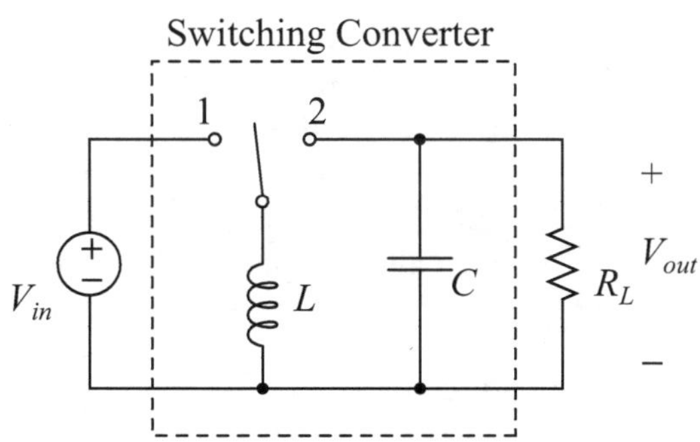

Figure 9.5 Schematic of a buck–boost converter.

9.2.2 Operation of a DC–DC Converter

From the energy point of view, the inductor stores and supplies energy as a current. Thus the inductor current is continuous, and its variation is very important to consider the operation of the DC–DC switching converter. The operation of the DC–DC switching converter is classified by the continuity of the inductor current. When the inductor current is always greater than zero and varies constantly, it is called continuous conduction mode (CCM) operation. If the inductor current has an interval of time staying at zero, it is working in discontinuous conduction mode (DCM) operation. Another operation is PCCM, which will be described in the next subsection in detail.

9.2.2.1 Continuous conduction mode

Figure 9.6a illustrates the inductor current waveform of CCM. The CCM boost and buck–boost converters have a right half-plane zero (RHPZ) in their control to the output transfer function. This means the design of the control circuit is complex compared to DCM. With boost and buck–boost converters operating in CCM, the peak and root-mean-square (RMS) currents of the inductor are low, resulting in less loss and ringing.

9.2.2.2 Discontinuous conduction mode

Figure 9.6b illustrates the inductor current waveform of DCM. The inductor current ripple of the converter operating in DCM becomes large at a light load. DCM occurs when converters operate with their loads removed. In the buck, boost, and buck–boost converters operating in DCM, input to output and control to the output transfer function of the converters have a single pole, which means that the control circuit can easily compensate its stability compared to converters operating in CCM. With boost and buck–boost converters operating in DCM, the peak and RMS currents of the inductor are high, resulting in more loss and ringing compared to CCM.

9.2.2.3 Pseudo-continuous conduction mode

Figure 9.6c illustrates the inductor current waveform of PCCM. PCCM was proposed in Ref. [13] in 2003. The inductor current always stays above zero and has a constant interval of time. This mode is suitable for a SIMO DC–DC converter, which is discussed in the next section. PCCM can realize both less ripple and peak of the inductor current.

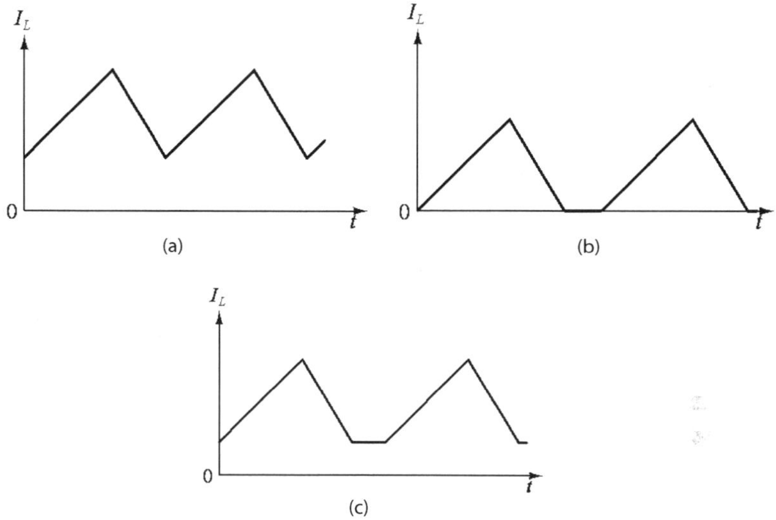

Figure 9.6 Inductor current waveforms. (a) Inductor current waveform of CCM. (b) Inductor current waveform of DCM. (c) Inductor current waveform of PCCM.

9.3 What Is a SIMO DC–DC Converter?

As mentioned in Section 9.1, a SIMO DC–DC converter can support more than one output, while requiring only one off-chip inductor. In this section, first, the basic operation of the SIMO converter is described and next the topology of the SIMO converter is classified and each operation is described.

9.3.1 Basics of a SIMO Converter

Now we consider the basic topology of a SIMO converter using the buck converter of Fig. 9.2. Typically, the switch can be replaced, as shown in Fig. 9.7. Switch S2 is sometimes interchanged with a diode; however, in this section, an ideal switch is employed, as shown in Fig. 9.7. The left part of Fig. 9.8 illustrates two buck converters, and the right shows the waveforms of their inductor currents and timing diagram of switches S1–S4. By using this timing diagram, we can obtain two different output voltages V_{out1} and V_{out2} working in

DCM because the output voltages V_{out1} and V_{out2} are controlled by an exclusive period.

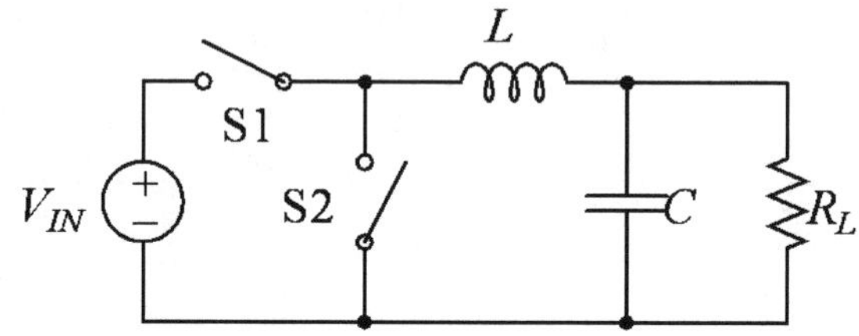

Figure 9.7 Schematic of a buck converter using an ideal switch.

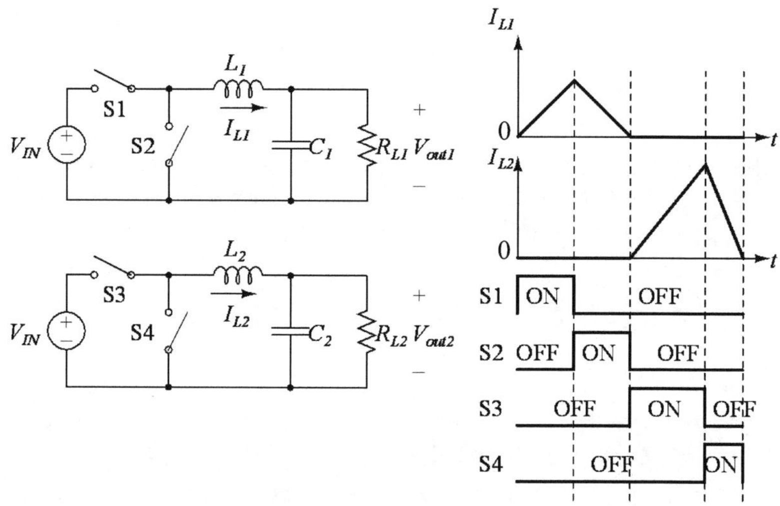

Figure 9.8 Timing diagram of S1–S4 and inductor currents of L1 and L2.

Now to realize a SIMO converter, we'd like to share the two inductors, as shown in Fig. 9.9. The left part of Fig. 9.9 shows the schematic of a basic SIMO converter using two buck converters, and the right shows the waveform of the inductor current and the timing diagram of switches S1–S4. Now we consider the period from T_1 to T_3. Because switch S3 turns on and switch S4 turns off, the inductor current is supplied to only the load resistor R_{L1}. In period T_1, the energy is stored from V_{in} and the stored energy in period T_1 is supplied to only R_{L1} in period T_2. In the period T_4–T_6, the inductor current is supplied to only the load resistor R_{L2}. In this period, the stored energy in period T_4 is supplied to only R_{L2} in the period

T_5. Because the switches S3 and S4 close exclusively, the inductor can be shared and the SIMO buck/buck converter is realized. Note that only four switches are required, which is the same number as two buck converters. By adding switches and capacitors, we can extend the SIMO converter to multiple outputs. However, increasing the outputs decreases a time slot of T/N for charging and discharging the inductor, which causes a problem of load regulation. Method to solve this problem is proposed in [14].

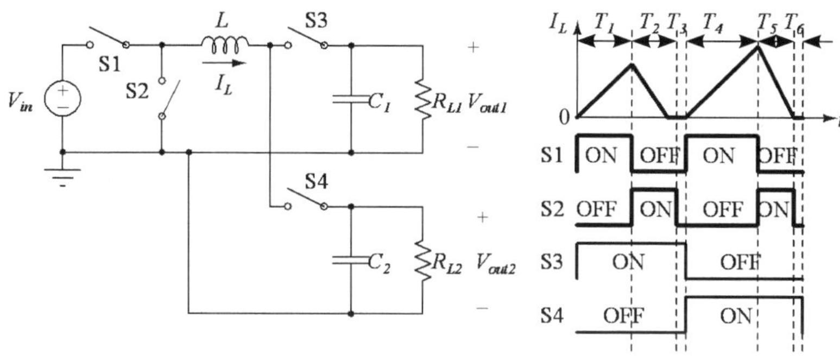

Figure 9.9 Schematic of a basic SIMO converter using two buck converters.

9.3.2 Topologies of a SIMO DC–DC Converter

In the previous section, a schematic example of a SIMO converter using a buck/buck converter is shown. As mentioned in Section 9.2, a DC–DC converter has three basic topologies, that is, buck, boost, and buck–boost. Additionally, a charge pump circuit can be utilized to realize a SIMO converter. Thus we can realize SIMO topologies as the number of combinations of these converters. In the next subsections, these SIMO topologies are described.

9.3.2.1 Buck/buck combination

This topology is described in the previous section. The output voltages V_{out1} and V_{out2} are obtained as

$$V_{\text{out1}} = \frac{T_1}{T_1 + T_2} V_{\text{in}} \tag{9.2}$$

$$V_{\text{out2}} = \frac{T_4}{T_4 + T_5} V_{\text{in}} \tag{9.3}$$

From these equations, the two output voltages can be controlled independently by changing the periods T_1 and T_4, and both output voltages are positive.

9.3.2.2 Buck/boost combination

The left part of Fig. 9.10 shows the schematic of a SIMO converter using buck and boost converters, and the right side shows its inductor current variation and the timing diagram of switches. By controlling the switches as this timing diagram, the output voltages V_{out1} and V_{out2} become

$$V_{\text{out1}} = \frac{T_1}{T_1 + T_2} V_{\text{in}} \tag{9.4}$$

$$V_{\text{out2}} = \frac{T_4 + T_5}{T_5} V_{\text{in}} \tag{9.5}$$

The feature of this topology is that both output voltages can be controlled independently and are positive.

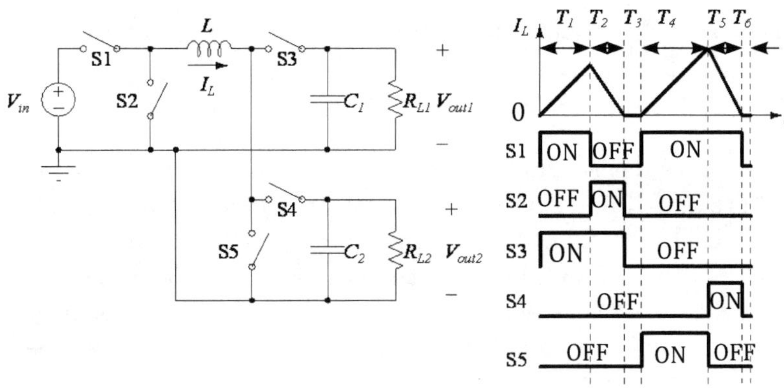

Figure 9.10 Schematic of a SIMO converter using buck and boost converters.

9.3.2.3 Boost/boost combination

The left side of Fig. 9.11 shows the schematic of a SIMO converter using two boost converters, and the right side shows its inductor current and the timing diagram of switches.

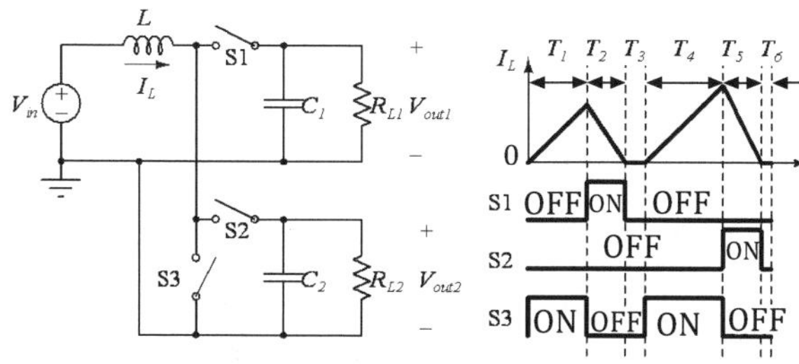

Figure 9.11 Schematic of a SIMO converter using two boost converters.

The output voltages V_{out1} and V_{out2} become

$$V_{\text{out1}} = \frac{T_1 + T_2}{T_2} V_{\text{in}} \tag{9.6}$$

$$V_{\text{out2}} = \frac{T_4 + T_5}{T_5} V_{\text{in}} \tag{9.7}$$

The feature of this topology is that both output voltages are positive and the number of switches is minimum.

9.3.2.4 Buck/positive buck–boost combination

The topology of a SIMO converter using buck and positive buck–boost converters is the same as in Fig. 9.10, but the timing diagram is different. The timing diagram of this SIMO converter is illustrated in the right part of Fig. 9.12.

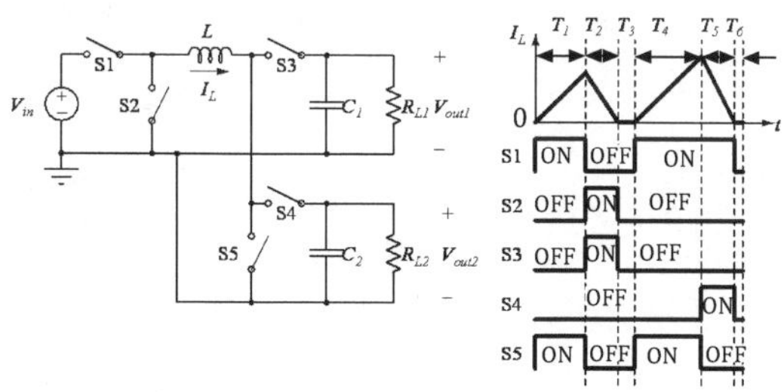

Figure 9.12 Schematic of a SIMO converter using buck and positive buck–boost converters.

The output voltages V_{out1} and V_{out2} are obtained as

$$V_{\text{out1}} = \frac{T_1}{T_1 + T_2} V_{\text{in}} \tag{9.8}$$

$$V_{\text{out}2} = \frac{T_4}{T_5} V_{\text{in}} \tag{9.9}$$

Note that both output voltages are positive.

9.3.2.5 Boost/positive buck–boost combination

The topology of a SIMO converter using boost and positive buck–boost converters is also the same as in Fig. 9.10, but the timing diagram is different. The timing diagram of this converter is illustrated in the right part of Fig. 9.13. The output voltages V_{out1} and V_{out2} are given as

$$V_{\text{out}1} = \frac{T_4}{T_5} V_{\text{in}} \tag{9.10}$$

$$V_{\text{out}2} = \frac{T_4 + T_5}{T_5} V_{\text{in}} \tag{9.11}$$

Note that both output voltages are positive.

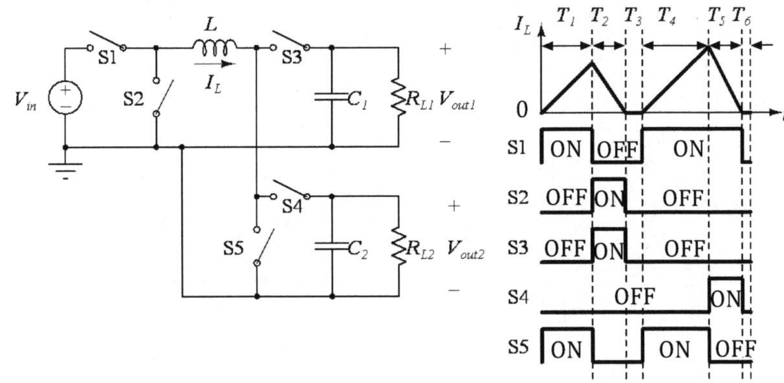

Figure 9.13 Schematic of a SIMO converter using boost and positive buck–boost converters.

9.3.2.6 Buck/negative buck–boost combination

Figure 9.14 shows the schematic of a SIMO converter using a buck and a negative buck–boost converter and the timing diagram of its switches. The output voltages V_{out1} and V_{out2} using the regions T_1, T_2, T_4, and T_5 are given as

$$V_{\text{out1}} = \frac{T_1}{T_1 + T_2} V_{\text{in}} \tag{9.12}$$

$$V_{\text{out2}} = -\frac{T_4}{T_5} V_{\text{in}} \tag{9.13}$$

The feature of this topology is that this converter can realize bipolar outputs. In many applications such as organic electroluminescence and charge-coupled devices (CCDs), both positive and negative power supplies, that is, a bipolar output converter, are required. Thus this combination is also useful for many applications.

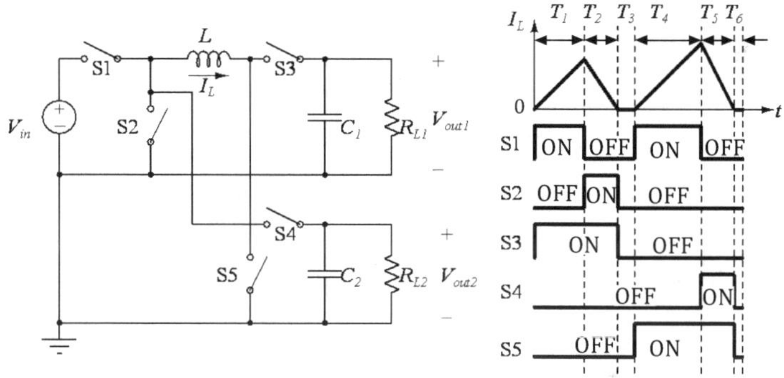

Figure 9.14 Schematic of a SIMO converter using buck and negative buck–boost converters.

9.3.2.7 Boost/negative buck–boost combination

The left part of Fig. 9.15 shows the schematic of a SIMO converter using boost and negative buck–boost converters, and the timing diagram of the switches is shown in the right part of Fig. 9.15. The output voltages V_{out1} and V_{out2} using the periods T_1, T_2, T_4, and T_5 are given as

$$V_{\text{out1}} = \frac{T_1 + T_2}{T_2} V_{\text{in}} \tag{9.14}$$

$$V_{\text{out2}} = -\frac{T_4}{T_5} V_{\text{in}} \tag{9.15}$$

Note that this converter can also realize bipolar outputs.

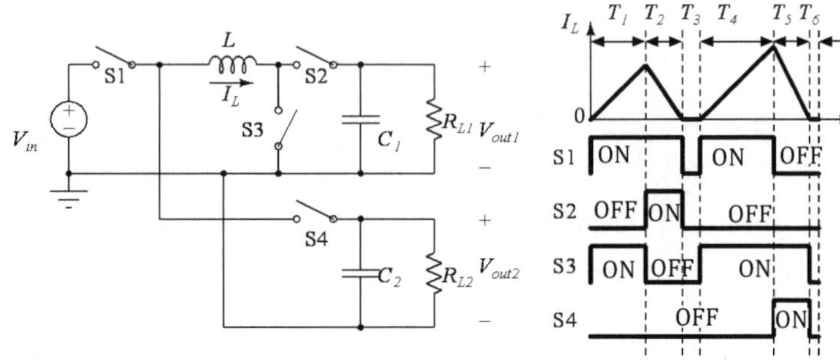

Figure 9.15 Schematic of a SIMO converter using boost and negative buck–boost converters.

9.3.2.8 Positive buck–boost/positive buck–boost combination

The topology of a SIMO converter using two positive buck–boost converters is the same as in Fig. 9.10, but the timing diagram is different. The timing diagram of this SIMO converter is illustrated in the right part of Fig. 9.16. The output voltages V_{out1} and V_{out2} are given as

$$V_{\text{out1}} = \frac{T_4}{T_5} V_{\text{in}} \tag{9.16}$$

Note that both output voltages are positive.

$$V_{\text{out2}} = \frac{T_4 + T_5}{T_5} V_{\text{in}} \tag{9.17}$$

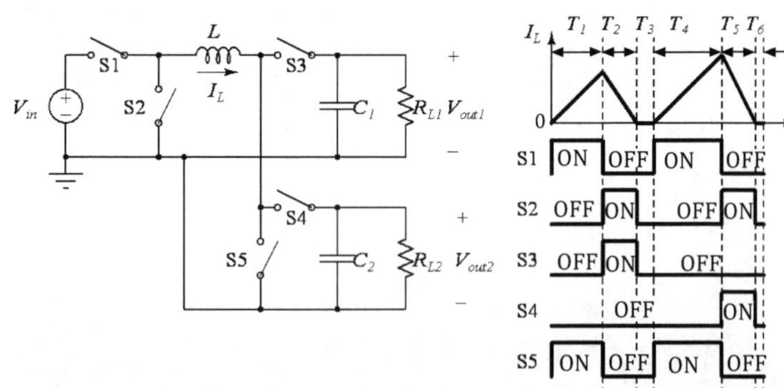

Figure 9.16 Schematic of a SIMO converter using two positive buck–boost converters.

9.3.2.9 Negative buck–boost/negative buck–boost combination

The left part of Fig. 9.17 shows the schematic of a SIMO converter using two negative buck–boost converters, and the right part shows the timing diagram of its switches. The output voltages V_{out1} and V_{out2} using the periods T_1, T_2, T_4, and T_5 are given as

$$V_{\text{out1}} = -\frac{T_1}{T_2} V_{\text{in}} \tag{9.18}$$

$$V_{\text{out2}} = -\frac{T_4}{T_5} V_{\text{in}} \tag{9.19}$$

The feature of this topology is that both output voltages are negative and the number of switches is minimum.

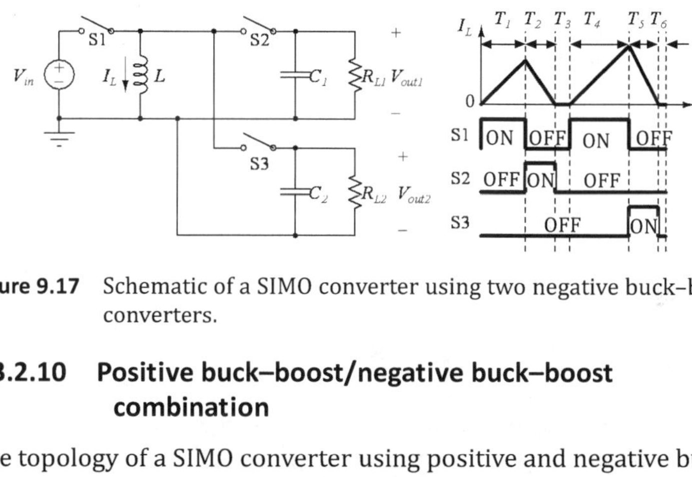

Figure 9.17 Schematic of a SIMO converter using two negative buck–boost converters.

9.3.2.10 Positive buck–boost/negative buck–boost combination

The topology of a SIMO converter using positive and negative buck-boost converters is the same as in Fig. 9.14, but the timing diagram is different. The timing diagram of this SIMO converter is illustrated in the right part of Fig. 9.18. The output voltages V_{out1} and V_{out2} are given as

$$V_{\text{out1}} = \frac{T_1}{T_2} V_{\text{in}} \tag{9.20}$$

$$V_{\text{out2}} = -\frac{T_4}{T_5} V_{\text{in}} \tag{9.21}$$

The feature of this topology is that this converter can realize bipolar outputs.

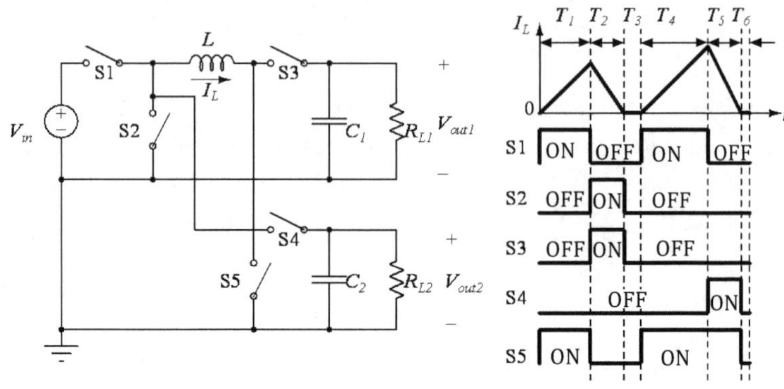

;

Figure 9.18 Schematic of a SIMO converter using positive and negative buck–boost converters.

9.3.2.11 SIMO using a boost converter and a charge pump circuit

The left part of Fig. 9.19 shows the schematic of a SIMO converter using a boost converter and a charge pump circuit, and the timing diagram of its switches is indicated in the right part of Fig. 9.19. The output voltages V_{out1} and V_{out2} are given as

$$V_{\text{out1}} = \frac{T_1 + T_2}{T_2} V_{\text{in}} \tag{9.22}$$

$$V_{\text{out2}} = -\frac{T_4 + T_5}{T_5} V_{\text{in}} + V_{\text{F}} \tag{9.23}$$

where V_{F} is the diode voltage drop. Note that the converter can realize bipolar outputs, and the number of switches is minimum.

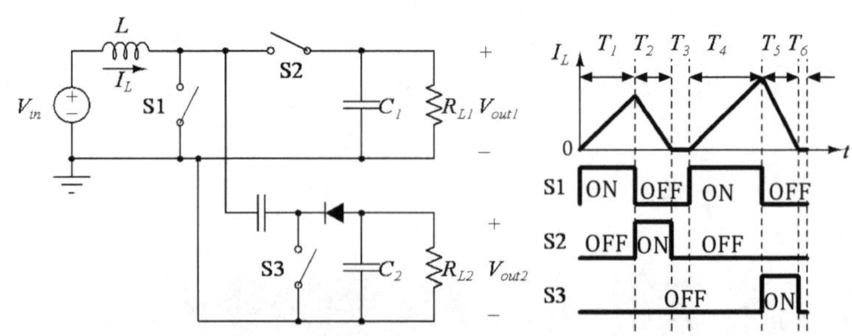

Figure 9.19 Schematic of a SIMO converter using a boost converter and a charge pump circuit.

9.3.3 Freewheel Technique for PCCM

When we try to realize a SIMO DC–DC converter, the inductor current is shared for each output terminal. Because two outputs share one inductor in turn, we must take the effect of current variation of each output terminal into account. When CCM is employed for control of the inductor current, the current variation of one output terminal affects the current of the other terminal because the inductor current is continuous in CCM. This effect is called cross regulation, and cross regulation is not good if CCM is employed for control of the inductor current. DCM has good cross-regulation characteristics because the inductor current is not continuous. However, the inductor current ripple is not small. To solve these problems, PCCM is proposed in Ref. [13]. Again, Fig. 9.20b indicates the waveform of the inductor current using PCCM. In PCCM, the floor of the inductor current is raised by a DC level of I_B. PCCM can realize a small inductor current ripple as in the DCM case. Compared to the CCM case, the inductor current alternately resets and stays constant at I_B, which successfully isolates the two output current variations. Individual output current variation can be adjusted by changing the duty ratio of S_{on} and S_{off} of the corresponding output terminal, which does not affect the other. To achieve PCCM, we must realize a constant inductor current. Switch Sf in Fig.9.20a keeps the inductor current constant, which is called the freewheel switch.

Figure 9.21 shows a SIMO converter applying the freewheel technique to Fig. 9.19 and the timing diagram of its switches. The output voltage is the same as in Fig. 9.19. Note that the cross regulation and inductor current ripple of this circuit are smaller than those of Fig. 9.19.

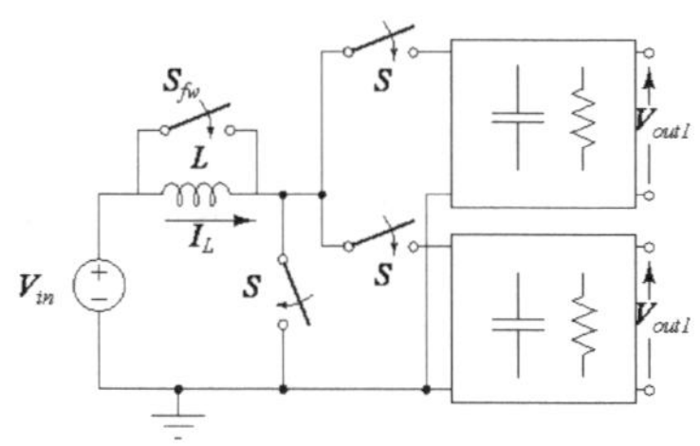

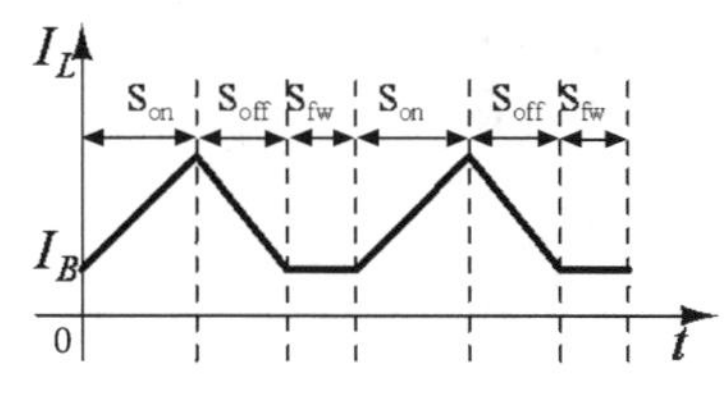

Figure 9.20 Waveform of inductor current using PCCM. (a) SIMO circuit with a freewheel switch. (b) Waveform of inductor current using PCCM.

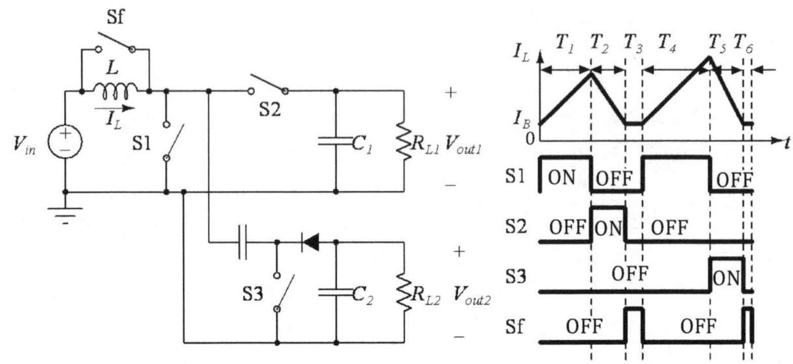

Figure 9.21 Schematic of a SIMO converter using a boost converter and a charge pump circuit.

9.4 Control Circuit

To realize the timing diagram of the switches described in the previous section, a control circuit is required. A control circuit is classified into three types: voltage-mode control, current-mode control, and hysteresis control. In this section, control circuits for the SIMO converter are described.

9.4.1 Voltage-Mode Control

Voltage-mode control is a basic, widely used control. This control method feeds only the output voltage back to the input through a feedback loop. The detailed operation is as follows. The output voltage is compared with the reference voltage using an error amplifier (EA). The output voltage of the EA is compared with a sawtooth wave using a comparator, which results in a pulse width of the pulse width modulation (PWM) signal. The PWM signal is applied to switches of the converter so as to control the output voltage. The feature of this mode is simple control, good EMI tolerance, and short on time. On the other hand, the demerit of this mode is complex phase compensation of the loop.

Figure 9.22 shows the schematic of the voltage-mode control of Fig. 9.21. To obtain the timing diagram of Fig. 9.21, a control circuit that consists of three components is required. One component is to divide the period with respect to positive and negative output voltages. The second component is to detect the inductor current, I_L, and limit the inductor current to I_B. The final component is a logic circuit to avoid not overlapping the timing of all switches.

The control circuit blocks are composed of resistors for dividing the output voltages, EAs, a reference voltage source, two ramp wave generators, CMPs (comparators), a logic circuit, and a current sensor.

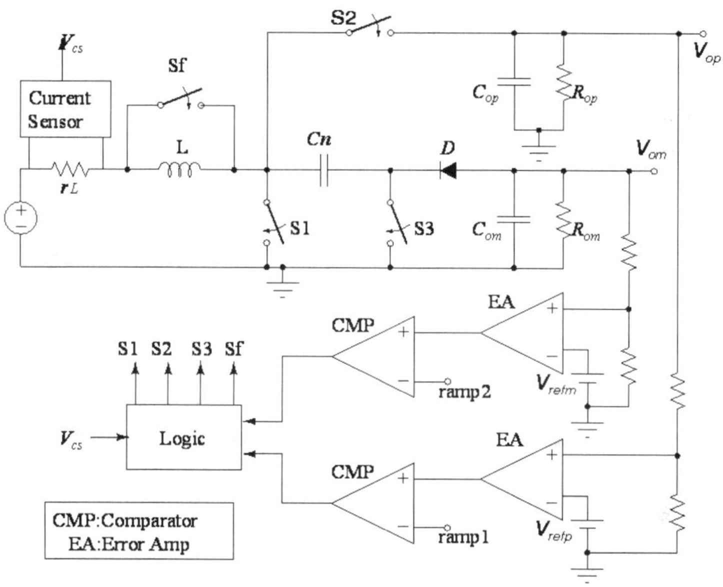

Figure 9.22 Schematic of the voltage-mode control of Fig. 9.21.

9.4.2 Current-Mode Control

Current-mode control is a modified control method of voltage-mode control. While the PWM signal is generated by using a sawtooth wave in voltage mode, an inductor current substitutes for the sawtooth wave in current mode. Thus current mode has a voltage loop and a current loop. The control becomes complex, but phase compensation becomes widely easy. Moreover, other merits of this mode are high stability of the feedback loop, improvement of the transient response, good line regulation, and an included overcurrent protection circuit.

Figure 9.23 shows the SIMO converter of Fig. 9.21 using current-mode control circuit. The current-mode control circuits are connected to each output terminal. A current sense circuit detects the amount of inductor current, IL. When the inductor current, I_L, becomes I_B, the output voltage of the current sense circuit becomes "high," and the logic and driver circuit makes switch Sf turn on. This realizes the freewheel operation. Also the logic and driver circuit prevents the switches from turning on at the same time. As mentioned before, the inductor shares the positive and negative output terminals. Thus inductor current detection is also separated for the positive and negative output periods.

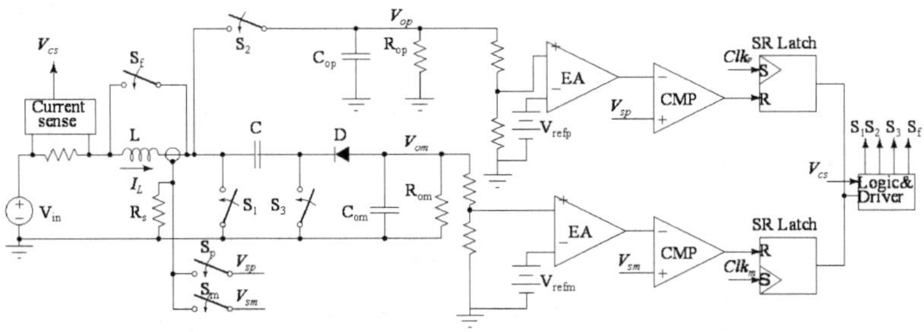

Figure 9.23 Schematic of the current-mode control of Fig. 9.21.

9.4.3 Ripple Control and Hysteresis Control

The voltage and current modes have the demerit that performance for the response time for quick variation of the load current is not good. This is caused by the delay time of the voltage loop and phase compensation circuit, which limits the frequency response.

In ripple control, the output voltage is compared with the reference voltage, and when the output voltage becomes lower than the reference voltage, the switches of the converter are controlled. This control is called fixed on time with bottom-detection control. If the switches of the converter start the control when the output voltage becomes more than the reference voltage, this control is called fixed off time with upper-detection control. Mixed control of the fixed on-time and off-time control is called hysteresis control.

In these controls, the control circuit consists of only comparator and reference voltages and a simple logic circuit.

Figure 9.24 shows the schematic of a boost converter using hysteresis control. The control circuit is composed of two comparators, an AND circuit, and a reference voltage. We can apply this circuit to a SIMO converter. Figure 9.25 shows the schematic of a SIMO boost–boost converter.

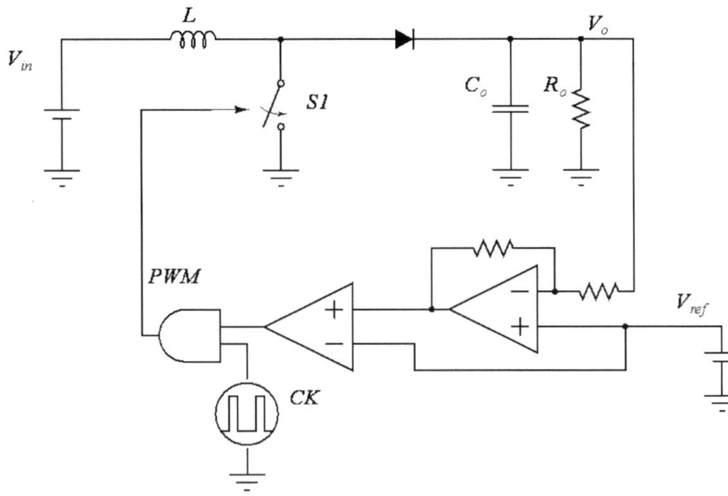

Figure 9.24 Schematic of a boost converter using hysteresis control.

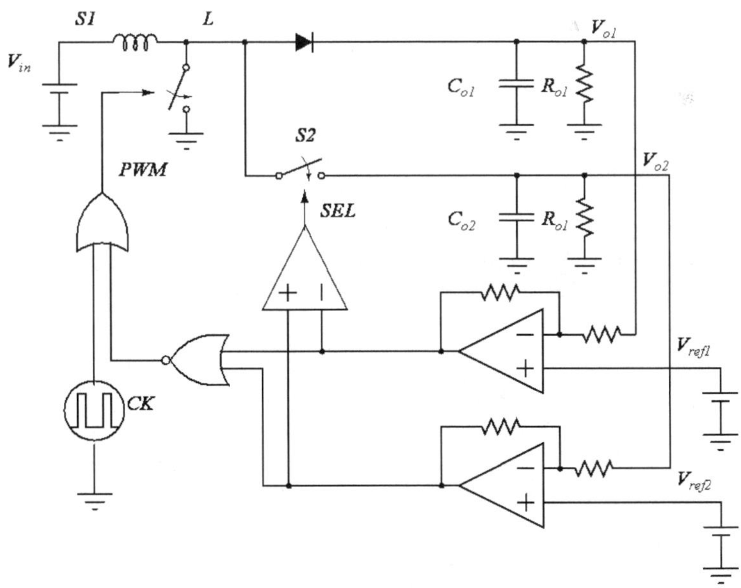

Figure 9.25 Schematic of a SIMO boost–boost converter using hysteresis control.

9.5 Conclusion

In this chapter, 11 types of SIMO DC–DC converters and their output voltages were described. A SIMO converter can supply positive outputs, negative outputs, and bipolar outputs. Moreover, three types of control methods and their schematic examples were indicated.

References

1. H.-P. Le, C.-S. Chae, K.-C. Lee, G.-H. Cho, S.-W. Wang, G.-H. Cho, and S. il Kim, A single-inductor switching DC-DC converter with 5 outputs and ordered power- distributive control, *Proc. ISSCC*, **29.9**, 534–535, 2007.
2. C.-S. Chae, H.-P. Le, K.-C. Lee, M.-C. Lee, G.-H. Cho, and G.-H. Cho, A single-inductor step-up DC-DC switching converter with bipolar outputs for active matrix OLED mobile display panels, *Proc. ISSCC*, 136–137, 2007.
3. S.-C. Koon, Y.-H. Lam, and W.-H. Ki, Integrated charge-control single-inductor dual-output step-up/step-down converter, *Proc. ISCAS*, 3071–3074, 2005.
4. W.-H. Ki and D. Ma, Single-inductor multiple-output switching converters, *Proc. IEEE Power Electron. Specialist Conf.*, **1**, 226 - 231, 2001.
5. Y. Kobori, F. Zhao, Q. Li, M. Li, S. Wu, Z. Nosker, S. N. Mohyar, N. Takai, H. Kobayashi, T. Odaguchi, I. Nakanishi, K. Ueda, and J. Matsuda, Single inductor dual output DC-DC converter design with exclusive control, *Proc. APCCAS*, 436–439, 2012.
6. N. Takai, K. Takahashi, H. Yokoo, S. Miwa, K. Tsushida, H. Iwase, K. Murakami, H. Kobayashi, T. Odaguchi, S. Takayama, T. Oomori, I. Nakanishi, K. Nemoto, and J. Matsuda, Single inductor DC-DC converter with independent bipolar outputs using charge pump, *Key Eng. Mater.*, **497**, 285–295, 2012.
7. K. Takahashi, H. Yokoo, S. Miwa, K. Tsushida, H. Iwase, K. Murakami, N. Takai, H. Kobayashi, T. Odaguchi, S. Takayama, I. Fukai, and J. Matsuda, Single inductor DC-DC converter with bipolar outputs using charge pump, *Proc. APCCAS*, 460–463, 2010.
8. N. Takai, T. Okada, K. Takahashi, H. Yokoo, S. Miwa, K. Tsushida, H. Iwase, K. Murakami, H. Kobayashi, T. Odaguchi, S. Takayama, T. Oomori, I. Nakanishi, K. Nemoto, and J. Matsuda, Single inductor bipolar outputs

DC-DC converter with current mode control circuit, *Key Eng. Mater.*, **534**, 220–226, 2013.

9. N. Takai, H. Iwase, T. Okada, T. Sakai, Y. Kobori, H. Kobayashi, T. Omori, T. Odaguchi, I. Nakanishi, K. Nemoto, and J. Matsuda, Low power consumption control circuit for sibo DC-DC converter, *Proc. Int. Conf. Analog VLSI Circuits*, 2012.
10. Y. Kobori, S. Tanaka, T. Nagashima, Z. Nosker, N. Takai, H. Kobayashi, T. Odaguchi, I. Nakanishi, K. Ueda, and J. Matsuda, High-speed response single inductor multi output DC-DC converter with hysteresis control, *Proc. Int. Conf. Power, Energy Electrical Eng.*, 2013.
11. Z. Hu and D. Ma, A pseudo-CCM buck converter with freewheel switching control, *Proc. ISCAS*, 3083–3086, 2005.
12. Y.-J. Woo, H.-P. Le, G.-H. Cho, G.-H. Cho, and S.-I. Kim, Load-independent control of switching DC-DC converters with freewheeling current feedback, *IEEE J. Solid State Circuit*, **43**(12), 2798–2808, 2008.
13. D. Ma, W.-H. Ki, and C.-Y. Tsui, A pseudo-ccm/dcm simo switching converter with freewheel switching, *IEEE J. Solid State Circuit*, **38**(6), 1007–1014, 2003.
14. Y. Kobori, Q. Zhu, M. Li, F. Zhao, Z. Nosker, S. Wu, S. N. Mohyar, M. Onozawa, H. Kobayashi, N. Takai, K. Niitsu, T. Odaguchi, I. Nakanishi, K. Nemoto, and J. Matsuda, Single inductor dual output DC-DC converter design with exclusive control, *IEEE Asia Pacific Conference on Circuits and Systems*, 436–439, Kaohsiung, Taiwan, 2012.

Chapter 10

A Small, Low-Power Boost Regulator Optimized for Energy-Harvesting Applications

Zachary Nosker
Division of Electronics and Informatics, Gunma University,
1—5—1 Tenjin-cho, Kiryu, Gunma 376-8515, Japan
t11802471@gunma-u.ac.jp

A small, low-power bootstrapped boost regulator is introduced that can start up with an input voltage of 240 mV and achieve a maximum efficiency of 97%. The proposed circuit uses two separate control schemes for start-up and steady-state operation. A fixed-frequency oscillator is used to initially start up the circuit and raise the output voltage. Once the output voltage has reached a level adequate to bias the internal circuitry, a constant-on-time-style hysteretic control scheme is used, which helps increase system efficiency compared to using a conventional pulse-width-modulated control scheme. While maintaining high efficiency, the proposed circuit only requires three external components: two capacitors (input and output) and an inductor. The effectiveness of this approach is shown through Spectre® simulation results.

Handbook of Power Management Circuits
Edited by Haruo Kobayashi and Takashi Nabeshima

ISBN 978-981-4613-15-6 (Hardcover), 978-981-4613-16-3 (eBook)
www.panstanford.com

10.1 Introduction

Recently there have been great advances in the realm of energy harvesting. Using vibrational, thermal, and solar transducers, small amounts of power can be captured by latent energy sources [1–8]. While previous works have shown that capturing power from these micropower sources is feasible, each approach has drawbacks. Some approaches require an external battery [1] or a storage capacitor [2] with a voltage of several volts to bias the control circuitry. Others have designed their circuits with external bias circuitry [3] or require a mechanical switch to start up the converter [4]. Furthermore, previous works also are not suitable to drive realistic loads, as their maximum load capability is in the microwatt range [1–6]. In real-world applications, where energy harvesters are used in hard-to-reach places, it would be impractical to replace the batteries used in these energy-harvesting systems. Additionally, since a typical low-power microcontroller requires several megawatts of power to operate, conventional circuits are unsuitable for real-world applications.

To address these issues, we have created a low-power boost circuit that is only operational when its input source is large enough to bootstrap the system. Our circuit is also optimized to provide several micro-watts of power, which would be suitable to drive a low-power microcontroller unit (MCU). Additionally, as previous papers only briefly mention total system efficiency, our circuit is designed to maximize efficiency, which, in a real-world application, would increase the maximum amount of energy that is available to the load. While there are a number of boost regulators on the market that can start up with a low input voltage [9, 10] these parts have very low efficiency (<40%) in the 1 mW output load range, and some require a large number of external components [10]. Since energy-harvesting inputs like thermal transducers only output a small amount of power, it is critical to optimize energy-harvesting power electronics at a lower load range.

This paper introduces a regulator that is able to start up with an input voltage of 240 mV and can supply up to 6.2 mW of power with a maximum efficiency of 97%. With our approach, these results are achieved with only three external components: an input capacitor, an output capacitor, and an inductor. The feasibility of our approach is shown with Spectre® simulation results.

10.2 Proposed Circuit Operation

10.2.1 Ideal Boost Regulator Operation

A boost or step-up regulator is a commonly used DC–DC converter that has an output voltage higher than the input voltage. Figure 10.1 shows the schematic of an ideal synchronous boost converter, where the two switches S_1 and S_2 are switched out of phase with a fixed duty cycle D. When S_1 is on, the current in inductor L increases linearly by the function $\Delta I/\Delta t = V_I/L$. Next, turning off S_1 (and hence turning on S_2) transfers the current built up in the inductor to the output capacitor and load. By operating this circuit at a constant D, it has been shown [11] that the steady-state output voltage of the boost regulator is

$$\frac{V_O}{V_I} = \frac{1}{1-D} \tag{10.1}$$

While Eq. 10.1 is a decent approximation, it suggests that at a 100% duty cycle, the output voltage would rise to infinity. However, in this situation (i.e., S_1 is always on), the current in the inductor would rise unencumbered until any component fails, never delivering its current to the output capacitor nor the output load, and hence the output voltage would be 0 V. Because of this effect, switch S_1 must turn off for some amount of time (and hence S_2 must turn on), guaranteeing that the output voltage will rise. Note that Fig. 10.1 contains a constant current source as an output. This is used to model the load current being delivered by the regulator.

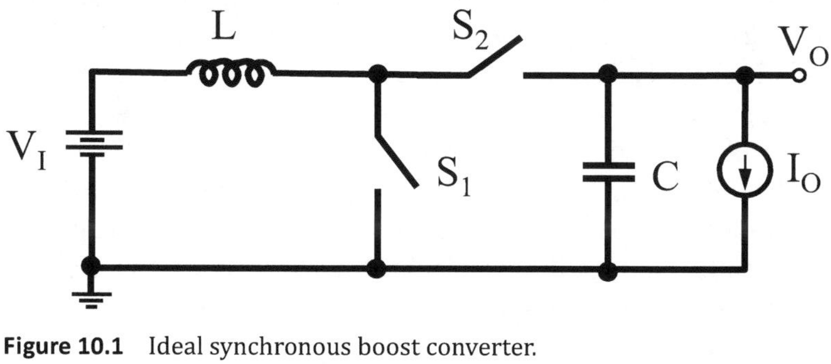

Figure 10.1 Ideal synchronous boost converter.

10.2.2 Design Methodology

The proposed circuit uses a novel combination of commonly used circuits to achieve a low input voltage, high efficiency, and a load range that is realistic for energy-harvesting systems. During start-up, a charge pump circuit (Section 10.2.4) is used to create a voltage high enough to begin switching the main power transistors. Once the output voltage has risen to an acceptable value, the system changes to a constant-on-time hysteretic control mode (Section 10.2.7) to regulate the output voltage, while maintaining high efficiency. This is accomplished by using only a single-input voltage supply (i.e., there is no extra gate biasing) and with only three external components: an input capacitor, an output capacitor, and an inductor.

All individual circuit blocks have been optimized to lower input voltage and increase efficiency, offering exceptional performance at load ranges suitable for energy-harvesting applications (~1 mW).

10.2.3 Block Diagram

A block diagram of the proposed circuit is shown in Fig. 10.2. The components L, M_N, M_P, and C_O form the core of a basic boost converter, similar to that in Fig. 10.1. Since the output voltage V_O will be well regulated and larger than the input, it is suitable for supplying power to the internal control circuitry, namely, the voltage loop comparator, one shot and voltage reference. To initially raise the output voltage, however, we are using a charge pump circuit that can operate from a very low input voltage, which will be described in Section 10.2.4. The output of this charge pump is then fed to a fixed-frequency start-up oscillator and to the main drivers that commutate the switches and raise the output voltage to bias the internal circuitry. Once the output voltage has reached an acceptable level, the charge pump and start-up oscillator are stopped and the driver power source switches to the output voltage. When this handoff is complete, the circuit begins running in a constant-on-time hysteretic mode, as described in Section 10.2.7. The input voltage V_I, output voltage V_O, and charge pump output voltage V_{CP} are highlighted in different colors to help the reader easily understand the voltage domains present in the proposed circuit.

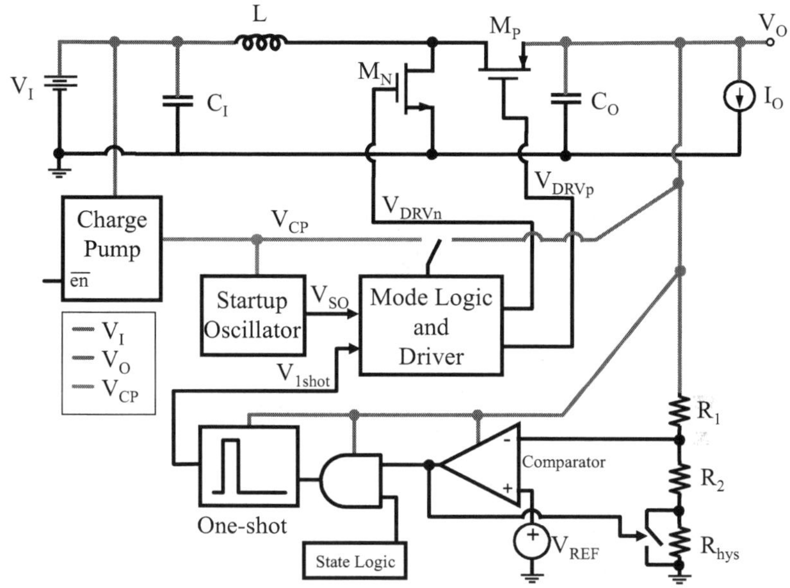

Figure 10.2 Bootstrapped boost converter block diagram.

10.2.4 Start-Up Charge Pump

The proposed circuit is being designed in a 0.18 μm complementary metal–oxide–semiconductor (CMOS) process that supports n-channel metal–oxide–semiconductor (NMOS) native-V_t and low-V_t transistors (threshold voltages for all available 1.8 V transistors are shown in Table 10.1). To reduce leakage current, however, nominal V_t devices are used as the output drivers and switches. Because the threshold voltage of the nominal devices is greater than 400 mV, an on-chip charge pump is used to ensure that the system can operate with V_I less than the switch threshold voltage. Figure 10.3 shows the oscillator used to operate the charge pump circuit. By using low-V_t transistors and a 20 fF capacitor, this circuit operates at 40 MHz with an input voltage of 300 mV. Since the charge pump is only used during the initial start-up mode of the circuit, this oscillator needs to be disabled when the constant-on-time loop takes over. To this end, the final p-channel metal–oxide–semiconductor (PMOS) MP3 in the oscillator of Fig. 10.3 is designed with a W/L size of 2 μm/5 μm to minimize the current consumed when this cell is disabled. The other

PMOS transistors are 5 μm/250 nm, and the NMOS transistors are 2 μm/300 nm, where 250 nm and 300 nm are the minimum channel length for low-V_t PMOS and NMOS transistors, respectively. It should be noted that this circuit is the limiting step in terms of minimum start-up voltage. As long as the charge pump oscillator operates, the charge pump will create a high enough output voltage to operate the main switches and operate the circuit (as described later). Current data shows that the minimum input voltage required to start up the charge pump oscillator, and hence the full circuit, is 240 mV.

Table 10.1 Transistor threshold voltage V_{t0}

Transistor type	Model V_{t0}
Native NMOS	–0.02 V
Low-V_t PMOS	–0.13 V
Low-V_t NMOS	0.27 V
Nominal V_t PMOS	–0.44 V
Nominal V_t NMOS	0.44 V

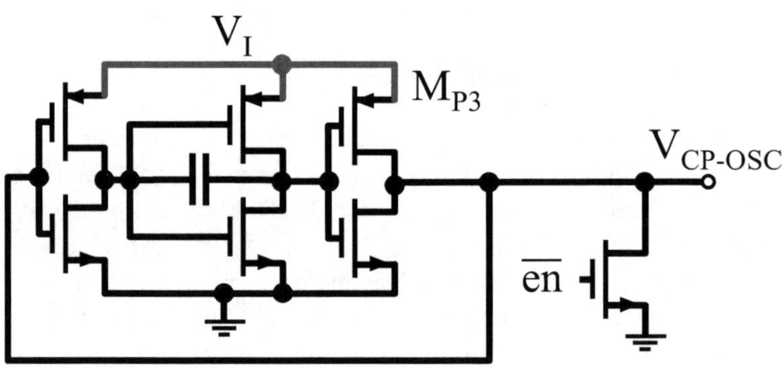

Figure 10.3 Start-up charge pump oscillator.

Figure 10.4 shows the complete charge pump schematic. The oscillator from Fig. 10.3 is fed through a buffer before connecting to a diode-connected native-V_t NMOS charge pump, similar to the approach in Ref. [12]. Unlike the previous work, our system relies on an oscillator signal generated on-chip and is connected to the output switch drivers, which require a large amount of power to operate. Because of these drawbacks, while using a 10-stage design, our charge pump achieves a peak voltage of only 950 mV from a

300 mV source. In our implementation of this charge pump, all native-V_t NMOS transistors are designed with a size of 40 μm/500 nm and the capacitors are 10 pF metal-metal capacitors. Note that in the CMOS process we are using, 500 nm is the minimum channel length for native-V_t NMOS transistors.

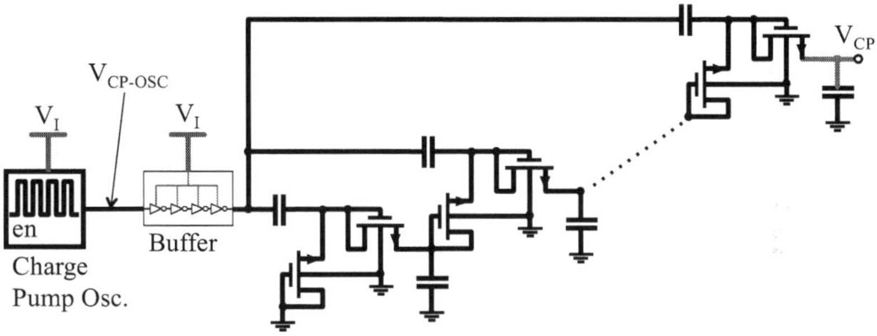

Figure 10.4 Full charge pump schematic.

10.2.5 Start-Up Oscillator and Driver

To raise the output voltage to an appreciable value to bias the hysteretic control circuitry, a high-duty-cycle oscillator is applied to the gate driver circuitry for the switch transistors, powered by the charge pump described in Section 10.2.4.

Using an oscillator similar to that in Fig. 10.3, but running at a frequency of 100 kHz, this low-frequency clock is modified to run at a very high duty cycle (83%). Figure 10.5 shows the high-duty-cycle transformation circuit. As long as either of the inputs to the NAND gate is low, the output of this circuit will be high. Since the input comes from a clock running at about a 50% duty cycle, the output is only low during the delay when the oscillator goes from LO to HI. During this time, the output V_{SO} will be low until the input signal propagates through the three-inverter, two-capacitor delay. As was described in Section 10.2.1, the duty cycle of the converter must be less than 100% to ensure that current from the inductor is transferred to the output capacitor, and hence the output voltage can rise. The schematic of this high-duty-cycle system connected to the drivers and output switches is shown in Fig. 10.6.

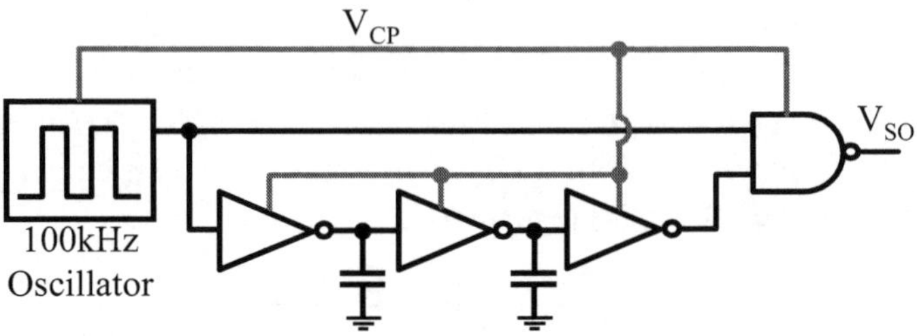

Figure 10.5 Start-up oscillator and high-duty-cycle circuitry.

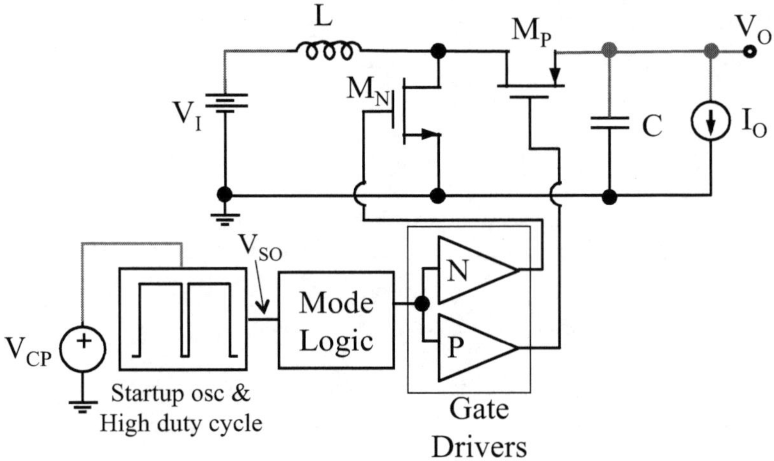

Figure 10.6 Start-up circuitry connected to drivers and output switches.

10.2.6 Voltage Reference

Since the proposed circuit operates with an output voltage of around 1 V, which is used to drive the control circuitry, the reference voltage (labeled V_{REF} in Fig. 10.2) must also run from a very low input voltage. Due to its low-input-voltage requirement and the ability to create a reference voltage less than a typical 1.2 V bandgap, we have modified the Banba bandgap reference described in Ref. [13].

To save space, we have incorporated an area mismatch of 8× to 1× between the two PNP bipolar junction transistors (BJTs). This has an added benefit that the transistors can be easily laid out in a 3 × 3 grid.

Additionally, we have adjusted the reference voltage output to 250 mV, which is better suited to our application. However, since these parameters have been changed, the values of the pertinent resistors must also be changed, as explained in the next section.

10.2.6.1 Voltage reference design equations

Like any bandgap voltage reference, the Banba bandgap operates by adding a voltage component that is complementary to absolute temperature (CTAT) to a voltage component that is proportional to absolute temperature (PTAT). In the circuit of Fig. 10.7, both substrate PNP devices and the output resistor R_3 share the same current:

$$I_1 = v_T \frac{\ln(A)}{R_1} + \frac{V_{BE}}{R_2} \quad (10.2)$$

where V_{BE} is the emitter-base voltage of the 1× transistor, $v_T = kT/q$ is the thermal voltage of silicon (which is directly proportional to temperature T), and A is the area ratio between the two PNP transistors. Since $V_{OUT} = I_1 R_3$:

$$V_{REF} = \frac{R_3}{R_1} v_T \ln(A) + \frac{R_3}{R_2} V_{BE} \quad (10.3)$$

Equation 10.3 shows that the reference voltage is a function of a PTAT component and a CTAT component; hence the resistor ratios and area ratio A can be adjusted to ensure that the reference voltage remains constant over temperature.

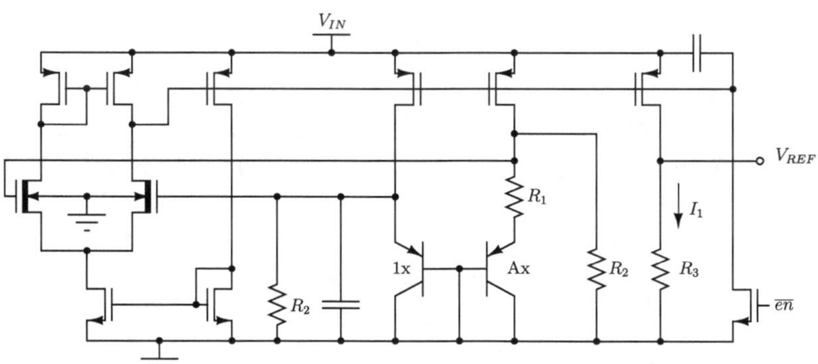

Figure 10.7 Banba bandgap voltage reference.

As described in Ref. [14], the base-emitter voltage of a BJT can be written as

$$V_{BE} = E_{GE} - H\,(E_{GE} - V_{BEN}) + V_{TN}\,H\ln\,(I_C/I_N) - \eta v_{TN}\,H\ln\,(H) \tag{10.4}$$

where E_{GE} is the bandgap voltage of silicon, H is the "hotness" factor ($H = T/T_N$), T_N is the normalizing temperature, V_{BEN} is the value of V_{BE} at a normalized point, I_C/I_N is the BJT collector current divided by the normalized current, v_{TN} is the value of the thermal voltage at nominal temperature ($v_{TN} = kT_N/q$), and η is a curvature factor. Assuming that the collector current of the transistor remains constant, V_{REF} can be rewritten as

$$V_{REF} = \frac{R_3}{R_1} H v_{TN} \ln(A) + \frac{R_3}{R_2}\left[E_{GE} - H(E_{GE} - V_{BEN}) - \eta v_{TN} H \ln(H)\right] \tag{10.5}$$

To make V_{REF} a proper reference voltage, it should remain constant over temperature; hence

$$\frac{\partial V_{REF}}{\partial H} = 0 \tag{10.6}$$

Substituting Eq. 10.5 into 10.6 and expanding, an expression can be developed for which V_{REF} does not change with temperature:

$$\frac{R_1}{R_2} = \frac{v_T \ln(A)}{E_{GE} - V_{BEN} + \eta V_{TN}} \tag{10.7}$$

As long as the condition in Eq. 10.7 is met, a closed-form expression can be gained for V_{REF} that does not change with temperature:

$$V_{REF} = v_T \ln(A) \frac{R_3}{R_1}\left(1 + \frac{V_{BEN}}{E_{GE} - V_{BEN}}\right) \tag{10.8}$$

To finalize the values for the resistors, we start by setting the area mismatch between the two PNP transistors to 8:1 ($A = 8$). Since the difference in the two V_{BE} voltages will appear across resistor R_1, and setting the current in each PNP to 300 nA, we have $R_1 = 54$ mV/300 nA = 180 kΩ. By using the result of Eq. 10.8 and then Eq. 10.7 and setting $V_{REF} = 250$ mV, it follows that $R_3 = 371$ kΩ and $R_2 = 2.07$ MΩ.

10.2.7 Hysteretic Control

10.2.7.1 Overview

The main control loop for the proposed circuit is implemented using the constant-on-time control scheme shown in Fig. 10.8 and similar to what is described in Ref. [15]. If the feedback voltage is below the reference voltage, the one-shot circuit will trigger, causing the NMOS switch to turn on for a fixed period of time. At the same time, the switch across R_{hys} will close, effectively setting the voltage threshold to a higher level and adding a predictable amount of positive hysteresis. Once the one shot times out, the NMOS turns off and the PMOS turns on, until the current through the inductor reaches zero. At this point, the same process repeats until V_O reaches the comparator threshold, at which point the circuit goes into a low-power coasting mode and the R_{hys} switch turns off, returning the circuit to its lower threshold as it waits for the output voltage to fall again.

The block labeled "State logic" in Fig. 10.8 controls when the circuit enters coasting mode (i.e., the one shot is disabled and both switches are OFF) as well as resetting the one shot after every zero current detect when the NMOS is OFF. A state diagram of the proposed control scheme is shown in Fig. 10.9.

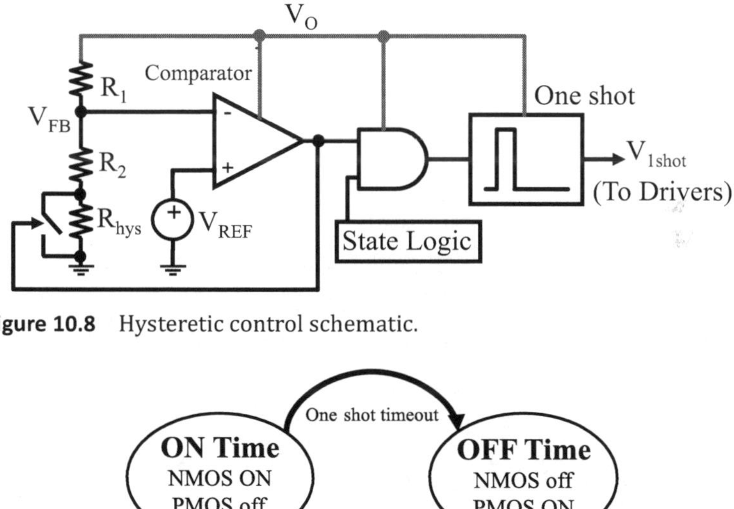

Figure 10.8 Hysteretic control schematic.

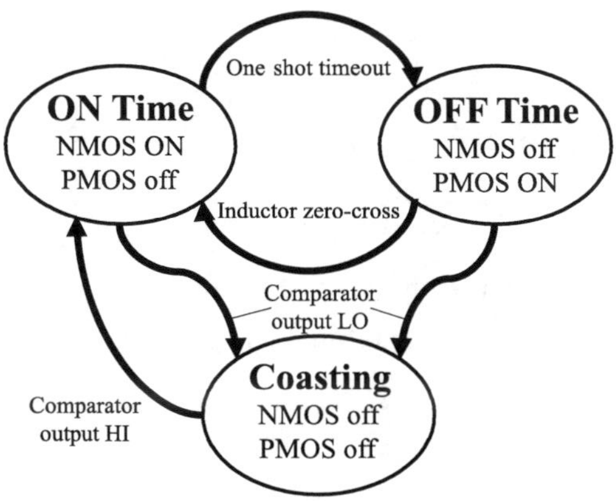

Figure 10.9 Hysteretic control state diagram.

10.2.7.2 Output voltage ripple

By examining Fig. 10.10 and assuming the output current I_O is zero, a closed-form expression can be developed for the output voltage ripple, Δv_O. Note that Fig. 10.10b shows the on-time system, while the NMOS and PMOS are commutating ON and OFF; hence here is no "Coasting" time. In the zero-load condition, the output voltage of the constant on-time regulator only increases when the inductor is discharging, hence, transferring its energy to the output capacitor. During this time (t_1 to t_2), switch S_1 is OFF, while S_2 is ON, as shown in Fig. 10.10a. The current in the inductor over an entire switching period is shown in Eqs. 10.9 and 10.10.

$$i_L(t) = \frac{V_I}{L}t, \; t_0 < t \le t_1 \tag{10.9}$$

$$i_L(t) = I_P - \frac{V_O - V_I}{L}(t - t_1), \; t_1 < t \le t_2 \tag{10.10}$$

And hence

$$t_{off} = t_2 - t_1 = \frac{I_P L}{V_O - V_I} \tag{10.11}$$

During t_{off}, when the current from the inductor flows to the output capacitor and load, the output voltage will increase for each cycle. Assuming the load current, I_O is zero, this change in output voltage, Δv_O, can be derived as shown in Eq. 10.12:

$$\Delta v_O = \frac{1}{C_O}\int i_L(t)dt = \frac{1}{C_O}\int_{t_1}^{t_2}\left[I_P - \frac{V_O - V_I}{L}(t - t_1)\right]dt = \frac{I_P^2 L}{2C_O(V_O - V_I)} \tag{10.12}$$

From Eq. 10.9, it can be shown that $I_P = V_I t_{on}/L$, and thus Eq. 10.12 can be simplified to the more readable expression shown in Eq. 10.13:

$$\Delta v_O = t_{on}^2 \frac{V_I^2}{2LC_O(V_O - V_I)} \tag{10.13}$$

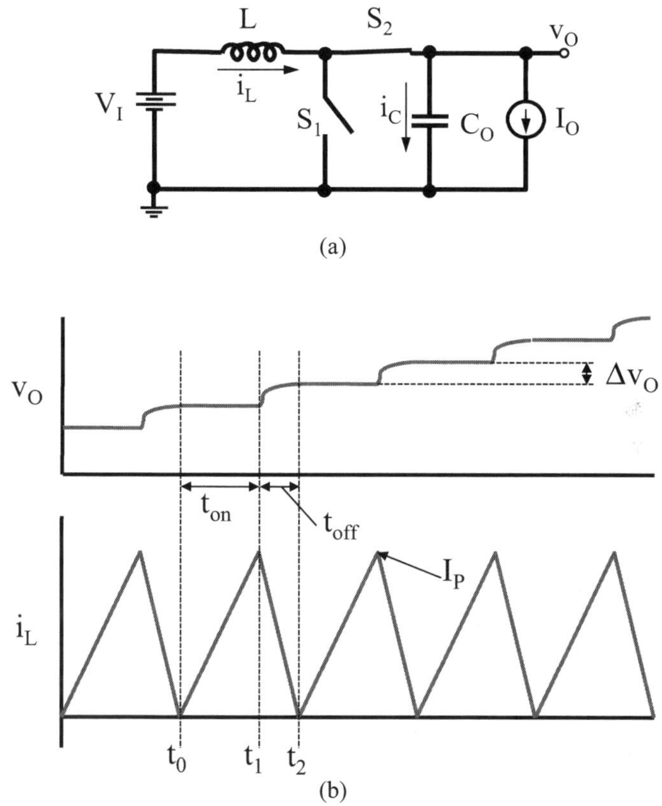

Figure 10.10 Constant on-time voltage ripple analysis. (a) Off-time schematic. (b) Switching inductor current and output voltage.

10.2.7.3 Maximum load current

Using the superposition principle, the output voltage ripple, including effects from the load current I_O, can be written as

$$\Delta v_O = t_{on}^2 \frac{V_I^2}{2LC_O(V_O - V_I)} - \frac{I_O(t_{on} + t_{off})}{C_O} \quad (10.14)$$

And since the output voltage ripple will be zero at the maximum output current (i.e., the output voltage cannot increase), the maximum load is as shown in Eq. 10.15. Note that the result from Eq. 10.1 was substituted to make this result more compact.

$$I_{O(max)} = t_{on} \frac{V_I^2}{2LV_O} \quad (10.15)$$

While increasing the on time or reducing the value of the inductor increases the available load, the peak inductor current also increases. While this is not a problem in the ideal mathematical case, larger peak currents are a problem in our practical circuit, where the switch resistance $r_{DS(on)}$ is fairly high. Because of this limitation, simulations have shown that the peak inductor current $I_P = 45$ mA.

10.2.7.4 Voltage hysteresis value

By analyzing the circuit of Fig. 10.8, and ignoring nonidealities in the comparator, it can easily be shown that the value of the voltage hysteresis, ΔV_O is

$$\Delta V_O = V_{REF} \frac{R_1}{R_2 + R_{hys}} \frac{R_{hys}}{R_2} \tag{10.16}$$

10.2.8 Ideal Components

In the currently described circuit, the inductor current sensor is modeled using ideal devices. To compensate for the current required to power this component, a 5 μA load is added to the output, which has a negative effect on the circuit efficiency but effectively models the impact this circuit will have on the complete system.

10.3 Simulation Results

10.3.1 Simulation Schematic

Simulations have been performed on the circuit shown in Fig. 10.11. The NMOS and PMOS have a size of 5 mm/0.18 μm and 10 mm/0.18 μm, respectively, V_{REF} is 250 mV, and t_{on} is 1.5 μs. Since circuit parasitics have a large effect on the operation of a boost regulator, parasitic resistances have been included, wherever possible. Note that the dashed box in the middle of Fig. 10.11 shows the internal (transistor-level) circuitry, while everything outside this box is an external component.

The load of the simulation circuit is modeled by a constant current source, which is commonly used to simulate power supply circuits. Note that since this circuit operates with a hysteretic control scheme, frequency compensation is not needed, and hence there are no stability advantages to using a current source rather than a resistive load [16].

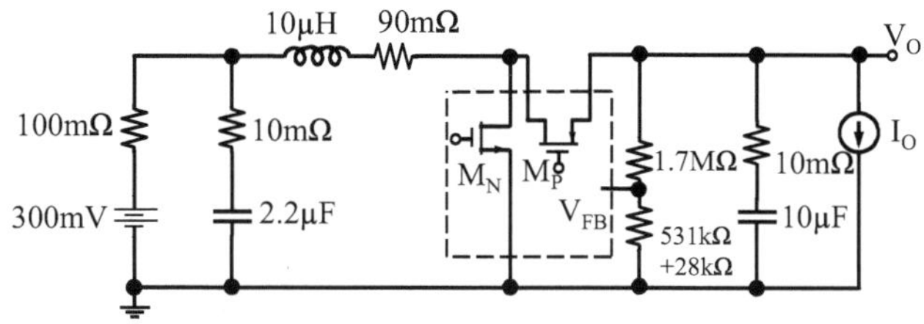

Figure 10.11 Simulation schematic including parasitic resistances.

10.3.2 Start-Up Results

The initial start-up of the regulator, using the charge pump circuitry and oscillator described in Section 10.2.4, is shown in Fig. 10.12. Once the charge pump oscillator starts switching, the output of the charge pump gradually rises (Fig. 10.12a). Once the charge pump output is large enough to run the start-up oscillator and drivers (shown previously in Fig. 10.6), the main switches start commutating, and the output voltage V_O slowly rises, as shown in Fig. 10.12b. Since the charge pump is not able to deliver a large amount of current, and since it is loaded by the drivers and output switches, the output of the charge pump reaches a maximum value of only 0.8 V, even with the 10-stage configuration described in Fig. 10.4. Once V_O has risen to the reference point, the constant-on-time controller takes over and the circuit switches to the steady-state, constant-on-time mode.

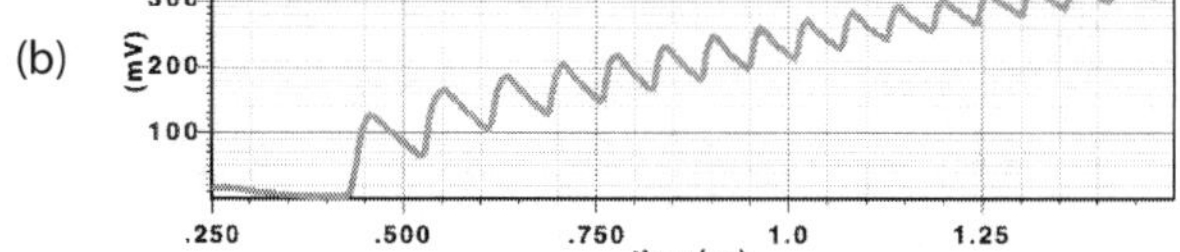

Figure 10.12 V_{CP} start-up.

10.3.3 Steady-State Operation

Figure 10.13 shows the steady-state operation of the proposed circuit. Starting from the far left of the figure, both switches are off and the output voltage is slowly coasting downward, with a slope determined by the output load current (i.e., this is the "Coasting" time shown in Fig. 10.9). During the coasting time, however, the hysteretic comparator is still watching V_O, and as soon as it reaches the lower threshold of the comparator, the constant-on-time control is enabled. Until the comparator reaches its upper threshold, the circuit turns on the NMOS (and hence charges the inductor) for a fixed amount of time and then turns off the NMOS (and hence turns on the PMOS) until the inductor current reaches zero. When the upper threshold of the comparator is finally reached, both switches are turned off and the circuit reenters the coasting mode, as was previously shown in the state diagram of Fig. 10.9.

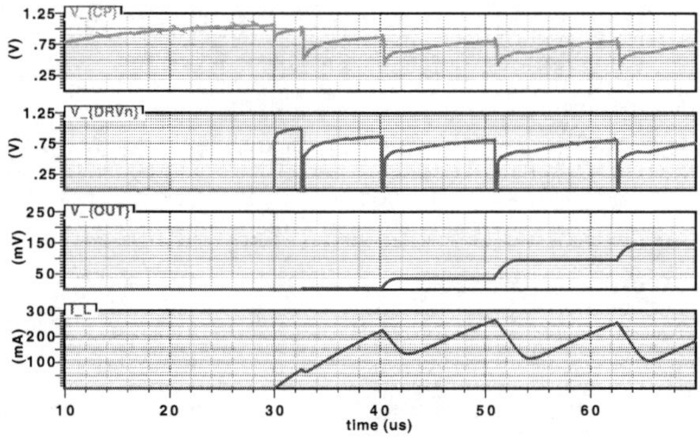

Figure 10.13 V_{CP} driving switch transistors.

10.3.4 Calculation and Simulation Comparison

Using the analysis in Section 10.2.7, we notice that the maximum load should be $I_{O(max)}$ = 6.7 mA with a peak-to-peak voltage ripple of ΔV_O = 40 mV (using the results from Eqs. 10.15 and 10.16, respectively). As can be seen in Fig. 10.14, the maximum load is just past P_O = 6.2 mW, and hence this agrees well with the predicted value. From Fig. 10.13, we can see that the voltage ripple is just about 40 mV.

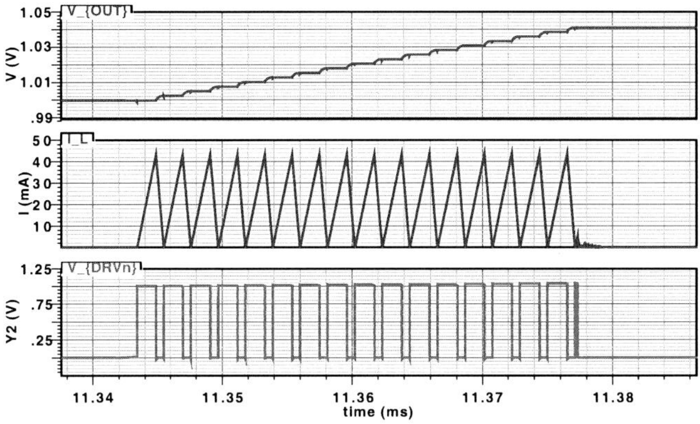

Figure 10.14 Steady-state operation simulation.

10.3.5 Efficiency

The total end-to-end efficiency of the proposed circuit is shown in Fig. 10.15. While the efficiency drops off slightly at a lower load, it remains above 95% for most of the load range, reaching a maximum of 97% at a load of 6.2 mW. The high efficiency in the proposed circuit is a virtue of the fact that switching losses are very low and the control circuitry consumes very little quiescent current. Note that this data comes from simulation results, and the efficiency values for an actual circuit will likely be lower. This is because there are various parasitic components that are difficult to model in simulation.

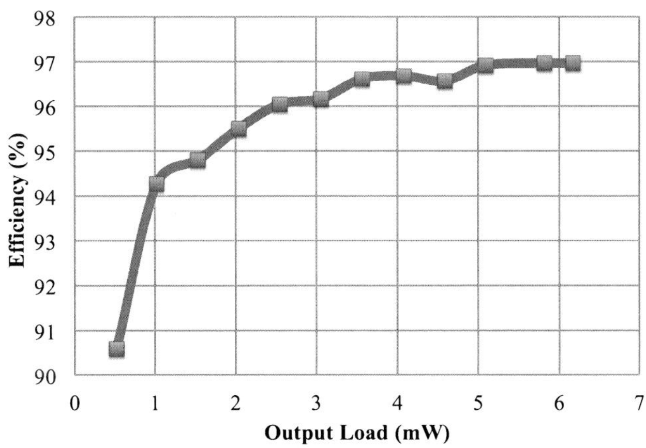

Figure 10.15 Efficiency over a load range.

A comparison between our peak efficiency and maximum load with recently published papers in the field of energy harvesting is shown in Fig. 10.16.

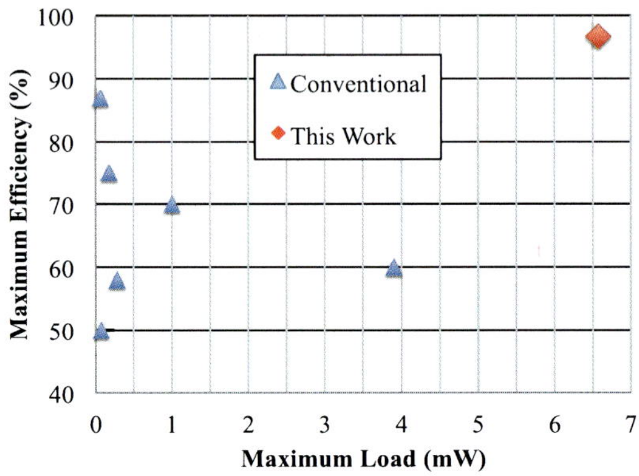

Figure 10.16 Comparison of maximum efficiency and output power with recent papers.

10.4 Test Chip

10.4.1 Chip Photomicrograph

A test chip comprising all the components of the charge pump start-up circuit (i.e., those shown in Fig. 10.4) were taped out in a 0.18 μ process from TSMC. A photomicrograph of the test chip is shown in Fig. 10.17. The 10-charge pump gain stages, star connected on top of the oscillator and buffer, can be clearly seen in the test chip image.

Figure 10.17 Charge pump test chip.

10.4.2 Chip Packaging

Due to packaging limitations, the chip was packaged in a QFP64 package, even though the test chip only uses five pins. This package was then soldered down to a simple breakout board for lab verification. A photograph of the packaged chip on the breakout board is shown in Fig. 10.18, and the pinout of the test chip is shown in Table 10.2.

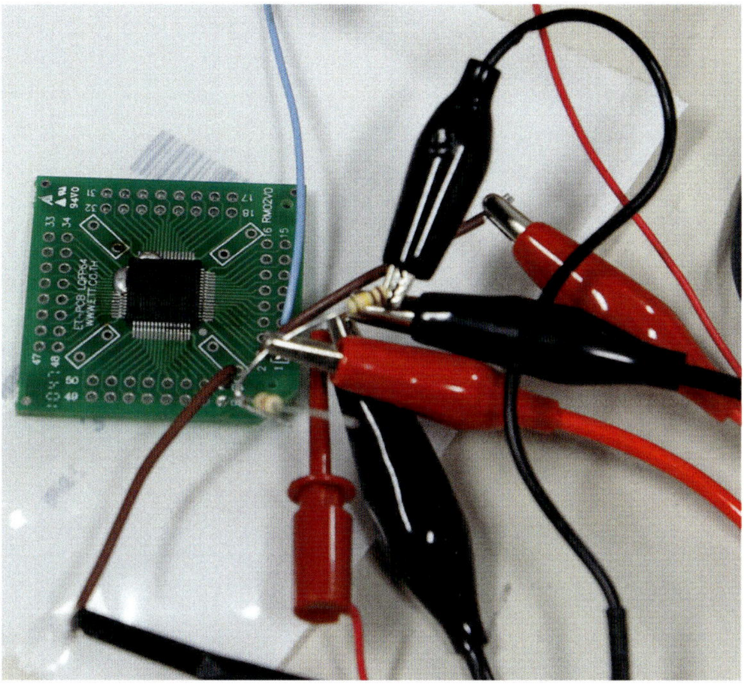

Figure 10.18 Packaged charge pump chip.

Table 10.2 Charge pump test chip pinout

Pin number	Function
1	VN
2	VP
3	_en
63	VOUT
64	Driver output
All others	NC

10.4.3 Bench Results

Using the lab setup shown in Fig. 10.19, bench measurements were performed with the packaged part on the breakout board shown in Fig. 10.18.

With an input voltage of 400 mV, the charge pump oscillator of the part switches, as shown in Fig. 10.20. While this plot proves that the oscillator switches and the charge pump works, adding an oscilloscope probe to the buffer output adds a very large loading capacitance to this node and ruins the gain capability of this circuit. This data point is very important to prove that the part is switching properly, but later data was taken without the oscilloscope connected to minimize loading effects.

As expected, with an input voltage of 400 mV, the charge pump oscillator switches between 0 V and 400 mV. Including the effect of loading the buffer with a large capacitance oscilloscope probe (as described above), the circuit operates at a frequency of about 30 MHz in this condition. The shape of the switching frequency and the operating frequency is similar to what was predicted in simulations.

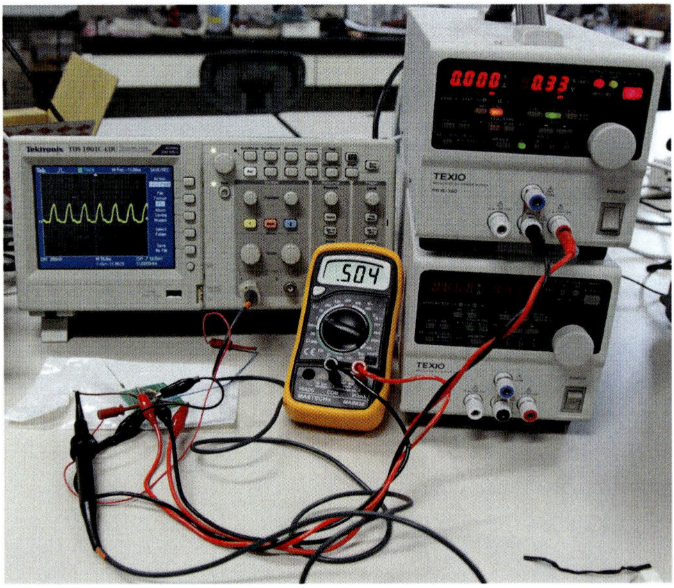

Figure 10.19 Charge pump lab setup.

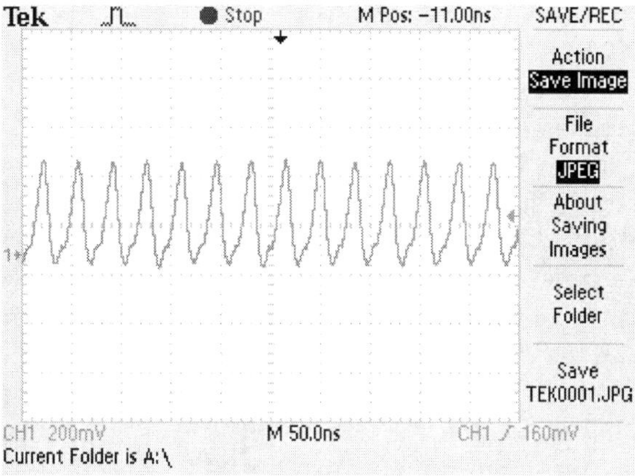

Figure 10.20 Charge pump oscillator buffer output at 400 mV.

10.4.3.1 Charge pump transfer function

To test the real-world performance of the charge pump, the chip was tested both with no load and with a 100 kΩ resistive load. Results from these experiments are shown in Fig. 10.21.

As expected, with no load there is a decent amount of gain, and the output voltage rises to almost 2 V with an input voltage of 340 mV, corresponding to a gain of about 6 V/V. With a resistive load, however, the output voltage rises slower for a given input voltage, but it still seems that the output voltage would rise high enough to start switching the part.

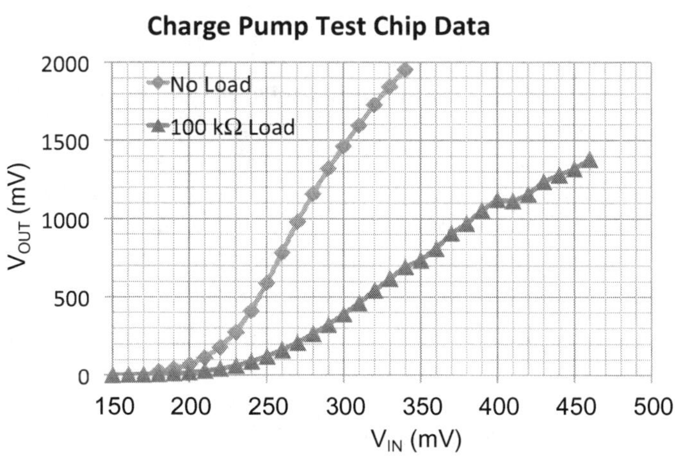

Figure 10.21 Test chip transfer function with and without a load.

10.4.4 Bench and Simulation Comparison

Keeping in mind the loading effect from the capacitor probe, the shape of the ring oscillator signal in the test chip is the same as the simulated result, shown previously in Fig. 10.12. The frequency of the oscillator is different from the simulated value, but this can be explained by two different phenomena. First, capacitive loading of the driver output will cause the oscillator to run slower. Second, the ring oscillator architecture of the circuit is very sensitive to the input voltage and the threshold voltage of the transistors—any small variation from the simulated values will have a large effect on the actual oscillation frequency.

In the actual application, this charge pump output would be applied to an oscillator and the main switch drivers, as shown previously in Fig. 10.6. Unfortunately it is difficult to model this condition without the actual circuit that the charge pump would be powering. But data from this test chip shows that the charge pump oscillator functions and that the diode-connected native NMOS charge pump stages can work together to create a high enough output voltage to start up the circuit.

10.5 Conclusion

We have introduced a low-power boost regulator that is optimized for energy-harvesting applications. While previously presented papers have shown that it is possible to convert energy from extremely low-power input sources, these papers have not designed their circuits to work with a realistic load nor to maximize the system efficiency. Compared to previous works, our circuit can handle enough output load to power a typical microcontroller system, while maximizing the end-to-end efficiency over the entire load range. Finally, the circuit we have introduced accomplishes all of this, while only requiring a single input voltage source and three external components: an input capacitor, an output capacitor, and an inductor.

References

1. D. Kwon and G. Rincon-Mora, A single-inductor AC-DC piezoelectric energy- harvester/battery-charger ic converting ±(0.35 to 1.2v) to (2.7

to 4.5v), *IEEE ISSCC Dig. Tech. Papers*, 494–495, 2010. DOI:10.1109/ISSCC.2010.5433867

2. Y. Ramadass and A. Chandrakasan, An efficient piezoelectric energy-harvesting interface circuit using a bias-flip rectifier and shared inductor, *IEEE ISSCC Dig. Tech. Papers*, 296–297, 297a, 2009. DOI:10.1109/ISSCC.2009.4977425
3. E. Carlson, K. Strunz, and B. Otis, A 20 mv input boost converter with efficient digital control for thermoelectric energy harvesting, *IEEE J. Solid-State Circuits*, **45**(4), 741–750, 2010. DOI:10.1109/JSSC.2010.2042251
4. Y. Ramadass and A. Chandrakasan, A batteryless thermoelectric energy-harvesting interface circuit with 35mv startup voltage, *IEEE ISSCC Dig. Tech. Papers*, 486–487, 2010. DOI:10.1109/ISSCC.2010.5433835
5. I. Doms, P. Merken, R. Mertens, and C. Van Hoof, Integrated capacitive power- management circuit for thermal harvesters with output power 10 to 1000µw, *IEEE ISSCC Dig. Tech. Papers*, 300–301, 301a, 2009. DOI:10.1109/ISSCC.2009.4977427
6. R. Dayal and L. Parsa, Low power implementation of maximum energy harvesting scheme for vibration-based electromagnetic microgenerators, *Proc. 26th IEEE APEC*, 1949–1953, 2011. DOI:10.1109/APEC.2011.5744863
7. P. H. Chen, K. Ishida, X. Zhang, Y. Okuma, Y. Ryu, M. Takamiya, and T. Sakurai, 0.18-v input charge pump with forward body biasing in startup circuit using 65nm cmos, *Proc. IEEE Custom Integr. Circuits Conf.*, 1–4, 2010. DOI:10.1109/CICC.2010. 5617444
8. Y. Rao and D. Arnold, Input-powered energy harvesting interface circuits with zero standby power, *Proc. 26th IEEE APEC*, 1992–1999, 2011. DOI:10.1109/APEC. 2011.5744870
9. Texas Instruments, Low input voltage synchronous boost converter with 1.3-a switches (rev. b), 2008, TPS61200 Datasheet.
10. Linear Technology, Ultralow voltage step-up converter and power manager, LTC3108 Datasheet.
11. R. W. Erickson and D. Maksimovic, *Fundamentals of Power Electronics*, 2nd ed., 24–45, Kluwer Academic, Norwell, MA, 2000.
12. Y. H. Kao, C. C. Liu, and H. C. Kuo, Study of front end of CMOS RFID tag with inductively- coupled broadband antenna, *Asia-Pacific Microwave Conf.*, 1–4, 2007. DOI:10.1109/APMC.2007.4555011

13. H. Banba, H. Shiga, A. Umezawa, T. Miyaba, T. Tanzawa, S. Atsumi, and K. Sakui, A CMOS bandgap reference circuit with sub-1-v operation, *IEEE J. Solid-State Circuits*, **34**(5), 670–674, 1999. DOI:10.1109/4.760378

14. B. Gilbert, *Monolithic Voltage and Current References: Theme and Variations*, 269– 352. Kluwer Academic, Norwell, MA, 1996.

15. Texas Instruments, Adaptive constant on-time (D-CAPTM) control study in notebook applications, 2007, www.ti.com/lit/an/slva281b/slva281b.pdf (accessed July 23, 2012).

16. S. Davis, Constant on-time buck regulator ICs, electronic design, 2006, http://www.electronicdesign.com/files/29/12253/12253.pdf (accessed July 26, 2012).

Chapter 11

Wireless Power Delivery

Kiichi Niitsu
Department of Electrical Engineering and Computer Science,
Graduate School of Engineering, Nagoya University,
Nagoya 464-8603 Japan
niitsu@nuee.nagoya-u.ac.jp

11.1 Introduction

Wireless power delivery technology has become a great concern in recent years. Since wireless power delivery does not require electrical contact, high reliability and low cost can be realized. One of the most practical applications is noncontact power delivery for 3D system integration and wireless wafer-level testing.

The design challenges for wireless power delivery are as follows:

- High-efficiency power conversion
- Long-distance power delivery
- High-reliability power delivery

In this chapter, the literature on the development of wireless power delivery is described in Section 11.2. The practical designs

Handbook of Power Management Circuits
Edited by Haruo Kobayashi and Takashi Nabeshima

ISBN 978-981-4613-15-6 (Hardcover), 978-981-4613-16-3 (eBook)
www.panstanford.com

of wireless power delivery are demonstrated in the subsequent two sections—wireless power delivery for 3D system integration in Section 11.3 and wireless power delivery design for noncontact wafer-level testing in Section 11.4. Section 11.5 demonstrates the efficiency improvement technique using thin-film magnetic material. A part of this chapter (from Section 11.2 to Section 11.7) is based on the previous report [17].

11.2 Literature on Wireless Power Delivery

In this section, the literature on the development of wireless power delivery is summarized. The literature was started by Nikola Tesla, who demonstrated the world's first wireless power delivery system [1]. After that, progress, such as radio-frequency identification (RFID), was reported several times. The turning point happened in 2007. A Massachusetts Institute of Technology (MIT) group reported wireless power delivery that lightened a 60 W light bulb at a 2 m distance using 60 cm coils with 40% efficiency [2]. This high-impact report motivated many researchers to develop wireless power delivery.

Wireless power delivery from IC to IC was demonstrated in 2007 for the first time [3]. The target application is 3D system integration. After that, some techniques for performance improvement were reported [4, 5]. Not only complementary metal–oxide–semiconductor (CMOS) implementation but organic MOS implementation was also demonstrated [6]. Application for a permanent memory system was developed in 2010 [7]. In 2011, 6 W power transfer for noncontact wafer-level testing was demonstrated in the International Solid-State Circuits Conference (ISSCC) 2011 [8].

11.3 Wireless Power Delivery for 3D System Integration

It becomes more and more difficult to exploit the benefits of device scaling. This difficulty motivates us to develop 3D system integration. Large-scale integration (LSI) chips are three-dimensionally stacked and connected with vertical interchip links in 3D system integration. It is important to develop high-performance vertical interchip links since the total system performance of 3D system integration deeply depends on the performance of interchip links. So far, several kinds of vertical interchip links have been developed and reported. As typical developments, microbump technology [9], through-silicon via (TSV) technology [10], a capacitive-coupling interchip link [11], and an inductive-coupling interchip link [12] have been proposed and investigated. The microbumps and TSV technology are in practical use; however, they require an additional mechanical process, which leads to high cost. On the other hand, capacitive-coupling links can be developed by a normal CMOS process and do not require an additional mechanical process. However, capacitive-coupling links have limitations in implementation. Capacitive-coupling links can be applied only in face-to-face implementation. Contrarily, inductive coupling has no limitations in implementation style. Inductive-coupling links enable communication between more than three stacked chips.

Figure 11.1 shows a conceptual image of wirelessly connected 3D system integration. The data, clock, and power are delivered using inductive-coupling links. Thus, noncontact 3D system integration can be realized.

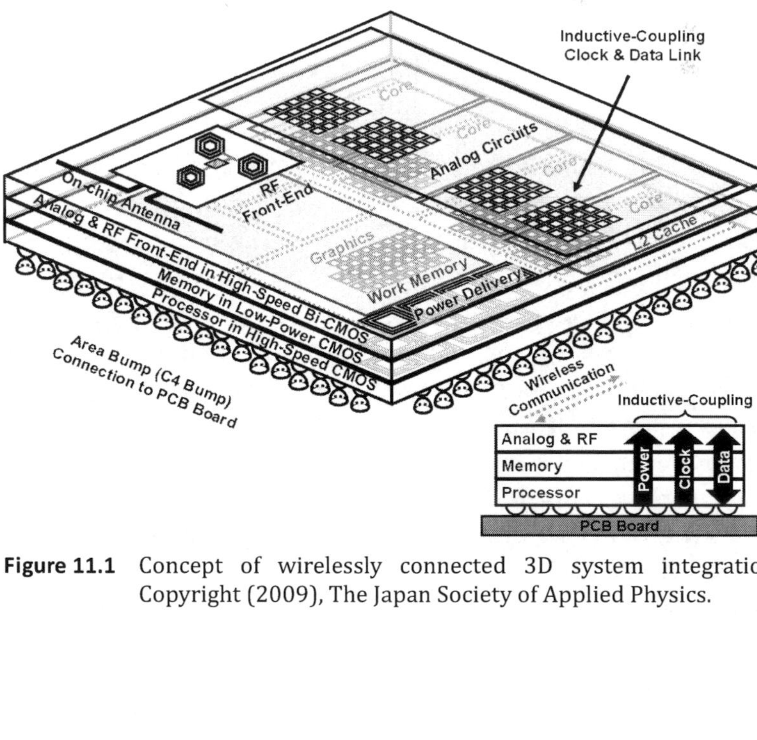

Figure 11.1 Concept of wirelessly connected 3D system integration. Copyright (2009), The Japan Society of Applied Physics.

11.4 Wireless Power Delivery for Noncontact Wafer-Level Testing

The wafer-level test is another attractive application of wireless power delivery. By removing the test pin for power delivery, the test cost can be dramatically reduced. As an approach for wireless wafer-level testing, three candidates are reported. The first is near-field electromagnetic communication that is developed by Scanimetrics [13]. The second is that using capacitive-coupling links [14]. The third utilizes inductive-coupling links. In a typical situation, wafer-level testing requires three kinds of signals (AC, DC, and power). An AC signal can be transferred by inductive-coupling data links. On the other hand, a DC signal can be delivered with pulse-based communication [15]. However, power cannot be transferred by conventional inductive-coupling pulse-based communication. To solve this issue, 6 W power transfer for noncontact wafer-level testing was reported [8]. In this report, several power transfer links collaborate to minimize losses and maximize efficiency.

11.5 Efficiency Improvement Using Thin-Film Magnetic Material

The transfer efficiency of wireless power delivery can be improved by introducing thin-film magnetic material [16]. The reason is that high permeability enhances magnetic flux. Figure 11.2 illustrates the mechanism of improving efficiency in power delivery by thin-film magnetic material. The existence of magnetic material with a horizontal hard magnetization axis enhances the magnetic flux density. With this change, mutual inductance of on-chip inductors increases and the efficiency can be improved. Figure 11.3 shows the magnetic flux that is calculated by a 3D electromagnetic solver. The hard magnetization axis was set to be horizontal. Simulation results without and with the thin-film magnetic material are depicted in the left and the right part, respectively. Strong magnetic flux is visualized by a bright color. As shown in the figure, the magnetic flux generated by on-chip inductors is enhanced by thin-film magnetic material.

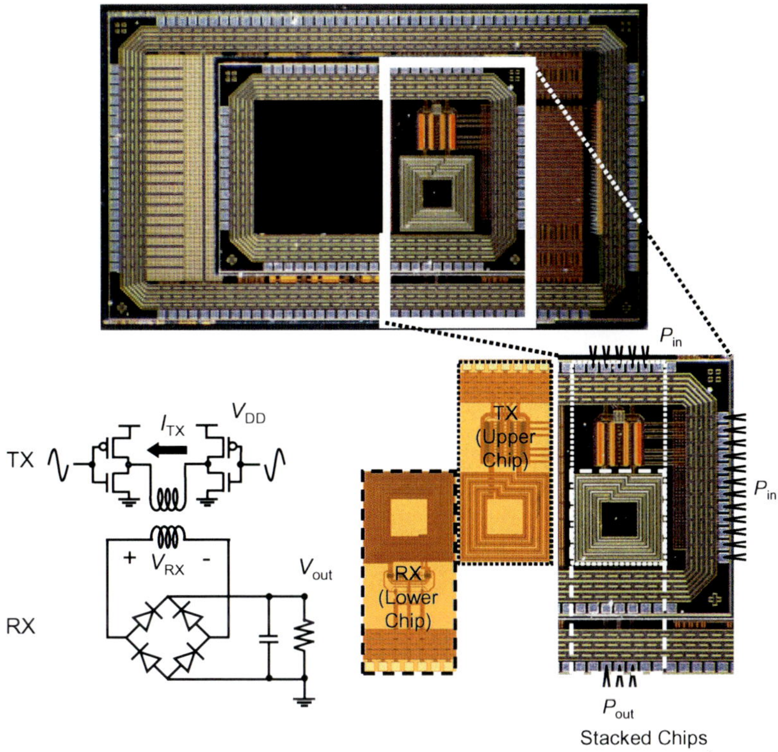

Figure 11.2 Microphotograph of stacked chips and schematic of wireless interchip power delivery. Copyright (2009), The Japan Society of Applied Physics.

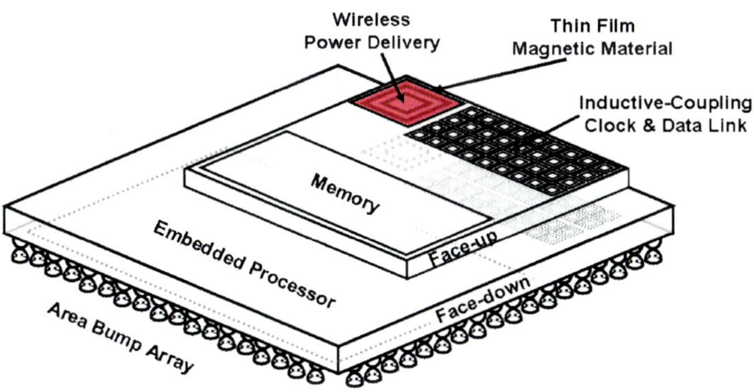

Figure 11.3 Concept of 3D system integration with wireless power delivery using thin-film magnetic material. Copyright (2009), The Japan Society of Applied Physics.

The stacked chip microphotograph and schematic of the circuitry in wireless interchip power delivery for this study are shown in Fig. 11.4. The design of the on-chip inductors and circuits is precisely described in the previous work [17]. The test chips were designed and fabricated using normal 0.18 μm CMOS technology. The transmitter chip is stacked on the receiver chip, and both chips are face-up (face-to-down stacking implementation). The transmitter chip is polished to have a thickness of 10 μm. The communication distance is 15 μm, including 5 μm of glue layer. The diameters of transmitter and receiver inductors are 700 μm.

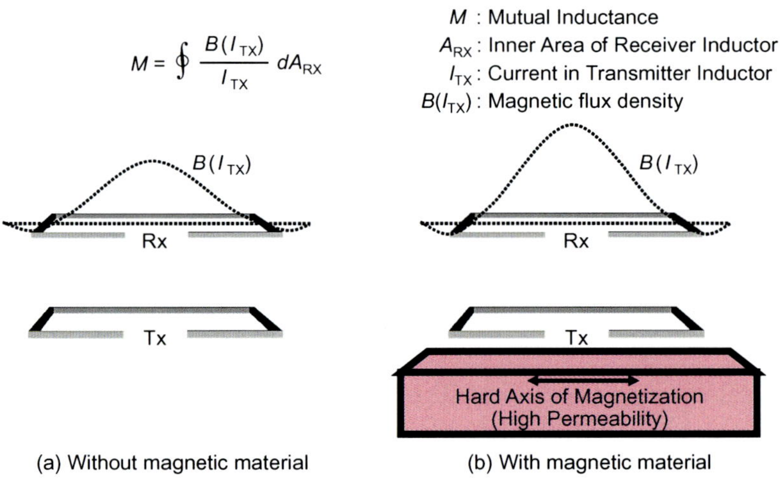

Figure 11.4 Improvement of efficiency using thin-film magnetic material (a) without magnetic material and (b) with magnetic material. Copyright (2009), The Japan Society of Applied Physics.

Figure 11.5 depicts the measurement setup and properties of the thin-film magnetic material utilized in this study. The thin-film magnetic material was diced from the wafer and attached to the stacked chips. Properties of the thin-film magnetic material are as follows: The composition is cobaltiferous, and the film-forming method is spattering. Permeability in the hard axis is approximately 470, where the operational frequency is 1 MHz. The ferromagnetic resonance frequency of this thin-film magnetic material is more than 500 MHz, and it is enough to apply the power delivery since the required operational frequency is less than 200 MHz (in this study, 140 MHz). The saturated magnetic flux density is 1.44 T, which is much larger than that in inductive-coupling power delivery. The Curie temperature is more than 800°C, which is high enough to utilize even in the automobile application.

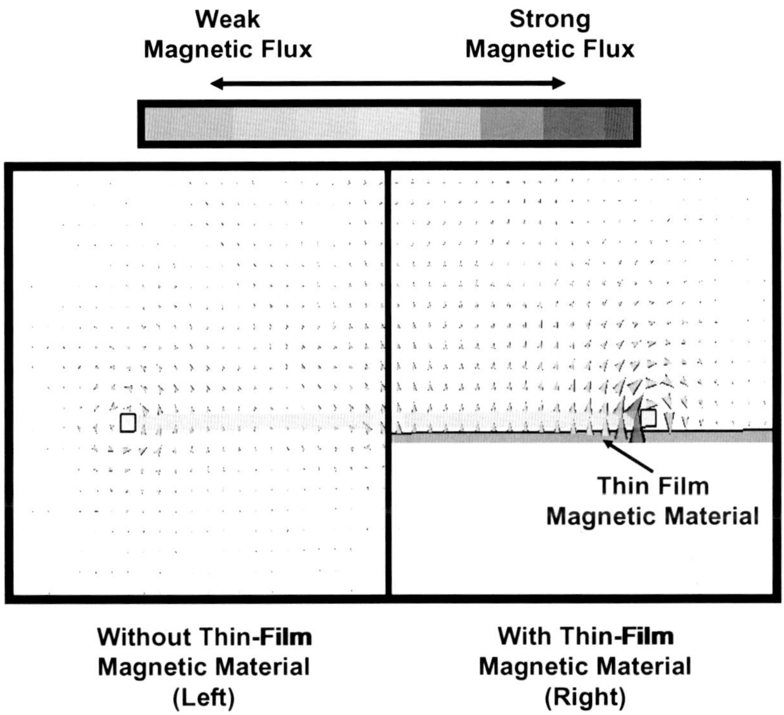

Figure 11.5 Simulated results of magnetic flux (a) with and (b) without thin-film magnetic material. Copyright (2009), The Japan Society of Applied Physics.

Figure 11.6 demonstrates the measurement results. The curve shown in the figure depicts the power efficiency as a function of the operational frequency of the power delivery circuit. The transferred power is maximized where operational frequency is approximately 140 MHz. At this point, the transferred power was improved from 36 mW to 48 mW. The power delivery efficiency was improved by 33%. The achieved efficiency of 13.3% is the highest among the previously reported results when published (in 2009).

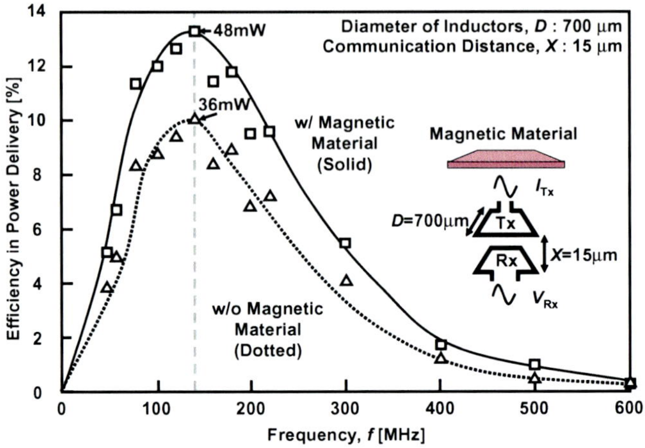

Figure 11.6 Measured efficiency dependence on operational frequency. Copyright (2009), The Japan Society of Applied Physics.

Figure 11.7 presents the improvement in efficiency as a function of operational frequency. The measured result matches well with the simulated result calculated by the 3D electromagnetic solver. The measured improvement in power delivery efficiency is 33% where operational frequency is 140 MHz. The result curve decreases rapidly as it approaches the ferromagnetic resonance frequency of the thin-film magnetic material (500 MHz) utilized in this work.

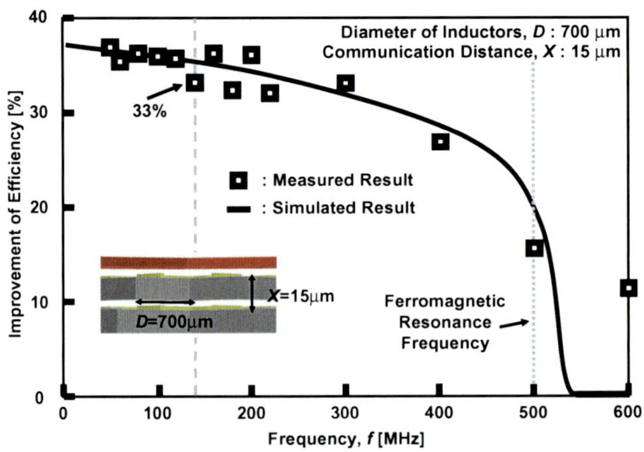

Figure 11.7 Measured improvement of efficiency dependence on operational frequency. Copyright (2009), The Japan Society of Applied Physics.

References

1. S. Cohen, Dr. Nikola Tesla and his achievements, in *The Electrical Experimenter*, 712–713, 777, 1917.
2. A. Kurs, A. Karalis, R. Moffatt, J. D. Joannopoulos, P. Fisher, and M. Soljacic, Wireless power transfer via strongly wireless power transfer via strongly coupled magnetic resonances, *Science*, **317**(5834), 83–86, 2007.
3. K. Onizuka, H. Kawaguchi, M. Takamiya, T. Kuroda, and T. Sakurai, Chip-to-chip inductive wireless power transmission system for SiP applications, *Proc. IEEE Custom Integr. Circuits Conf.*, 575–578, 2006.
4. Y. Yuan, Y. Yoshida, and T. Kuroda, Non-contact 10% efficient 36mW power delivery using on-chip inductor in 0.18-um CMOS, *Proc. IEEE Asian Solid-State Circuits Conf.*, 115–118, 2007.
5. Y. Yuan, Y. Yoshida, N. Yamagishi, and T. Kuroda, Chip-to-chip power delivery by inductive coupling with ripple canceling scheme," *Proc. Int. Conf. on Solid-State Devices Mater.*, 502–503, 2007.
6. M. Takamiya, T. Sekitani, Y. Miyamoto, Y. Noguchi, H. Kawaguchi, T. Someya, and T. Sakurai, Designsolutions for a multi-object wireless power transmission sheet based on plastic switches, *Proc. IEEE Int. Solid-State Circuits Conf.*, 362–363, 2007.
7. Y. Yuan, N. Miura, S. Imai, H. Ochi, and T. Kuroda, Digital Rosetta Stone: a sealed permanent memory with inductive-coupling power and data link, *Proc. IEEE Symp. VLSI Circuits*, 26–27, 2009.
8. A. Radecki, H. Chung, Y. Yoshida, N. Miura, T. Shidei, H. Ishikuro, and T. Kuroda, 6W/25mm2 inductive power transfer for non-contact wafer-level testing, *Proc. IEEE Int. Solid-State Circuits Conf.*, 230–232, 2011.
9. T. Ezaki, K. Kondo, H. Ozaki, N. Sasaki, H. Yonemura, M. Kitano, S. Tanaka, and T. Hirayama, A 160Gb/s interface design configuration for multichip LSI, *Proc. IEEE Int. Solid-State Circuits Conf.*, 140–141, 2004.
10. H. Yoshikawa, A. Kawasaki, T. Iiduka, Y. Nishimura, K. Tanida, K. Akiyama, M. Sekiguchi, M. Matsuo, S. Fukuchi, and K. Takahashi, Chip scale camera module (CSCM) using through-silicon-via (TSV), *Proc. IEEE Int. Solid-State Circuits Conf.*, 476–477, 2009.
11. A. Fazzi, R. Canegallo, L. Ciccarelli, L. Magagni, F. Natali, E. Jung, P. L. Rolandi, and R. Guerrieri, 3Dcapacitive interconnections with mono- and bi-directional capabilities, *Proc. IEEE Int. Solid-State Circuits Conf.*, 356–357, 2007.

12. K. Niitsu, Y. Shimazaki, Y. Sugimori, Y. Kohama, K. Kasuga, I. Nonomura, M. Saen, S. Komatsu, K. Osada, N. Irie, T. Hattori, A. Hasegawa, and T. Kuroda, An inductive-coupling link for 3D integration of a 90nm CMOS processor and a 65nm CMOS SRAM, *Proc. IEEE Int. Solid-State Circuits Conf.*, 480–481, 2009.

13. C. V. Sellathamby, M. M. Reja, F. Lin, B. Bai, E. Reid, S. H. Slupsky, I. M. Filanovsky, and K. Iniewski, Non-contact wafer probe using wireless probe cards, *Proc. IEEE Int. Test Conf.*, 452–457, 2005.

14. M. Daito, Y. Nakata, S. Sasaki, H. Gomyo, H. Kusamitsu, Y. Komoto, K. Iizuka, K. Ikeuchi, G.S. Kim, M. Takamiya, and T. Sakurai, Capacitively coupled non-contact probing circuits for membrane-based wafer-level simultaneous testing, *Proc. IEEE Int. Solid-State Circuits Conf.*, 144–145, 2010.

15. Y. Yoshida, K. Nose, Y. Nakagawa, K. Noguchi, Y. Morita, M. Tago, T. Kuroda, and M. Mizuno, Wireless DC voltage transmission using inductive-coupling channel for highly-parallel wafer-level testing, *Proc. IEEE Int. Solid-State Circuits Conf.*, 470–472, 2009.

16. K. Niitsu, Y. Yuan, H. Ishikuro, and T. Kuroda, A 33% improvement in efficiency of wireless inter-chip power delivery by thin film magnetic material for three-dimensional system integration, *Jpn. J. Appl. Phys.*, **49**(4B), 04C073-1–04C073-5, 2009.

17. Y. Yuan, Y. Yoshida, and T. Kuroda, Analysis of inductive coupling and design of rectifier circuit for inter-chip wireless power link, *IEICE Trans. Electron.*, **E93-C**(2), 164–171, 2009.

Chapter 12

High-Power GaN HEMT for Cellular Base Stations

Norihiko Ui
Sumitomo Electric Device Innovations, Inc., Showa-cho, Yamanashi, Japan
ui-norihiko@gr.sei.co.jp

12.1 Introduction

Due to recent requirements of broadband and high-speed wireless systems for data and video communications, broadband modulated signals, such as multicarrier 2nd generation (2G) cellular phone system GSM, W-CDMA (3G) and LTE (4G) signals, have been employed. Modulation speed is expanding from a few megahertz GSM to 100 MHz LTE-Advance. Frequency bands for these systems are also expanding. In addition to conventional 700 MHz, 800 MHz, 900 MHz, 1.8 GHz, 1.9 GHz, and 2.1 GHz for GSM and W-CDMA, a 2.6 GHz band has been licensed for LTE, and a 3.5 GHz band will be done in the near future. The 2.3–2.7 GHz bands have been used for TDD systems, mainly WiMAX. These frequencies would be changed for high-speed data communications of TDD-LTE

Handbook of Power Management Circuits
Edited by Haruo Kobayashi and Takashi Nabeshima

ISBN 978-981-4613-15-6 (Hardcover), 978-981-4613-16-3 (eBook)
www.panstanford.com

systems. In these systems, not only broadband RF characteristics but also broad signal capabilities (i.e., broad video band width or low memory effects) are required for power amplifiers in base stations.

A Si laterally diffused metal–oxide–semiconductor (LDMOS) has been widely used for UHF, L-, and S-band base station power amplifiers. After the first commercialization of GaN high-electron-mobility transistor (HEMT) in 2005, the market of the LDMOS is gradually replaced by the GaN HEMT due to its superior high power, high efficiency, and broadband characteristics.

In this section, efficiency enhancement techniques for a base station amplifier of handy phone systems are described with high-power GaN HEMT technologies.

First, material properties of GaN and other semiconductors and basic DC and RF characteristics are introduced.

Second, high-efficiency switching-mode operations such as high-frequency class E and class F using high-voltage GaN HEMTs are introduced. And then the recent trend of the Doherty amplifier is also introduced with a simpler design method with some examples of class F GaN HEMTs.

Finally, some feasibility studies for future applications of GaN HEMTs will be shown in Section 12.6.

12.2 Basic Characteristics of GaN HEMTs

12.2.1 Material Properties

Table 12.1 shows material properties of major semiconductors—Si, GaAs, SiC, and GaN—for microwave applications [1].

GaN has a wider bandgap and higher critical breakdown voltage than Si and GaAs, and it enables higher voltage operation and higher-temperature operation. High current density is expected with high saturation velocity and high carrier density in the HEMT structure. Due to these superior material properties, a GaN HEMT shows high power and high efficiency with broadband characteristics, as described in later sections.

GaN HEMTs are typically fabricated on SiC substrates. Pure SiC has quite high thermal conductivity, exceeding Cu—398 W/mK. Such high heat dissipation characteristics contribute to a small die size with good thermal resistance for high-power devices.

Table 12.1 lists material properties of major semiconductors for microwave applications.

Table.12.1 Material properties

	Unit	Si	GaAs	GaN	SiC
Energy gap	eV	1.12	1.42	3.39	2.86
Electron drift mobility	cm^2/Vs	1400	8500	2000	400
Electron saturation velocity	cm/Vs	1.0×10^7	2.0×10^7	2.7×10^7	2.0×10^7
Breakdown field strength	V/cm	3.0×10^5	4.0×10^5	2.0×10^6	3.0×10^6
Thermal conductivity	W/cm.K	1.5	0.46	1.5	4.9

By the material point of view, a GaN HEMT on a SiC substrate is one of best material combinations for high-power devices.

12.2.2 Comparison Si, GaAs, and GaN for DC and RF Characteristics

Figure 12.1 shows typical drain current and drain voltage (I_{ds}–V_{ds}) characteristics of GaN HEMTs. High power devices have large gate width (gate periphery) to get high power. But the I_d is normalized by unit gate width(/mm) for easier comparison in Fig. 12.1 and Table 12.2. This device has more than 200 V of breakdown voltage with good pinch-off characteristics. As described in a later section, the peak drain voltage reaches three times the operation (biased) voltage in "inverse class F" and class E operations. Roughly a 65 V operation (65 V × 3 = 195 V < 200 V) can be applied for this device.

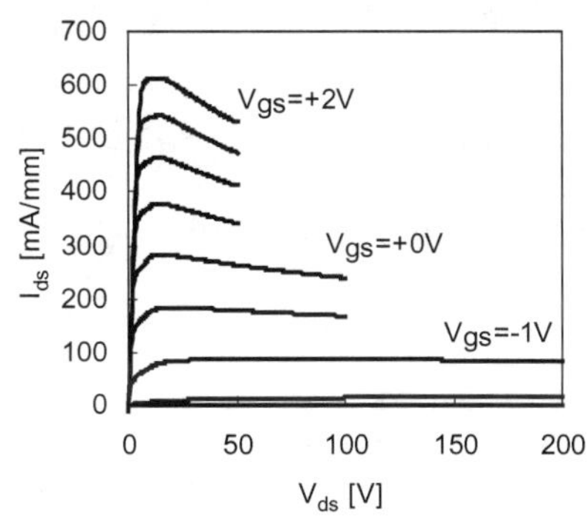

Figure 12.1 GaN HEMT I_d–V_d.

Table 12.2 shows several key parameters of typical semiconductors—Si LDMOS [2], GaAs field-effect transistor (FET), and GaN HEMT—for microwave applications.

A GaN HEMT has a large I_{max} (maximum current of the FET) compared to Si LDMOS and GaAs FET—three times and two times, respectively—in spite of a higher voltage operation of 50 V. By such a high-current and high-voltage operation, GaN has a seven times higher power density of 5W/mm. This power density is just one of production examples, but a much higher power density of 10–20 W/mm is reported in many articles. It means GaN still has much space to improve the performance with real products.

This higher power density means a small gate and drain and source electrodes; therefore GaN has a small capacitance between each electrode—C_{gs}, C_{gd}, and C_{ds}—with a similar factor of power density.

Table 12.2 Comparison of transistor parameters

Device	V_{ds}	I_{fmax}	C_{ds}		**Power**
Unit	**V**	**mA/mm**	**pF/mm**	**pF/W**	**W/mm**
Si LDMOS	28	200	0.26	0.37	0.7
10-GaAs	10	300	0.2	0.5	0.4
GaN HEMT	50	600	0.22	0.04	5

12.3 RF Operation and Load Line

12.3.1 Load Impedance and Operation Voltage

Optimum (maximum power) load impedance is simply described as

$$R_{opt} = \frac{V_{ds}}{I_{max}} \tag{12.1}$$

where V_{ds} is the operation voltage and I_{max} is the maximum current at class B bias, ignoring the on resistance (R_{on}) (Fig. 12.2).

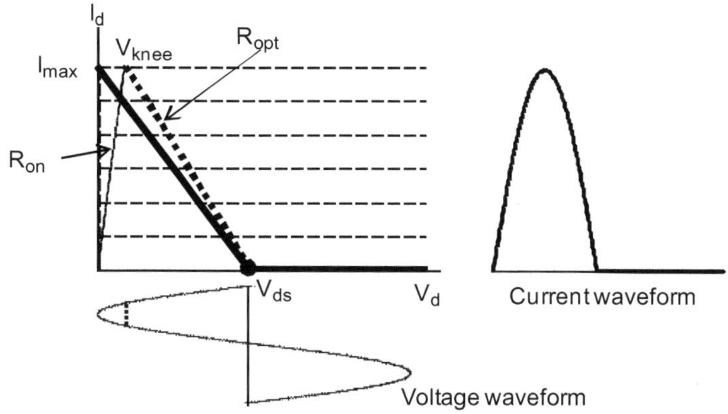

Figure 12.2 Load resistance and waveform at class B.

Typically RF devices are very deep class AB biased (a few percent of I_{max}, almost class B) to obtain higher efficiency. Voltage swing loss is described as

$$V_{loss} = V_{knee} = R_{on} \times I_{max} \tag{12.2}$$

Therefore

$$R_{opt} = \frac{V_{ds} - V_{loss}}{I_{max}} = \frac{V_{ds}}{I_{max}} - R_{on} \tag{12.3}$$

The maximum RF power is described as

$$P_{max} = \frac{1}{2} V_{rf} \cdot I_{rf} \tag{12.4}$$

where V_{rf} and I_{rf} are fundamental RF components, i.e., voltage swings; $V_{rf} \cdot \cos \omega t$ and current swing; $I_{rf} \cdot \sin \omega t$ respectively. In the class B operation, the voltage waveform shows sinusoidal form, ideally with short harmonic termination, and the current waveform shows a half sinusoidal form and consists of fundamental and even harmonic components (harmonic current can follow in the RF device with harmonic short circuit) with a peak current of I_{max} described by Fourier series analysis as

$$I(t) = \frac{I_{max}}{\pi} + \frac{I_{max}}{2} \sin \omega t + \frac{I_{max}}{\pi} \sum_{m=1}^{\infty} \frac{2}{1-4m^2} \cos 2m\omega t \tag{12.5}$$

The first component is a DC term, and the second one is a fundamental one with $I_{max}/2$ as a magnitude.

$$P_{max} = \frac{1}{2} V_{rf} \cdot I_{rf} = \frac{1}{2} V_{rf} \cdot \left(\frac{I_{max}}{2} \right) = \frac{V_{ds}^{2}}{4 \cdot R_{opt}} \tag{12.6}$$

$$\left(\text{Efficiency} = \frac{P_{RF}}{P_{DC}} = \frac{\pi}{4} = 78.5\% \right) \tag{12.7}$$

Ideally V_{rf} should be V_{ds}:

$$R_{opt} = \frac{V_{ds}^2}{4 \cdot P_{max}} \tag{12.8}$$

The optimum load is proportional to V_{ds}^2 in the case of the same maximum power.

In this relationship a two times operation voltage makes a four times load impedance with the same output power.

In a base station amplifier, more than 100 W peak output power is needed. For example, ignoring any parasitic component (i.e., at DC)

R_{opt} (28V) = 2.0 Ω

R_{opt} (50V) = 6.3 Ω

at 100 W output power.

Of course to make maximum output power at a system impedance of 50 Ω, a matching network is needed. Impedance converter ratios to 50 Ω in 28 V and 50 V operations are 25 and 8, respectively; high efficiency and broadband performance are expected by a small conversion ratio in high-voltage operation.

12.3.2 Load Impedance and C_{ds}

Figure 12.3 shows simplified small signal equivalent circuit for drain side. C_{ds} is connected to g_m in parallel. The conjugate impedance of the optimum load Z_{opt}^* is described in a similar schematic as a small signal replacing g_m with R_{opt}.

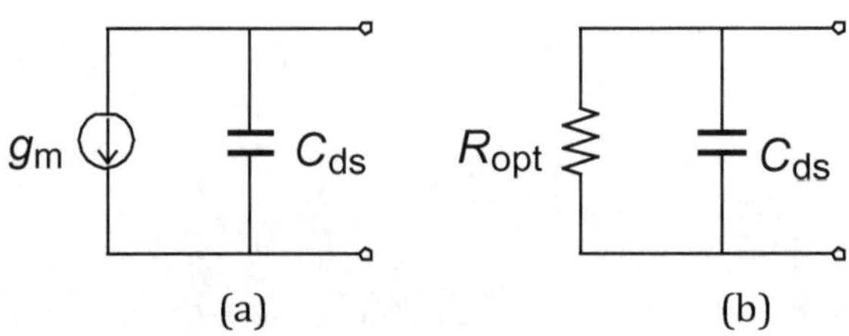

Figure 12.3 Simplified equivalent circuit at the drain side (a) and Z_{opt}^* (b).

The admittance of this circuit is

$$\frac{1}{Z_{\text{opt}}^{\ *}} = Y_{\text{opt}}^{\ *} = \frac{1}{R_{\text{opt}}} + j \cdot \omega C_{\text{ds}} \tag{12.9}$$

In this function, the frequency locus of Z_{opt}^* moves to "short" on the admittance circle through $1/R_{\text{opt}}$ (at DC), as shown in Fig. 12.4.

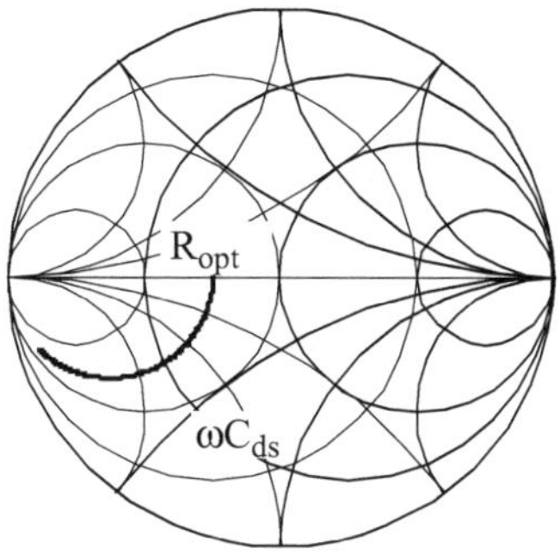

Figure 12.4 Frequency locus of Z_{opt}^*.

In the impedance expression

$$Z_{\text{opt}}^{\ *} = \frac{R_{\text{opt}} - j\omega C_{\text{ds}} R_{\text{opt}}^{\ 2}}{1 + (\omega C_{\text{ds}} R_{\text{opt}})^2} \tag{12.10}$$

$$R_{\text{opt}}{}' = R_{\text{opt}} \frac{1}{1 + (\omega C_{\text{ds}} R_{\text{opt}})^2} \tag{12.11}$$

The converted optimum load, R_{opt}', decreases with the factor $(\Omega\, C_{\text{ds}}\, R_{\text{ont}})^2$. Using Table 12.2 and each R_{opt}, Z_{opt}^* is calculated easily, as shown in Fig. 12.5.

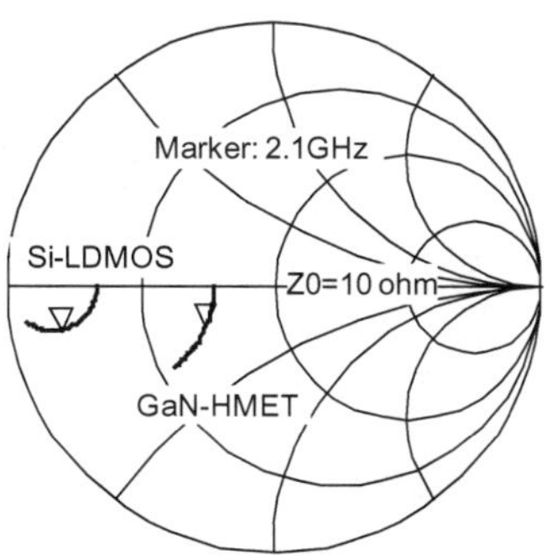

Figure 12.5 Z_{opt}^* for a GaN HEMT and Si LDMOS.

GaN has very high impedance with high voltage operation and small C_{ds}, and this small C_{ds} is very important not only for frequency band characteristics but also for efficiency characteristic.

A normally small (but not ignored at higher frequency) resistance R_s exists in a series of C_{ds}, as shown in Fig. 12.6. In this circuit, the maximum efficiency is calculated [3] as

$$\text{Efficiency} = 78.5\% \frac{1}{1+(\omega C_{ds})^2 R_s R_l} \tag{12.12}$$

Efficiency degrades with the factor $(\omega C_{ds})^2$.

This estimation is only for the fundamental component. Of course harmonics lose power by R_s through C_{ds}, and this loss is much larger than the fundamental one due to $(\omega C_{ds})^2$. A higher reflection of harmonics is most important for higher efficiency in class E and class F because voltage and current waveforms are shaped by reflection of the harmonic component. Therefore C_{ds} is very important for both fundamentals and harmonics.

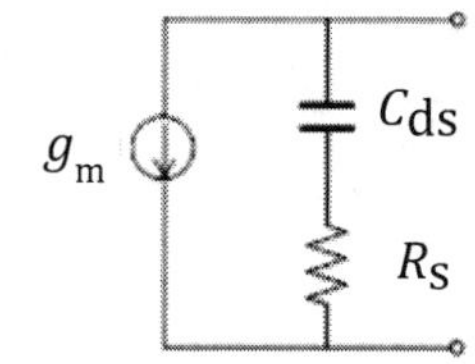

Figure 12.6 C_{ds} with R_s.

12.4 High-Efficiency Operation

At microwave frequency, high-efficiency operation of class F has been reported for many years. Another high-efficiency operation of class E (showing 100% efficiency ideally, the same as class F) has a frequency limitation by large C_{ds} in Si-LDMOS and GaAs FET. GaN HEMT proposes the class E in microwave frequency with very small C_{ds}.

12.4.1 Class E Operation

12.4.1.1 Principle of class E

A common class E amplifier consists of a (i) switch, (ii) parallel capacitance C_p, and (iii) series resonators L_s and C_s, as shown in

Fig. 12.7 [4]. The RF current flows through the switch during the switch's on state and the C_p during the off state. An RF transistor, GaN HEMT, is used as this switch, and the C_p should include GaN HEMT parasitic capacitance of C_{ds}. In the ideal class E operation, the synthesized current waveform of the drain current (I_d) and the C_p current (I_c) is perfectly sinusoidal because higher harmonics are filtered by the large L_s (which works as a choke for a higher frequency, and it is "open" for harmonics. The harmonic current in the load is not allowed; only the fundamental current can flow) of the L–C series resonator. RF power is calculated as the product of current and voltage components at each harmonic. Therefore this means no harmonic power dissipation (harmonic voltage terms must be exist, but there is no harmonic current and therefore no harmonic power), and only fundamental RF power is transmitted to the load. The I_d and drain voltage (V_d) have a zero value alternately in a cycle. Therefore no DC power dissipation is in the circuit. In other words, all the applied DC power converts to fundamental RF power; it means 100% efficiency ideally if there is no loss inside.

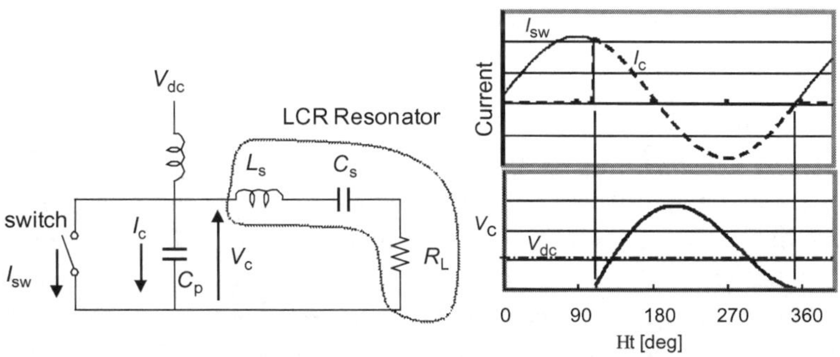

Figure 12.7 Common class E circuit and current and voltage waveforms.

A class E operation is basically a switching-mode amplifier that has simply two situations, $(V, I) = (1, 0)$ at off and $(0, 1)$ at on, as shown in Fig. 12.8. Therefore low R_{on} (to make better $V = 0$) and good pinch-off characteristics (to make better $I = 0$) are needed for the switching device. High breakdown voltage is also needed, as seen in a later section. The device indicates a small on resistance and very good pinch-off characteristics at more than 200 V.

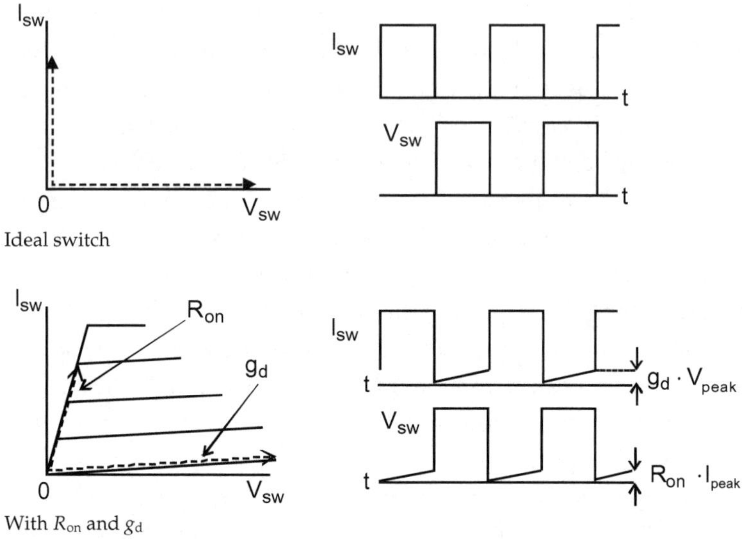

Figure 12.8 Switching operation.

12.4.1.2 Operational limitation of class E

The following function shows a relationship between V_{dc}, I_{pk}, and C_p for a class E amplifier.

$$V_{dc} = \frac{1}{\omega C_p} I_{pk} \cdot v_{dc}(\alpha) \tag{12.13}$$

V_{dc} is the operation voltage; C_p is the parallel capacitance, which should include the drain–source capacitance; C_{ds} is the package capacitance; and I_{pk} is the peak of the current waveform. The maximum I_{pk} is I_{fmax} of the actual device, and $v_{dc}(\alpha)$ is a normalized function of the conduction angle α.

The maximum operation frequency of class E can be calculated using the parameters in Table 12.2:

$$\text{freq\_max} = \frac{1}{2\pi C_{ds}} \cdot \frac{I_{max}}{V_{ds}} v_{dc}(\alpha) \tag{12.14}$$

freq_max is the frequency upper limit of a class E amplifier.
freq_max of Si LDMOS = 1.3 GHz
freq_max of GaN HEMT = 2.6 GHz

Therefore a GaN HEMT can cover mostly major frequency bands used cellar base stations with a class E operation.

12.4.1.3 Waveform simulation of class E

V_{ds}, I_{max}, and α (typically 113°) are selected in the initial design of class E. C_p and load impedance are automatically (a little complicated with numerical analysis) calculated using functions in Ref. [4].

To confirm the current and voltage waveforms, large signal simulation is performed for the I_d, I_c, and V_d waveforms, as shown in Fig. 12.9 [5], with a C_p of 0.6 pF, an L_s of 7.5 nH, a C_s of 1.8 pF, and a GaN HEMT gate width of 2.3 mm (10 W) at V_{ds} = 50 V and V_{gs} = −1.4 V at 2.1 GHz.

The C_p is the drain–source capacitance (C_{ds}) of the GaN HEMT itself. I_d and I_c are flowing alternately, and their synthesized current is nearly sinusoidal. I_d and V_d have around zero value alternately in a cycle. These simulated waveforms show remarkable similarity to the ideal class E waveform. Notice that the peak of V_{ds} indicates a fairly high value of 135 V, approximately 2.7 (theoretically 2.8) times as much as V_{ds} = 50 V. But the drain–source breakdown voltage of the GaN HEMT is high enough at 200 V to cope with the peak voltage.

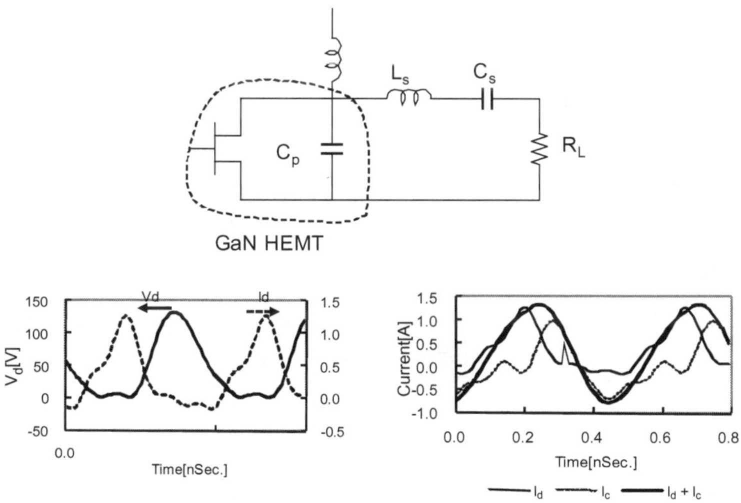

Figure 12.9 Simulated class E waveform.

To realize a sinusoidal waveform, the series inductance is needed to filter the higher harmonics. For that reason, an L–C low-pass output-matching network including enough high series inductance for the harmonics is expected to substitute for the series L–C resonator [6]. A low-pass-type class E amplifier configuration and simulation results are shown in Fig. 12.10 with an L_s of 6.5 nH and a C_m of 1.1 pF. The other conditions are the same as the series resonator type. All the waveforms are quite similar to the series resonator type. Therefore the low-pass-type class E amplifier is acceptable for a GaN HEMT at 2.1 GHz and 50 V operation.

An advantage of this low-pass type is that a single pole (L_s + C_m) low-pass network woks as not only a harmonic chock but also as fundamental matching.

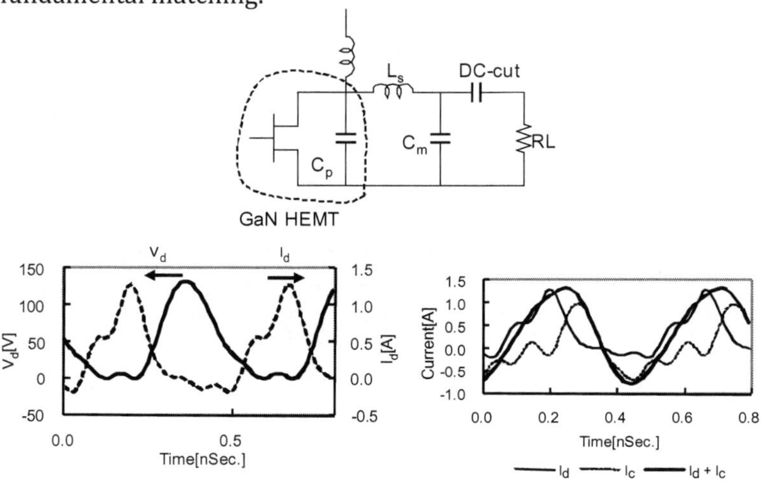

Figure 12.10 Low-pass-type class E circuit and waveform.

12.4.1.4 Circuit design of a 10 W class E

Figure 12.11 shows a packaged class E, which includes a 10 W GaN HEMT and an output low-pass network and its equivalent circuit [6].

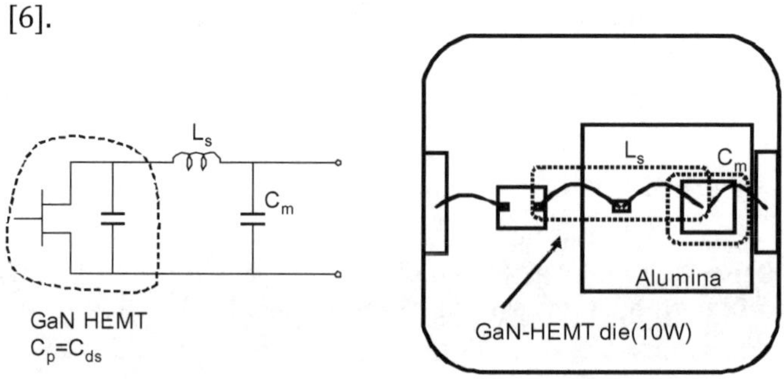

Figure 12.11 Designed class E amplifier.

The low-pass class E circuit composed of a parallel capacitor C_p (=C_{ds}), a series inductor L_s, and a matching capacitor C_m for the output is inside the package. The other input- and output-matching networks are on a printed circuit board (PCB) outside the package. A single-cell GaN HEMT with a gate width of 2.3 mm is mounted on the package with a Cu-based heat spreader. The L_s is composed of Au bonding wires with an interconnection alumina substrate for connecting the wires.

Advantages of this circuit topology are not only flexibility of a bias network connecting positions on the output circuit but also application to a higher-power multicell GaN HEMT easily by connecting GaN HEMT unit cells, drain bonding wires, and matching capacitors in parallel inside a package.

P_{in} and P_{out} characteristics of the 10 W GaN HEMT class E amplifier are shown in Fig. 12.12. An output power of 11 W, a drain efficiency of 82%, an associated power gain of 15.5 dB at an input power of 25 dBm, and a linear gain of 19.5 dB at a V_{ds} of 50 V and 2.1 GHz were achieved.

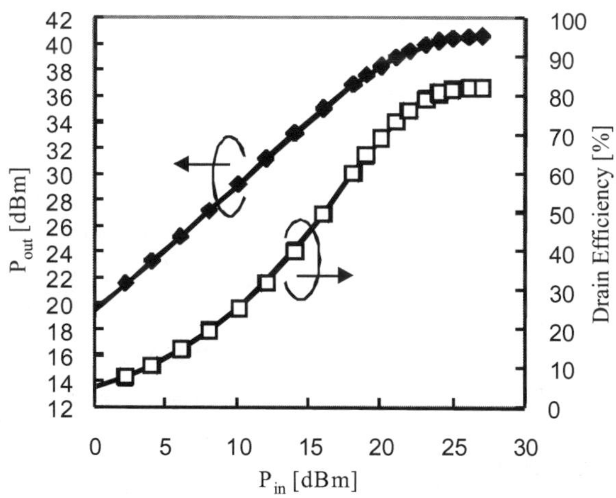

Figure 12.12 Power characteristics of a 10 W class E amplifier.

12.4.1.5 High-power class E

Typically a number of small cells are connected in parallel to get higher power, and a low-pass-type class E circuit can be easily expanded to a higher-power device, as shown Fig. 12.13. The condition of C_p, L_s, and C_m is already fixed in the previous section, and they are simply connected in parallel as the same number of cells.

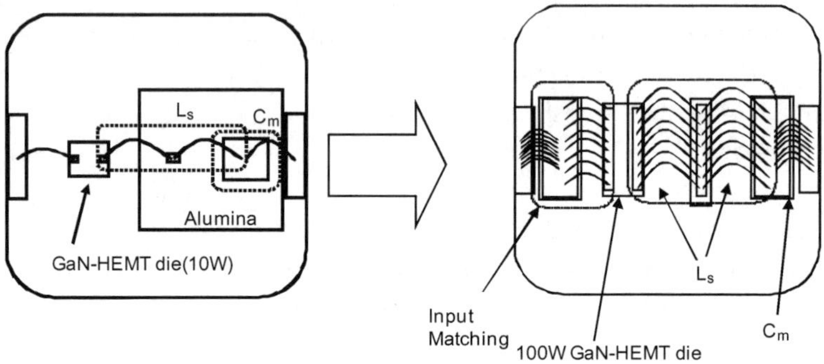

Figure 12.13 High-power GaN die and class E circuit.

But input and output impedances are smaller with $1/N$; therefore a matching network must be redesigned for proper frequency bandwidth.

A two-stage input-matching circuit of a quarter-wavelength impedance transformer inside the package and on a PCB are employed, and a low-pass class E circuit [6] composed of a parallel capacitor ($C_p \times N$), a series inductance (L_s/N), and a matching capacitor ($C_m \times N$) for the output. Other output-matching network elements are capacitors and open stubs on the test fixture. A GaN HEMT gate of width 20 mm is mounted on the package with a Cu-based heat spreader. C_p is the drain–source capacitance of the GaN HEMT itself.

P_{in} and P_{out} characteristics of the developed 100 W class E GaN HEMT are shown in Fig. 12.14. An output power of 100 W, a drain efficiency of 75%, and an associated power gain of 12 dB at P_{in} = 38 dBm were achieved. Flatness of the output power and the drain efficiency are 0.5 dB and 2 percentage points, respectively, between 2.11 GHz and 2.17 GHz. Figure 12.15 shows the output power spectrum of class C and class E operations at P_{out} = 100 W. The second harmonic component of class E is 20 dB smaller than class C. The third and fourth harmonics are less than –50 dBc of the fundamental component.

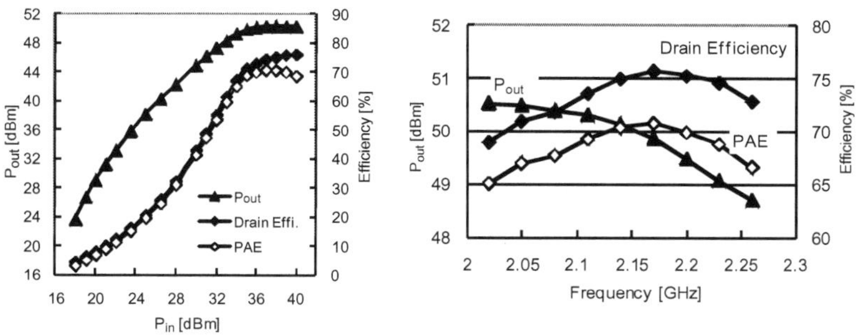

Figure 12.14 Power and frequency characteristics of 100 W class E GaN HEMT.

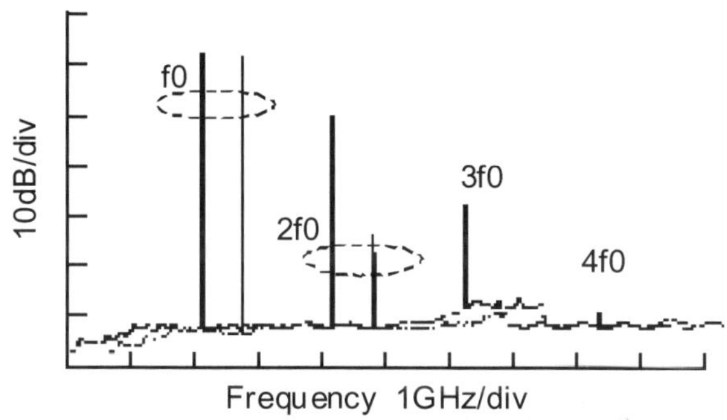

Figure 12.15 Spectrum of class C and class E operations.

12.4.2 Class F Operation

12.4.2.1 Principle of class F

A typical circuit and ideal current and voltage waveforms of class F and inverse class F are shown in Figs. 12.16 and 12.17, respectively.

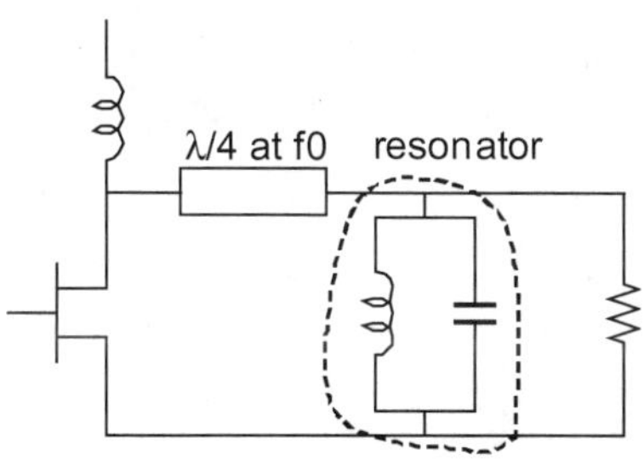

Figure 12.16 Typical circuit and waveforms of class F.

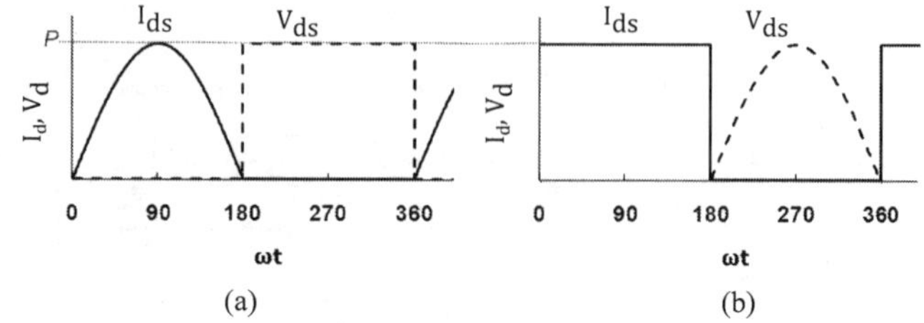

Figure 12.17 Waveforms of (a) class F and (b) inverse class F.

Harmonic terminations of class F are short for even harmonics and open for odd harmonics. And those of inverse class F are open for even and short for odd. The class F and inverse class F consist of a combination of a half sinusoidal and rectangle waveform for current or voltage as a function of

$$\text{Half sinusoidal } f(t)=\frac{P}{\pi}+\frac{P}{2}\sin\omega t+\frac{P}{\pi}\sum_{m=1}^{\infty}\frac{2}{1-4m^2}\cos 2m\omega t \tag{12.15}$$

$$\text{Rectangle } f(t)=\frac{P}{2}+\frac{2P}{\pi}\sum_{m=1}^{\infty}\frac{2}{1-4m^2}\sin(2m-1)\omega t \tag{12.16}$$

where P is the peak value of each waveform and the first term of the functions are DC or average value. In class F, current and voltage waveforms consist of even and odd harmonics, respectively, and this means no harmonic power dissipation in a cycle.

The relationships between peak value and DC (average) are

Class F: $I_{\text{peak}} = \pi \cdot I(\text{DC})$; $V_{\text{peak}} = 2 \cdot I(\text{DC})$

Inverse class F: $I_{\text{peak}} = 2 \cdot I(\text{DC})$; $V_{\text{peak}} = \pi \cdot I(\text{DC})$

The maximum voltage of class F is twice of the biased V_{ds} (DC), and the maximum current is three times the average I_{ds}.

In inverse class F, peak voltage reaches three times the biased V_{ds}. Therefore a very high breakdown voltage is needed in the inverse class F operation.

12.4.2.2 Harmonic load pull measurement and waveform simulation

To find the optimum load condition for the second harmonic, load pull measurement is performed. It is measured by fundamental load pull at each second harmonic angle with a high gamma greater than

0.9. The device is 2.3 mm gate width GaN HEMT which shows 10W output power at 50 V and 2.14 GHz.

Figure 12.18 shows a drain efficiency versus the second harmonic load angle. It shows the highest efficiency at around 30°. This point is very close to open. Compared to the fundamental load impedance, the second harmonic optimum load is much higher (even if C_{ds} de-embedded). This indicates inverse class F.

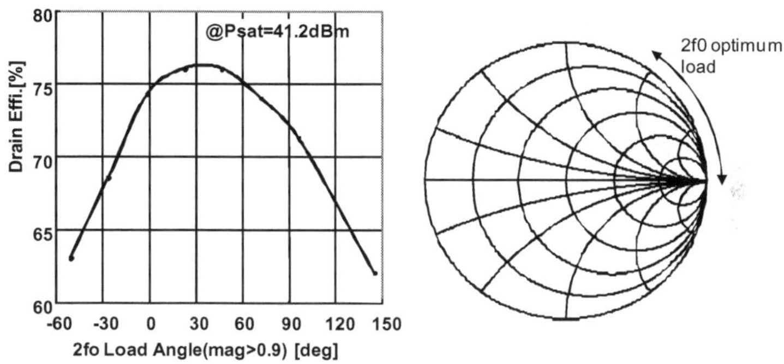

Figure 12.18 Load pull measurement results.

To analyze this result, waveform simulation is performed. Figure 12.19 shows the results of class F and inverse class F.

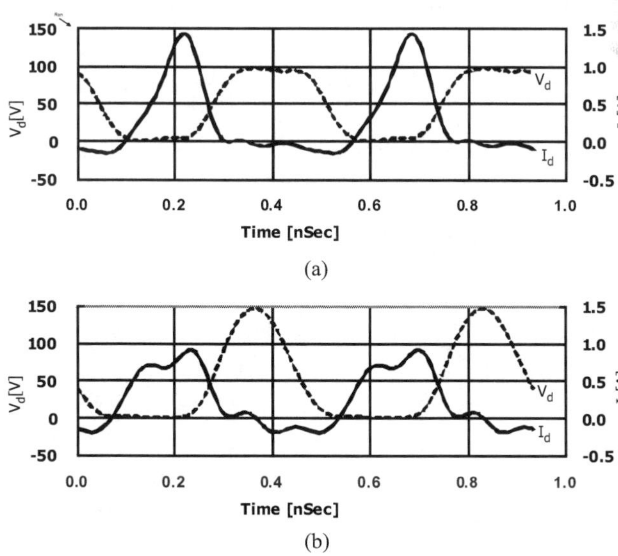

Figure 12.19 Simulated waveform of (a) class F and (b) inverse class F.

Both waveforms show significant similarity with ideals. In class F, the current waveform reaches 1.5 A at the peak, which is almost I_{max}, showing waveform distortion by on resistance (R_{on}). In inverse class F, the current peak is around half of I_{max}, but the voltage peak reaches 150 V, which is very closer to theoretical πn 50 V.

Figure 12.20 shows a simplified load line of class F and inverse class F.

- Class F: High I_d peak and low V_d peak with a large loss by on resistance and a distorted current waveform.
- Inverse F: Low I_d peak and high V_d peak with a small loss with a low I_d peak and good pinch-off characteristics with a small drain conductance of up to 200 V.

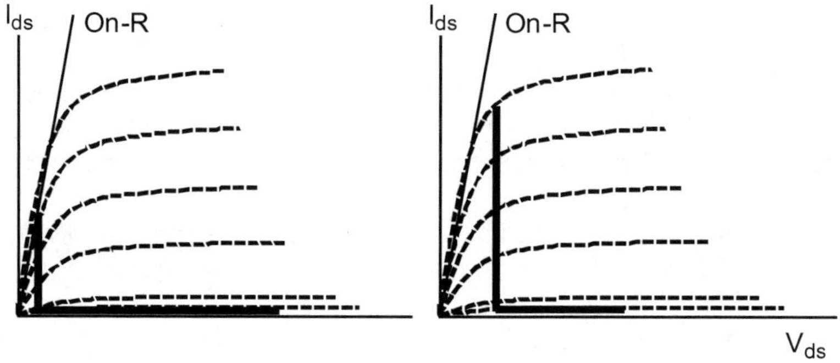

Figure 12.20 Switching load line for class F and inverse class F.

Therefore a high-breakdown-voltage GaN HEMT is suitable for inverse class F operation.

12.4.2.3 Circuit design of a 10 W class F

Figure 12.21 shows an internal view of a package inverse class F consisting of a 10 W GaN HEMT die and a second harmonic tuning circuit inside the package with a quarter-wavelength open stub for the second harmonic on a 50 Ω RF line, which gives a short plane for the second harmonic at the connection plane. The second harmonic angle is adjusted by the line length to the die plane.

The fundamental frequency is matched outside the package for both input and output.

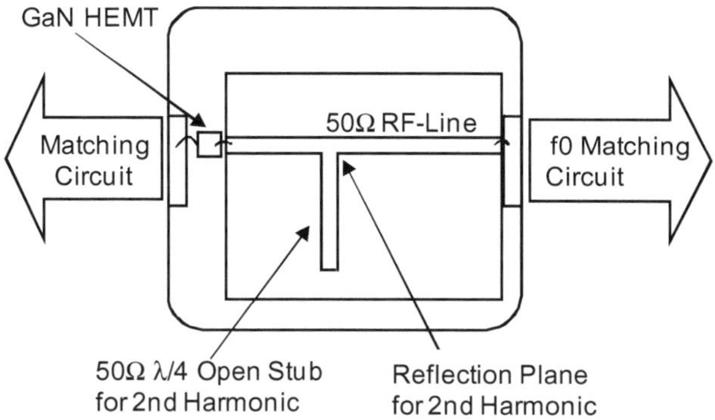

Figure 12.21 Designed circuit of class F.

As shown in Fig. 12.22 inverse class F shows high efficiency closer to 80%. But compared to class E, class F has slightly lower efficiency. In this circuit, only the second harmonic is terminated, but in class E, higher harmonics are automatically terminated by a large series inductance. This is one of the reasons for higher efficiency in class E. But such long wire bonding for large inductance are not proper for real products due to variation of the wire length in lot-lot or in bonding machines. Only second harmonic termination but high-power inverse class F with very good productivity is proposed in the next section.

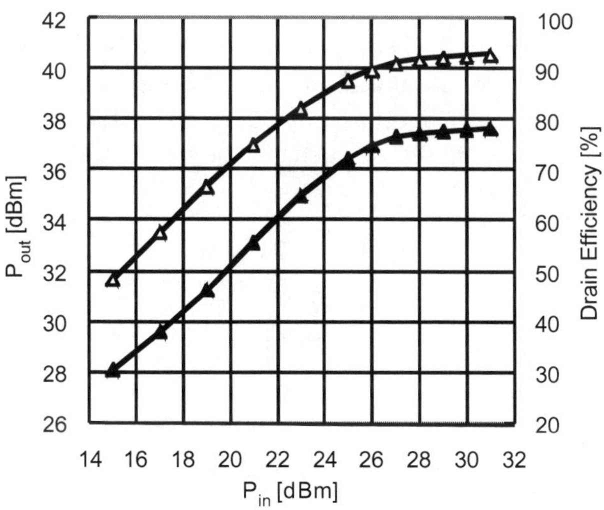

Figure 12.22 Power characteristics of a 10 W class F.

12.4.2.4 High-power class F

Figure 12.23 shows the concept of high-power inverse class F. An *L–C* low-pass-type network that gives a matching for the fundamental frequency and a higher impedance for the second harmonic is employed. The advantage of this network is that both fundamental and second harmonic impedances can be controlled with one pole-matching network.

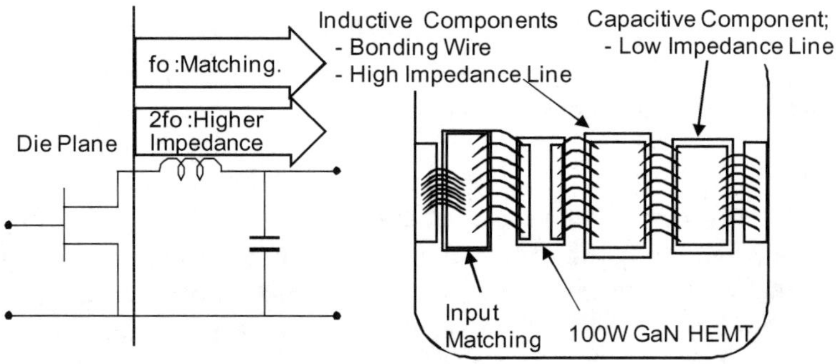

Figure 12.23 High-power class F concept.

Bonding wires and a high-impedance transmission line on a ceramic substrate as an inductive component and a low-impedance line on a high-dielectric material as a capacitive component are used.

S parameters in Fig. 12.24 show simulated load impedance at the die plane, including a fundamental frequency of 2.14 GHz and a second harmonic of 4.28 GHz. The fundamental and second harmonic are located on the optimum impedance, as shown in Fig. 12.18.

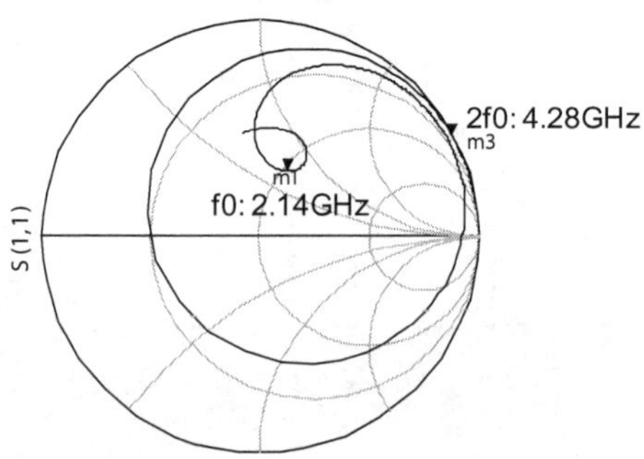

Figure 12.24 Simulated load impedance.

Measurement results of a 100 W class F are shown in Fig. 12.25. This inverse class F has higher efficiency over 70% with a 100 W output power in a real product.

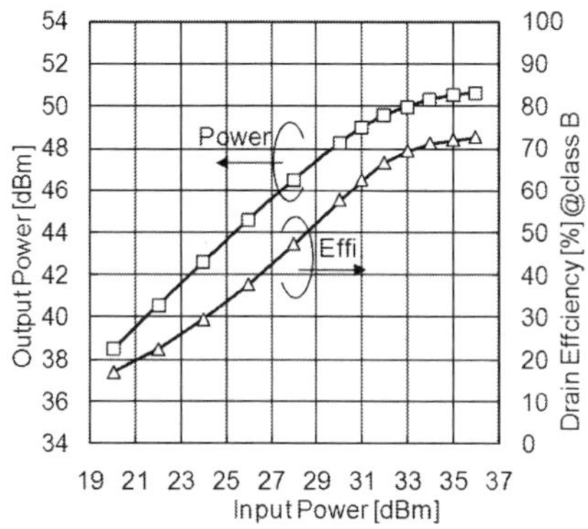

Figure 12.25 Power characteristics of a 100 W class F.

12.5 Doherty Amplifier

12.5.1 Principle of the Doherty Amplifier

The Doherty amplifier is one of the efficiency enhancement techniques for a large back-off region from saturation power [3].

Figures 12.26 and 12.27 show the well-known Doherty amplifier configuration and its efficiency characteristics. The Doherty amplifier has two amplifiers, main and peak. Typically main and peak amplifiers are biased to class B (or deep class AB closer to –B) and class C, respectively. The class C bias of the peak amplifier could be tuned at a large signal simulation and RF measurement; this bias condition determines the peak amplifier operation start and is very important for not only Doherty efficiency behavior but also adjacent channel leakage ratio (ACLR) and inter modulation distortion (IMD) characteristics, which is generated by non-linearity of power amplifier. If the main and peak amplifiers have the same power, the first efficiency peak is obtained at 6 dB back-off power, where the main amplifier saturates. The typical average power of 6–8 dB back-off for 3G W-CDMA and LTE is close to this first efficiency peak; therefore the saturation efficiency is very important for Doherty amplifier components.

Load impedance is set to half of R_{opt}. R_{opt} is described as the optimum load impedance for the maximum output power, as shown in Fig. 12.26, is simply defined as a linear line connecting class B bias (with V_{ds}) and I_{max}.

There are two regions in efficiency characteristics, as shown Fig. 12.26. In each region, impedance conditions are shown in schematics in Fig. 12.27. The most important thing in Doherty design is to realize the impedance condition in region 1 for the main amplifier. Open (or as high impedance as possible in reality) termination for the peak amplifier is realized easily because in class C bias on an FET, and g_m equals zero, meaning "open" at small-signal operation. Load impedance for the main amplifier must be twice the Z_{opt} by a quarter-wavelength impedance converter on the output side of the main amplifier. It is not easy to give twice the Z_{opt} for the main amplifier at the package plane, because amplifiers have parasitic components, mainly, C_{ds} and a matching network inside the package.

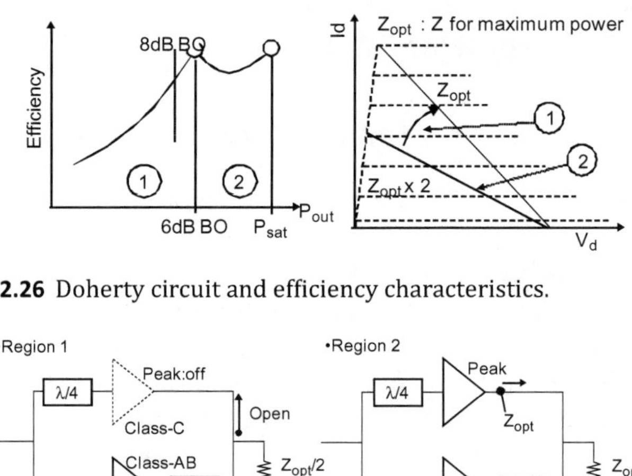

Figure 12.26 Doherty circuit and efficiency characteristics.

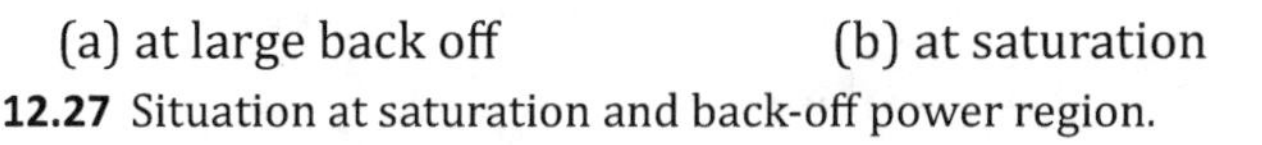

(a) at large back off (b) at saturation

Figure 12.27 Situation at saturation and back-off power region.

12.5.2 Load Pull Theory for the Doherty Design

Two reference planes for load pull are defined in Fig. 12.28, plane A at current source (g_m) plane in a GaN HEMT intrinsic area and plane

B at the package lead plane. Z_{opt} and $Z_{opt} \times 2$ must be given at plane A, but it is commonly said that people design a Doherty amplifier using information at plane B. Therefore Z_{opt} and $Z_{opt} \times 2$ must be found at plane B. Amplifier designers can find Z_{opt} easily by load pull measurement but cannot find $Z_{opt} \times 2$ directly.

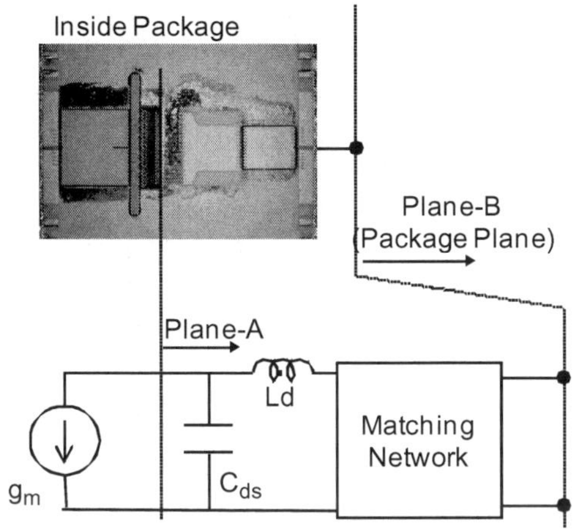

Figure 12.28 Definition of a reference plane.

To find $Z_{opt} \times 2$, the designers must understand load pull theory at plane A. R_{opt} is already determined in the previous discussion and is put on a Smith chart in Fig. 12.29 as an example.

R_1 and R_2 are R_{opt} divided and multipled by an arbitral coefficient, a. A power drop from the maximum power on both R_1 and R_2 is the same in the ideal case of no on resistance of the transistor, the voltage swing is $(/a)$, and current swing is $(\times$ a) for R_1 and R_2, respectively.

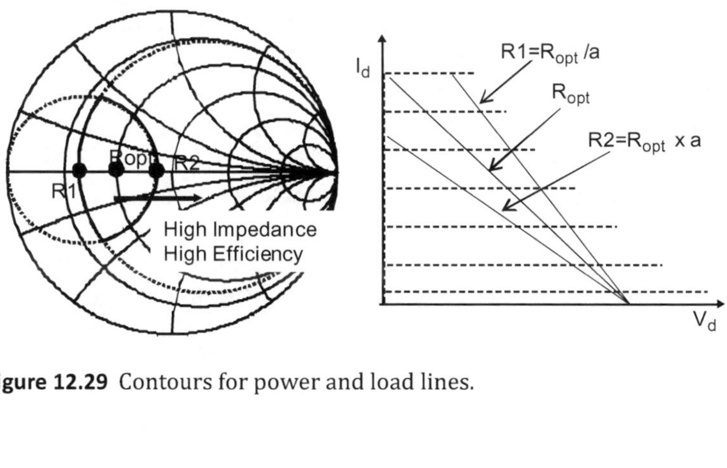

Figure 12.29 Contours for power and load lines.

A contour is described by two circles through R_1 and R_2. The same power is obtained as R_1 and R_2 on any impedance on the contour. Power on R_1 and R_2 is the same, but efficiency is different. Figure 12.30 shows ideal voltage waveforms. In R_2, the voltage reaches zero, but in R_1 the minimum voltage floats from zero by hitting I_{max}. This means DC loss and efficiency must be lower than R_{opt} and R_2. In the ideal case, R_{opt} and R_2 must have the same efficiency. But real transistors have unneglectable on resistance. In this case, the voltage must be floated, and the floating voltage at R_2 is smaller than R_{opt}. Therefore better efficiency is obtained at R_2 with small voltage loss.

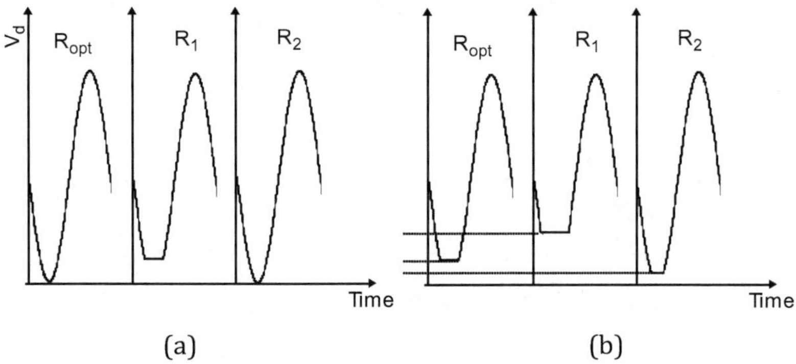

Figure 12.30 Voltage waveform (a) without and (b) with R_{on}.

Figure 12.31a shows load pull simulation at the current–source plane. The normalized impedance of this Smith chart is set to R_{opt}. This power and efficiency contours show higher efficiency in a higher-impedance region. On this Smith chart, the designer can find R_{opt} and $R_{opt} \times 2$ easily and notice that $R_{opt} \times 2$ locates in the high-efficiency direction. This means the high-efficiency direction is a high-impedance direction; therefore $Z_{opt} \times 2$ is in the high-efficiency direction at least.

Figure 12.31b shows load pull measurement results at plane B. The normalized impedance is 50 Ω. Z_{opt} and $Z_{opt} \times 2$ are converted by a transistor parasitic component, a matching network inside the package, and a package feed.

Z_{opt} is found easily at optimum power impedance, and $Z_{opt} \times 2$ should be on the line to the optimum-efficiency direction. The accurate impedance of $Z_{opt} \times 2$ cannot be found, but the direction from Z_{opt} is enough for Doherty design.

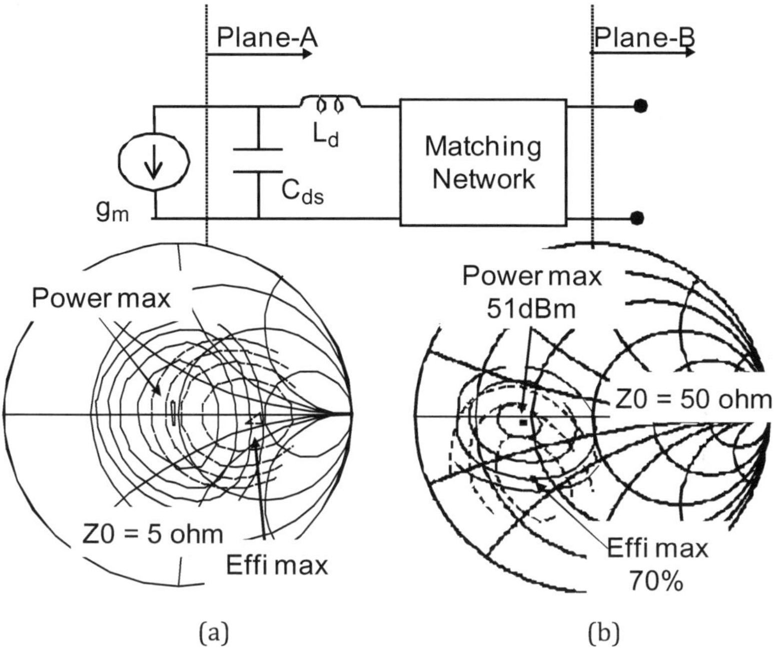

Figure 12.31 Load pull at (a) plane A (simulated) and (b) plane B (measurement).

12.5.3 Main Amplifier Design

There are just three steps for the main amplifier design, as shown in Fig. 12.32:

1. Make a matching network for a power tune at 50 Ω. (For an easier explanation, a 50 Ω match is used in this example but not limited. Designers can select any impedance.)
2. Add a 50 Ω delay line to realize 50 Ω for the power tune and 100 Ω for an efficiency tune.
3. Add a $\lambda/4$50 Ω impedance converter. At this plane the external power and efficiency optimum load locate at 50 Ω and 25 Ω, respectively.

All circuits were designed with linear simulation easily using load pull measurement at the package plane.

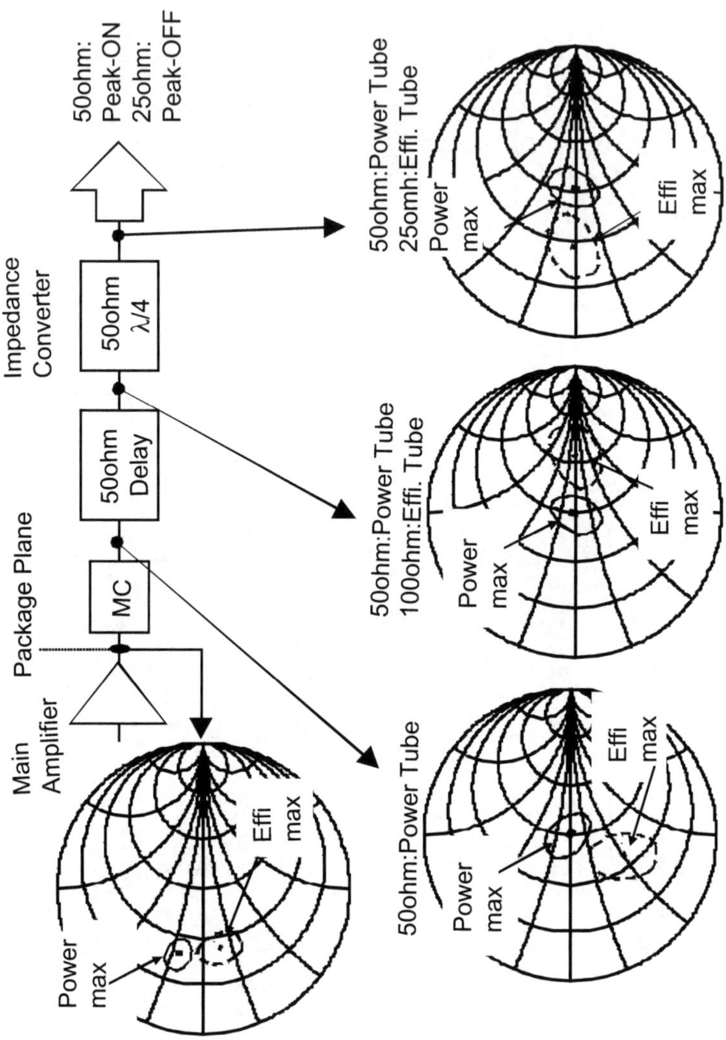

Figure 12.32 Design of the main amplifier.

12.5.4 Peak Amplifier Design

As previously discussed, a class C biased peak amplifier is "open" at the small-signal region because of $g_{\mathrm{m}} = 0(Z = 1/g_{\mathrm{m}})$. But the device has a parasitic component, and the packaged one has an internal circuit and a package feed, same as the main amplifier. But analysis is much easier than the main amplifier.

S parameters at the class C bias would show a much higher gamma (reflection coefficient) if there is not a large loss component inside the package. At this point, of course, it would not "open" by the parasitic component. To make "open" (or higher) impedance at the Doherty combiner plane, the designer just adds a proper length of 50 Ω with a proper matching network.

1. Make a matching network for a power tune at 50 Ω (it would be very similar to that of the main amplifier).
2. Add a 50 Ω delay line to make "open" (or higher) impedance at the Doherty combiner plane.

See Fig. 12.33.

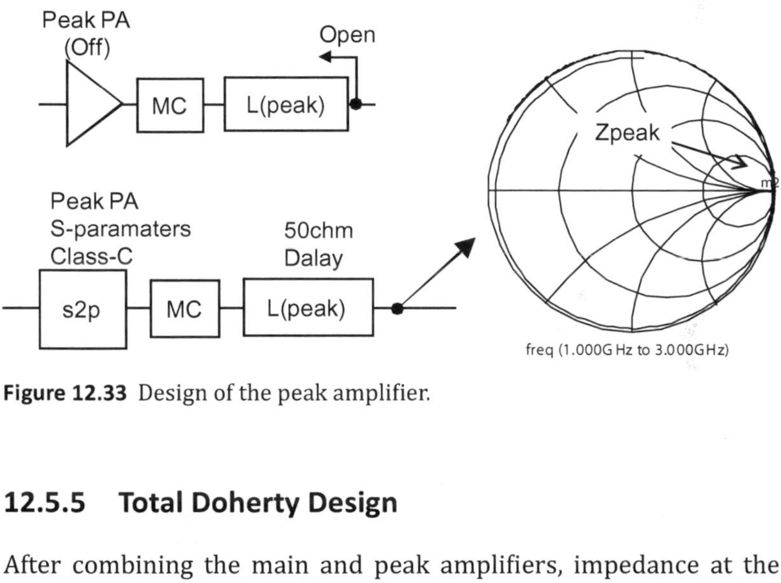

Figure 12.33 Design of the peak amplifier.

12.5.5 Total Doherty Design

After combining the main and peak amplifiers, impedance at the combining plane is 25 Ω; therefore another impedance transformer from 25 Ω to 50 Ω is needed.

For the input side, the same matching network as a single-ended amplifier (means gain matching) would be used. Note that the peak amplifier operation starting power is controlled by not only the bias circuit but also input matching of the peak amplifier (the main amplifier). A $\lambda/4$ delay line must be inserted to the peak amplifier side to compensate the output Doherty $\lambda/4$ transformer (Fig. 12.34).

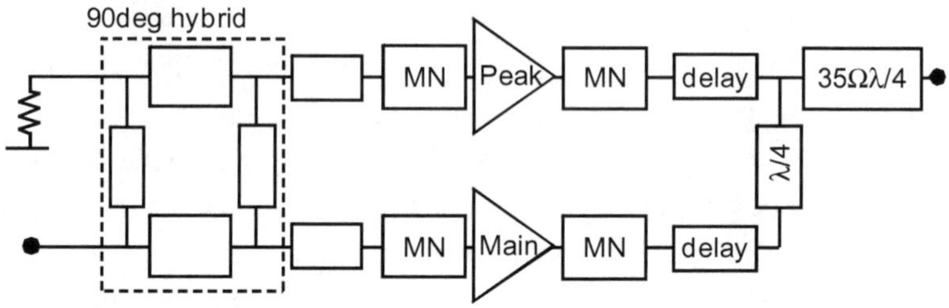

Figure 12.34 Total design of a Doherty amplifier.

To reduce interaction between the main and peak amplifiers, a 90° hybrid coupler is typically used with a function of a $\lambda/4$ delay line for the input side. At a large back-off power, the peak amplifier doesn't work; this means a 3 dB input loss by the input combiner. But in the ideal transistor case, a gain at $R_{\text{opt}} \times 2$ in the main amplifier is 3 dB higher than that at R_{opt}. The voltage swing at $R_{\text{opt}} \times 2$ is double of R_{opt} at the same g_{m} and the input voltage swing (input power), as described in Fig. 12.35. Therefore in the ideal situation, a 3 dB loss by the combiner is compensated by the output impedance converter. However, the designer cannot get a 3 dB higher gain at a high-impedance-load condition by loss of drain conductance and the parasitic component (experimentally the gain in Doherty is 1–2 dB lower than the single-ended case).

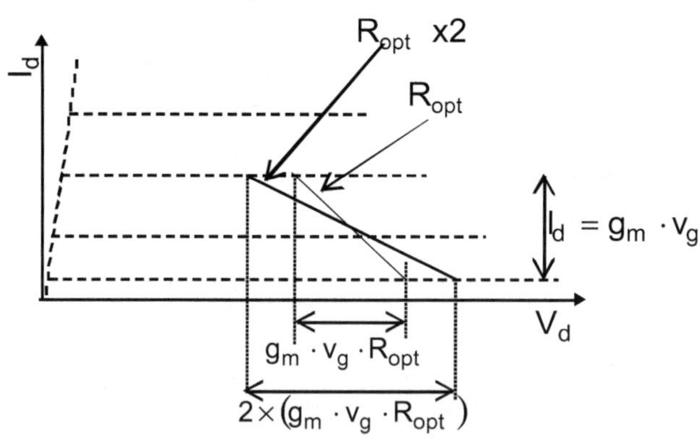

Figure 12.35 Voltage swing at R_{opt} and $R_{\text{opt}} \times 2$.

The photo in Fig. 12.36 shows a Doherty amplifier using an inverse class F 100 W GaN HEMT. Figure 12.37 show the power, efficiency, and drain current characteristics. Efficiency shows the first peak with 60% at 46 dBm; i.e. around 6 dB of back-off power from saturation of 52 dBm.

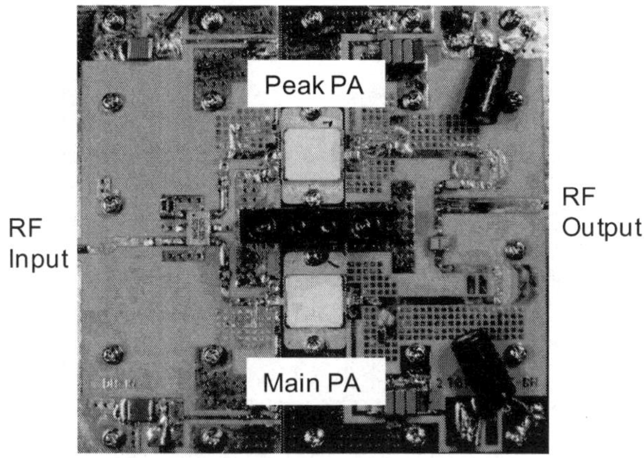

Figure 12.36 Photo of a Doherty amplifier.

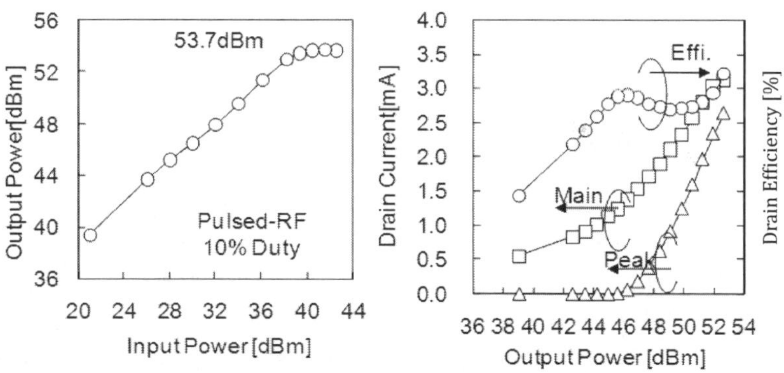

Figure 12.37 Power characteristics of a Doherty amplifier.

A Doherty amplifier generates larger signal distortion (larger IMD, AM–AM/AM–PM conversion and adjacent channel leakage power ratio [ACLR]) compared to a class AB single-ended amplifier due to its nonlinear operation. Typically digital predistortion (DPD) techniques are employed to compensate this large distortion. Figure 12.38 shows the ACLR with DPD operation. The efficiency doesn't show a peak at 6 dB back-off due to a large peak modulation signal of 7.8 dB peak-to-average-power ratio (PAPR) but shows 55% efficiency with a proper ACLR smaller than 52 dBc.

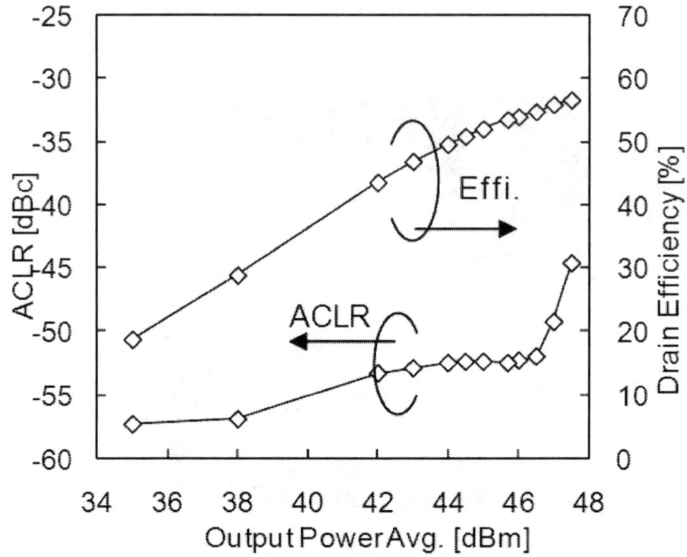

Figure 12.38 Efficiency and ACLR of a Doherty amplifier. V_{ds} = 50 V, $I_{ds\text{-}main}$ = 400 mA (V_{gs} = −1.24 V), $I_{ds\text{-}peak}$ = 0 mA (V_{gs} = −3 V), F = 2.14 GHz W-CDMA 2-carrier, 5 MHz spacing PAPR = 7.8 dB with a DPD evaluation test system.

12.5.6 Doherty Variations

A Doherty is one of the load modulation techniques, and it has several variations.

The simple design method of a Doherty that has the same output power for both main and peak amplifiers (called symmetrical Doherty) was introduced in previous sections. It has a peak efficiency at 6 dB back-off power from saturation. But due to recent higher data rate requirements, the PAPR in a modulated envelop signal is getting larger to around 8–12 dB. Therefore symmetrical Doherty is not enough, and typically "asymmetric Doherty" having a larger peak amplifier or "3-way Doherty" having two peak amplifiers is used for such higher-PAPR applications.

In addition to such classical Doherty configuration, an advanced N-way Doherty has been presented with (N + 1) ports Doherty combiner analysis in Ref. [7], as the configuration shown in Fig. 12.39.

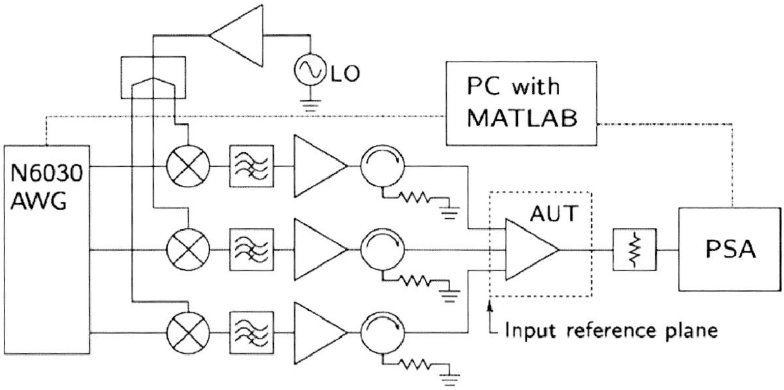

Figure 12.39 Example of an N-way Doherty with a mixed-signal approach.

In this configuration, input signals (magnitude and phase) are individually controlled by an arbitrary waveform generator (AWG) in coherence. In the classical way, the input signal is divided by an analog circuit (90° hybrid, Willkinson, etc.), and the designer must tune and find the best condition of input magnitude and phase for both main and peak amplifiers. But in this mixed-signal approach, the designer can control and optimize input signals for efficiency, power, and linearity by a computer digitally.

12.6 Other Efficiency Enhancement Techniques

Several efficiency enhancement techniques have been proposed, such as envelope tracking (ET), outphasing (or nonlinear amplification using nonlinear component [NLNC]), and switching amplifier (class S). Loads lined for each technique are shown in Fig. 12.40. All techniques intend to maximize voltage swing in any input level. Larger of the I_d–V_d curves is used in ET and outphasing compared to symmetrical Doherty. Therefore higher efficiency is expected in large back-off power.

In ET, operation voltage changes according to the input level, and nonlinearity (mainly of C_{ds}) for V_{ds} is very important. Typically the load condition in ET is fixed in any input level. If C_{ds} changes largely for V_{ds}, the load condition at the current source (g_m) plane inside the FET changes.

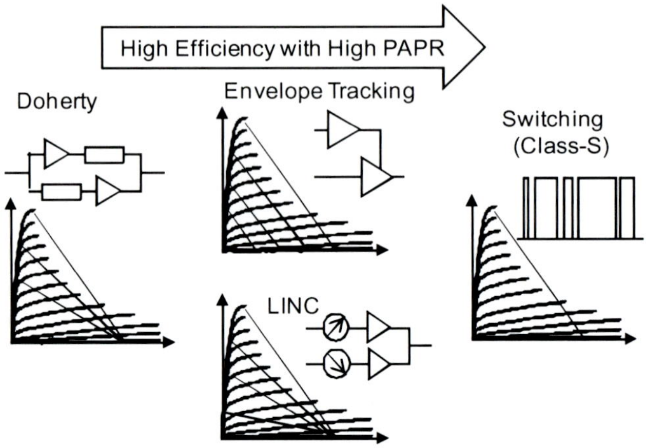

Figure 12.40 Efficiency enhancement techniques.

An improvement of C_{ds} nonlinearity is reported in Ref. [8]. Figure 12.41 shows its improvement and efficiency characteristics at several operation voltages (note that these are not real ET operations but are supposed to be ET). The thick line is the efficiency locus in ET operation and shows a quite high efficiency of 70% at 8 dB back-off power.

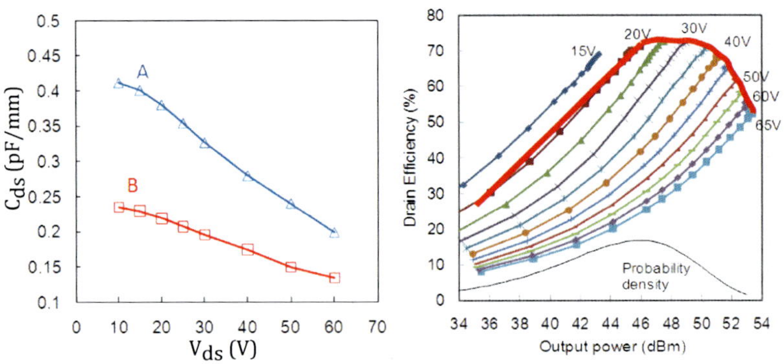

Figure 12.41 C_{ds} linearity improvement and efficiency for ET.

References

1. Y. Ohno, Current status and prospects of GaN-based high-frequency devices, *FED Rev.*, **1**(13), 2002 (in Japanese).

2. W. Burger, H. Brech, D.Burdeaux, C.Dragon, G.Formicone, M.Honan, B. Pryor, and X. Ren, A device technology for high power RF infrastructure applications, *CSIC2004*, 2004.

3. F. van Rijs and S. J. C. H. Theeuwen, Efficiency improvement of LDMOS transistors for base stations towards the theoretical limit, *IEDM2006*, 2006.

4. S. C. Cripps, *RF Power Amplifiers for Wireless Communications*, Artech House, 1999.

5. N. Ui and S. Sano,. A 45% drain efficiency,-50dBc ACLR GaN HEMT class-E amplifier with DPD for W-CDMA base station, *IMS2007*, 2006.

6. N. Ui and S. Sano, A 100W class-E GaN HEMT with 75% drain efficiency at 2GHz, *EuMIC2006*, 2006.

7. W. C. Edmund Neo, J. Qureshi, M. J. Pelk, J. R. Gajadharsing, and Leo C. N. de Vreede, A mixed-signal approach towards linearand efficient N-way Doherty amplifiers, *IEEE Trans. Microwave Theory Tech.*, **55**(5), 866–879, 2007.

8. F. Yamaki, K. Inoue, N. Ui, A. Kawano, and S. Sano, A 65 % drain efficiency GaN HEMT with 200 W peak power for 20 Vto 65 V envelope tracking base station amplifier, *IMS2011*, 2011.

Chapter 13

Understanding the Efficiency of Switched-Capacitor Power Supply Circuits

Haruo Kobayashi,[a] Daiki Oki,[a] Biswas Sumit Kumar,[a] and Keith Wilkinson[b]*

[a]*Division of Electronics and Informatics, Gunma University, 1-5-1 Tenjin-cho, Kiryu, Gumma 376-8515, Japan*

[b]*Technical Consultant, Kanagawa, Japan*

k_haruo@el.gunma-u.ac.jp

This chapter contains a tutorial explaining the energy loss and power efficiency of switched-capacitor switching power supply circuits. The objective of this tutorial is to understand how to realize high-efficiency charge pump circuits [1–3].

13.1 Basic Study of Switched-Capacitor Power Electronics

This section explains why switched-capacitor power supply circuits have a fundamental energy loss that limits their maximum possible efficiency, even if the loss in switches is negligible.

*Deceased.

Handbook of Power Management Circuits

Edited by Haruo Kobayashi and Takashi Nabeshima

ISBN 978-981-4613-15-6 (Hardcover), 978-981-4613-16-3 (eBook)

www.panstanford.com

Switching power supply circuits can achieve high theoretical efficiency because ideal switches do not consume power: their current is zero in the switched-off state, and their voltage drop is zero in the switched-on state (Fig. 13.1). The switching loss of real switches occurs during transitions between the off state and the on state, where there is a finite current through the switch and a finite voltage drop across the switch (Fig. 13.2). However, even if the on resistance of the switches is close to zero, there is still a theoretical limit to the efficiency of switched-capacitor power supply circuits, as explained below.

Assumption: Hereafter in this chapter, we assume that the switches are ideal and that their switching losses are zero.

Figure 13.1 Switch off state and on state.

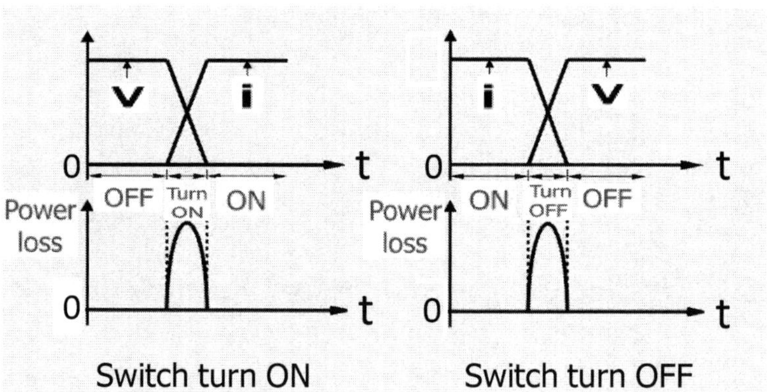

Figure 13.2 Switching loss (which is ignored in this chapter). (a) When the switch is turned on and (b) when the switch is turned off.

We would like to call the reader's attention to the following two theorems that explain power losses in switched-capacitor switching power supply circuits:

Theorem 1: Let us consider the charging of a capacitor C in the RC circuit of Fig. 13.3. Here the voltage across the capacitor is zero at time $t = 0$.Then the energy E_c stored in C in the steady state ($t = \infty$), the energy E_R dissipated in R from $t = 0$ to ∞, and the energy E_{supply} are, respectively, expressed by

$$E_C = \frac{1}{2}CV_{dd}^2,\ E_R = \frac{1}{2}CV_{dd}^2,\ E_{supply} = CV_{dd}^2$$

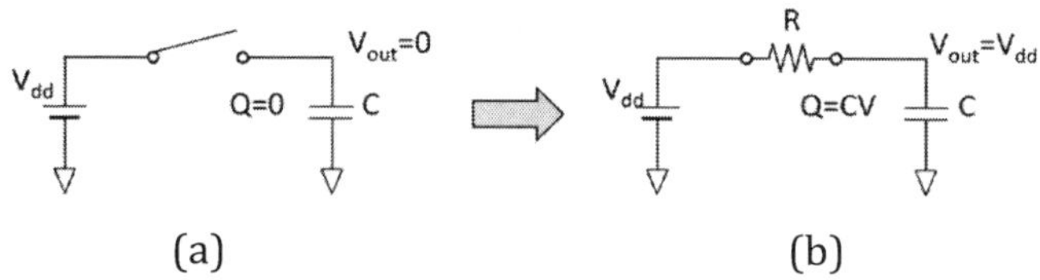

Figure 13.3 Charging a capacitor.

Proof: The differential equation for Fig. 13.3b is given by

$$RC\frac{d}{dt}V_{out}(t)+V_{out}(t)=V_{dd},V_{out}(0)=0$$

and $V_{out}(t)$ is obtained as follows:

$$V_{out}(t)=V_{dd}\left(1-\exp\left(-\frac{t}{RC}\right)\right)$$

The energy E_{supply} supplied from V_{dd} from $t = 0$ to ∞ is given by

$$E_{supply} = CV^2_{dd}$$

because the charge CV_{dd} flows from V_{dd}. The energy stored in C in the steady state ($t = \infty$) is given by

$$E_C = \frac{1}{2}CV_{dd}^2$$

Also the energy E_R dissipated in R from $t = 0$ to ∞ is obtained by

$$E_R = \frac{1}{R}\int_0^\infty (V_{dd} - V_{out}(t))^2 dt = \frac{1}{2}CV_{dd}^2 \quad \text{(QED)}$$

Remarks:

(i) Note that we have the following relationships:

$E_{supply} = E_c + E_R$, $E_C = E_R$

(ii) One half of the energy supplied from V_{dd} is stored in C and the other half is dissipated in R. E_R does not depend on the value of R.

Theorem 2: Let us consider the circuit in Fig. 13.4, which consists of capacitors C_1 and C_2 and a switch. At first the switch is off, the voltages across C_1 and C_2 are V_1 and V_2, respectively, and the total energy stored in both C_1 and C_2 is E. Next the switch turns on, and the total energy stored in both C_1 and C_2 is E'. Then we have the following:

When $V_1 \neq V_2$, then $E > E'$.

When $V_1 = V_2$, then $E = E'$.

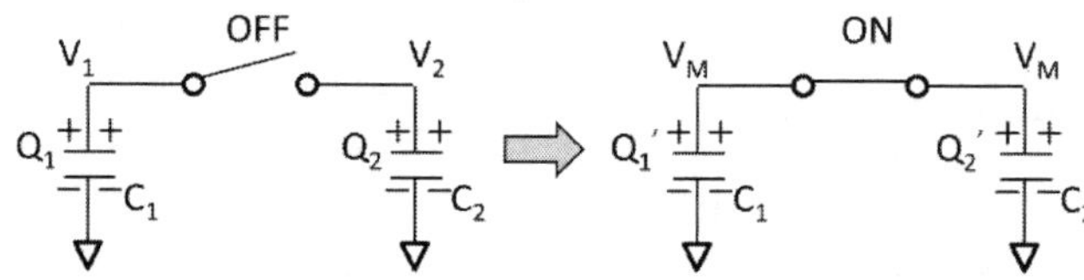

Figure 13.4 Two capacitors and a switch. (a) When the switch is off, the voltages across C_1 and C_2 are V_1 and V_2, respectively. (b) When the switch is on, the charges in C_1 and C_2 move so that the voltages across C_1 and C_2 become the same (V_M), and some energy is dissipated in the switch.

Proof: When the switch is off, and the voltages across C_1 and C_2 are V_1 and V_2, respectively, the charges Q_1 and Q_2 stored in C_1 and C_2, and the corresponding stored energies E_1 and E_2 are, respectively, given by

$$Q_1 = C_1 V_1, Q_2 = C_2 V_2,$$

$$E_1 = \frac{1}{2} C_1 V_2^2, \; E_2 = \frac{1}{2} C_2 V_2^2$$

Then we obtain the total energy E as follows:

$$E = E_1 + E_2 = \frac{1}{2} C_1 V_1^2 + \frac{1}{2} C_2 V_2^2 \quad (13.1)$$

Next let the switch turn on. The voltages across C_1 and C_2 become the same (V_M) in the steady state, which leads to the following:

$$Q'_1 = C_1 V_1, Q'_2 = C_2 V_2 \quad (13.2)$$

Since the charges are preserved between the two states in Fig. 13.4, we have the following relationship:

$$Q' = C_1 V_M, Q'_2 = C_2 V_2 \quad (13.3)$$

It follows from Eqs. 13.2 and 13.3 that

$$C_1 V_1 + C_2 V_2 = C_1 V_M + C_2 V_M$$

$$V_M = \frac{C_1V_1 + C_2V_2}{C_1 + C_2}$$

Hence the energies E_1 and E_2 stored in C_1 and C_2 in the switch's on state are given by

$$E_1' = \frac{1}{2}C_1V_M^2 = \frac{1}{2}\frac{C_1\left(C_1V_1 + C_2V_2\right)^2}{\left(C_1 + C_2\right)^2}$$

$$E_2' = \frac{1}{2}C_2V_M^2 = \frac{1}{2}\frac{C_2\left(C_1V_1 + C_2V_2\right)^2}{\left(C_1 + C_2\right)^2}$$

Thus the total energy E' in Fig. 13.3b reduces to

$$E' = E_1' + E_2' = \frac{1}{2}\frac{\left(C_1V_1 + C_2V_2\right)^2}{C_1 + C_2}. \tag{13.4}$$

It follows from Eqs. 13.1 and 13.4 that the energy difference ΔE between the two states in Fig. 13.4 is given by

$$\Delta E = E - E' = \frac{1}{2}\frac{C_1C_2}{C_1 + C_2}(V_1 - V_2)^2. \tag{13.5}$$

(QED)

Remark: When $V_1 \neq V_2$, then $\Delta E > 0$ and the energy ΔE is dissipated in the switch. On the other hand, when $V_1 = V_2$, ΔE is equal to zero and this is called zero-volt switching (ZVS).

13.2 Capacitor Size, Switching Frequency, Load Current, and Energy Loss

Let us consider the switched-capacitor circuit of Fig. 13.5, where the capacitor is charged to V_{dd} when the switch is on and the load current I_{load} flows from the capacitor C during period T in the switch's off state. Then the voltage drop ΔE across the capacitor is

$$\Delta V = \frac{I_{load}T}{C}. \tag{13.6}$$

The energy E_{load} provided to the load during period T in the switch's off state is

$$E_{load} = \frac{1}{2}\left[V_{dd} + (V_{dd} - \Delta V)\right]I_{load}T = \left(V_{dd} - \frac{1}{2}\Delta V\right)I_{load}T. \tag{13.7}$$

The energy provided from V_{dd} to the capacitor in the switch's on state is

$$E_{supply} = I_{load} T V_{dd} \tag{13.8}$$

Thus the energy loss during period T is given by

$$E_{loss} = E_{supply} - E_{load} = \frac{C(\Delta V)^2}{2} = \frac{(I_{load}T)^2}{2C}. \tag{13.9}$$

Thus we obtain the power loss as follows:

$$P_{loss} = \frac{E_{loss}}{T} = \frac{I_{load}^2 T}{2C}. \tag{13.10}$$

Also we have the efficiency

$$\eta = \frac{E_{load}}{E_{supply}} = \frac{\left(V_{dd} - \frac{1}{2}\Delta V\right) I_{load} T}{I_{load} T V_{dd}} = 1 - \frac{I_{load} T}{2CV_{dd}} \tag{13.11}$$

These are derived from Theorems 1 and 2. We note the following:

(1) The power loss decreases as the capacitor size increases.

(2) The power loss decreases as the switching frequency $f_s \left(= \frac{1}{T}\right)$ increases.

(3) The power loss increases as the load current increases.

We would like to emphasize that the efficiency of the switched-capacitor power supply circuit depends strongly on the capacitor size, the switching frequency, and the load current.

Remark: One might wrongly conclude from Theorem 1 that the efficiency of the switched-capacitor circuit cannot exceed 50%. This is true at start-up time but is NOT true in the steady state. The power loss in the steady state should be calculated with Eq. 13.10, and the efficiency can be greater than 50%.

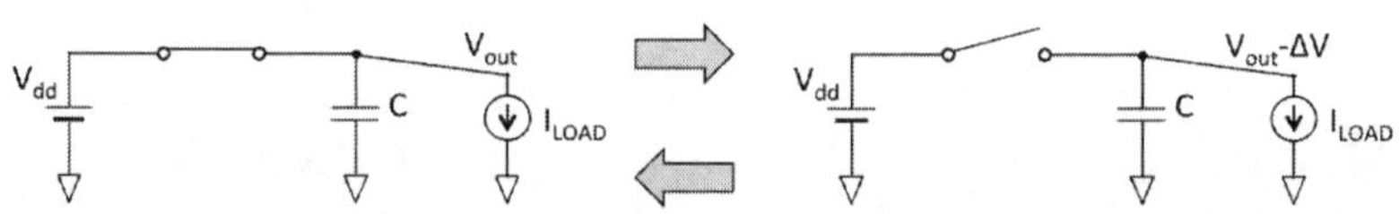

Figure 13.5 Output load current, capacitor size, sampling frequency, and power loss.

13.3 Parasitic Capacitance and Output Voltage

Parasitic capacitances degrade the efficiency of the switched-capacitor power supply circuit, and this section explains the reasons. Figure 13.6a shows the principle of a Dickson charge pump [1–4] and explains how the output voltage V_{out} can reach $2V_{dd}$ when the output node is connected to the pumping capacitor. In a real circuit, the output node has a large capacitor C_{out} to obtain $2V_{dd}$ at the steady state in both the left and right cases of Fig. 13.6b.

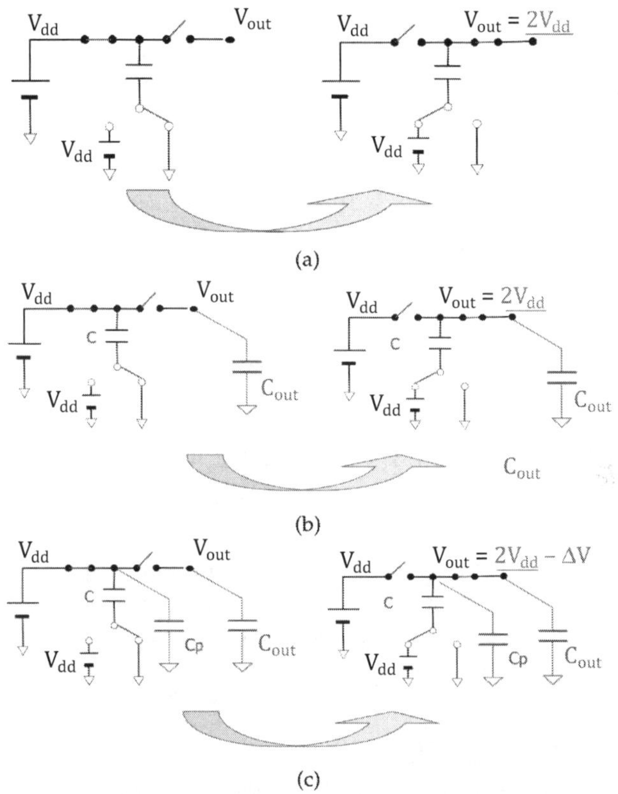

Figure 13.6 Charge pump circuit. (a) Principle of high voltage generation. (b) Charge pump circuit with output capacitor C_{out}. (c) Charge pump circuit with parasitic capacitor in parallel with output capacitor C_{out}.

Now let us consider the case that there is a parasitic capacitor C_p, as shown in Fig. 13.6c. In the steady state

$$CV_{dd} + C_pV_{dd} = C(V_{out} - V_{dd}) + C_pV_{out} \tag{13.12}$$

Then we have

$$V_{out} = \left(2 - \frac{C_p}{C + C_p}\right)V_{dd} \tag{13.13}$$

Thus the parasitic capacitor prevents the output voltage V_{out} from reaching $2V_{dd}$. The total energy E_1 stored in capacitors C and C_p in Fig. 13.6c (left) is given by

$$E_1 = \frac{1}{2}CV_{dd}^2 + \frac{1}{2}C_pV_{dd}^2. \tag{13.14}$$

Also the total energy E_2 stored in capacitors C and C_p in Fig. 13.6c (right) is given by

$$E_2 = \frac{1}{2}C(V_{out} - V_{dd})^2 + \frac{1}{2}C_pV_{out}^2 \tag{13.15}$$

Then the energy loss E_{loss} due to the parasitic capacitor C_p is given by

$$E_{loss} = E_2 - E_1 = \frac{1}{2}\frac{CC_p}{C + C_p}V_{dd}^2. \tag{13.16}$$

Here E_{loss} is provided from V_{dd}.

13.4 Dual Form of Theorem 2 Circuit: Two Inductors and a Switch

This section examines a circuit with two inductors and a switch—the dual form of the two capacitors and switch circuit of theorem 2—and defines zero-current switching (ZCS).

Theorem 3: Let us consider the circuit in Fig. 13.7, which consists of two inductors L_1 and L_2 and a switch. At first the switch is on, the currents that flow through L_1 and L_2 are I_1 and I_2, respectively (Fig. 13.7a), and the total energy stored in L_1 and L_2 is E. Next the switch turns off (Fig. 13.7b), and the total energy stored in L_1 and L_2 is E'. Then we have the following:

When $I_1 \neq I_2$, then $E > E'$.

When $I_1 = I_2$, then $E = E'$

Proof: When the switch is on and the currents that flow through L_1 and L_2 are I_1 and I_2, respectively, in Fig. 13.7, the flux linkage numbers

ϕ_1 and ϕ_2 in L_1 and L_2 and the corresponding stored energies E_1 and E_2 are given by

$$\phi_1 = N_1\varphi_1 = L_1 I_1,\ \phi_2 = N_2\varphi_2 = L_2 I_2$$

Here, N_1, N_2 are numbers of turns and φ_1, φ_2 are magnetic fluxes in L_1, L_2.

$$E_1 = \frac{1}{2}L_1 I_1^2,\ E_2 = \frac{1}{2}L_2 I_2^2.$$

Then the total energy E is given by

$$E = E_2 + E_1 = \frac{1}{2}L_1 I_1^2 + \frac{1}{2}L_2 I_2^2 \tag{13.17}$$

Next let the switch be turned off. The currents that flow through L_1 and L_2 become the same (I_M) in the steady state. Then we obtain the following:

$$\phi'_1 = L_1 I_M,\ \phi'_1 = L_2 I_M \tag{13.18}$$

Since the magnetic fluxes are preserved between the states in Figs. 13.7a and 13.7b, we have the following relationship:

$$\phi_1 + \phi_2 = \phi'_1 + \phi'_2 \tag{13.19}$$

Then it follows from Eqs. 13.18 and 13.19 that

$$L_1 I_1 + L_2 I_2 = L_1 I_M + L_2 I_M$$

$$I_M = \frac{L_1 I_1 + L_2 I_2}{L_1 + L_2}$$

Hence the energies E'_1 and E'_2 stored in L_1 and L_2 in Fig. 13.7b are given by

$$E'_1 = \frac{1}{2}L_1 I_M^2 = \frac{1}{2}\frac{L_1(L_1 I_1 + L_2 I_2)^2}{(L_1 + L_2)^2}$$

$$E'_2 = \frac{1}{2}L_2 I_M^2 = \frac{1}{2}\frac{L_2(L_1 I_1 + L_2 I_2)^2}{(L_1 + L_2)^2}.$$

Thus the total energy E'_2 in Fig. 13.7b reduces to

$$E' = E'_1 + E'_2 = \frac{1}{2}\frac{(L_1 I_1 + L_2 I_2)^2}{(L_1 + L_2)^2} \tag{13.20}$$

It follows from Eqs. 13.17 and 13.20 that the energy difference ΔE between the states in Fig. 13.17a,b is given by

$$\Delta E := E - E' = \frac{1}{2}\frac{L_1 L_2}{L_1 + L_2}(I_1 - I_2)^2. \tag{13.21}$$

(QED)

Remark: When $I_1 \neq I_2$, then $\Delta E > 0$ and the energy ΔE is dissipated in the switch. On the other hand, when $I_1 = I_2$, ΔE is equal to zero and this is called ZCS.

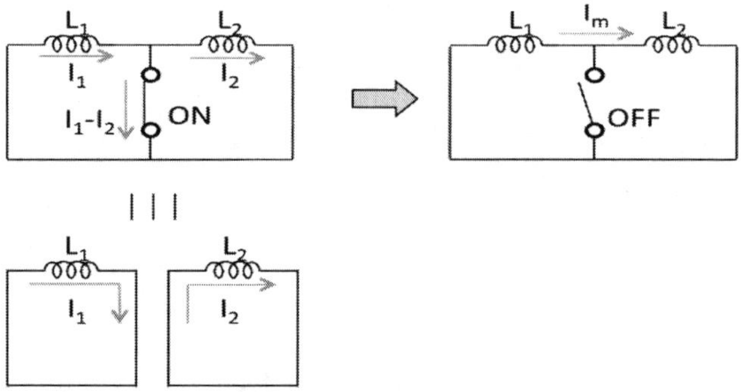

Figure 13.7 A circuit with two inductors and a switch and a dual circuit with two capacitors and a switch.

13.5 Efficient Capacitor-Charging Method

13.5.1 Problem 1

Let us consider the problem of charging a capacitor C from 0 to $2V_{\mathrm{dd}}$ with two voltage sources V_{dds}.

Method 1

Let us consider the case that one terminal of C is connected to ground and the other to $2V_{\mathrm{dd}}$. Then the stored energy is given by

$$E_{\mathrm{store1}} = \frac{1}{2}C(2V_{\mathrm{dd}})^2 = 2CV_{\mathrm{dd}}^2 \tag{13.22}$$

Then the dissipated energy is given by

$$E_{\mathrm{loss1}} = \frac{1}{2}C(2V_{\mathrm{dd}})^2 = 2CV_{\mathrm{dd}}^2. \tag{13.23}$$

Method 2

Next connect one terminal of C to ground and the other to V_{dd}. Then the stored energy is

$$E_{\text{store2a}} = \frac{1}{2}CV_{\text{dd}}^2 \tag{13.24}$$

Then the dissipated energy is given by

$$E_{\text{loss2a}} = \frac{1}{2}CV_{\text{dd}}^2. \tag{13.25}$$

Also the output voltage is V_{dd}. Then connect one terminal of C to V_{dd} and the other to $2V_{\text{dd}}$. The added stored energy is given by

$$E_{\text{store2b}} = \frac{1}{2}CV_{\text{dd}}^2. \tag{13.26}$$

The dissipated energy is given by

$$E_{\text{loss2b}} = \frac{1}{2}CV_{\text{dd}}^2. \tag{13.27}$$

The total stored energy E_{store2} is given by

$$E_{\text{store2}} = E_{\text{store2a}} + E_{\text{store2b}} = CV^2{}_{\text{dd}} \tag{13.28}$$

Also the total energy loss E_{loss2} is given by

$$E_{\text{loss2}} = E_{\text{loss2a}} + E_{\text{loss2b}} = CV^2{}_{\text{dd}} \tag{13.29}$$

The output voltage is $2V_{\text{dd}}$.

We see that the energy loss E_{loss2} of method 2 is half of E_{loss1}. If the input signal source rises from 0 to $2V_{\text{dd}}$ and the capacitor C is charged gradually, then the energy loss approaches zero: adiabatic complementary metal–oxide–semiconductor (CMOS) logic circuits use this principle, as shown later.

13.5.2 Problem 2

Let us consider the problem of charging two capacitors *Cs* from 0 to V_{dd} with one voltage source V_{dd}.

Method 1

Let us consider the case that one terminal of each C is connected to ground and the other to V_{dd} (Fig. 13.8a). Then the stored energy of each C is given by

$$E_{\text{store1}} = \frac{1}{2}CV_{\text{dd}}^2. \tag{13.30}$$

The dissipated energy for each C is given by

$$E_{\text{loss1}} = \frac{1}{2} C V_{\text{dd}}^2. \tag{13.31}$$

Then the total energy stored in the two capacitors is given by

$$E_{\text{store1total}} = 2E_{\text{store1}} = CV^2_{\text{dd}} \tag{13.32}$$

Also the total dissipated energy for the two capacitors is given by

$$E_{\text{loss1total}} = 2E_{\text{loss1}} = CV^2_{\text{dd}} \tag{13.33}$$

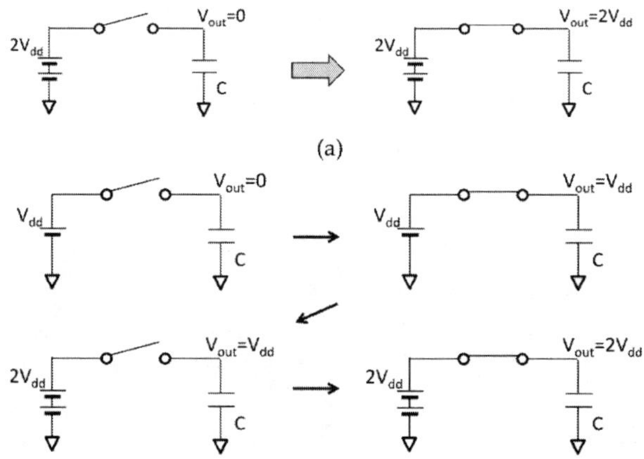

Figure 13.8 Capacitor charging to $2CV_{\text{dd}}$. (a) Method 1. (b) Method 2.

Method 2

Connect two capacitors, Cs, in series, as shown in the left part of Fig. 13.9b. Then the stored energy of each capacitor C is given by

$$E_{\text{store2a}} = \frac{1}{2} C \left(\frac{1}{2} V_{\text{dd}} \right)^2 \tag{13.34}$$

The dissipated energy for each capacitor is given by

$$E_{\text{loss2a}} = \frac{1}{2} C \left(\frac{1}{2} V_{\text{dd}} \right)^2 \tag{13.35}$$

Then connect the terminals of the two capacitors, as shown in the right part of Fig. 13.9b. The stored energy of each capacitor C is given by

$$E_{\text{store2b}} = \frac{1}{2} C V_{\text{dd}}^2 \tag{13.36}$$

The dissipated energy for each capacitor is given by

$$E_{\text{loss2b}} = \frac{1}{2}C\left(\frac{1}{2}V_{\text{dd}}\right)^2 \quad (13.37)$$

Then the total stored energy in the two capacitors is given by

$$E_{\text{store2total}} = 2E_{\text{store2b}} = CV^2_{\text{dd}} \quad (13.38)$$

Also the total dissipated energy for the two capacitors is given by

$$E_{\text{loss2total}} = 2(E_{\text{loss2a}} + E_{\text{loss2b}}) = \frac{1}{2}CV^2_{\text{dd}} \quad (13.39)$$

We see that the total stored energies of methods 1 and 2 are the same, but the energy loss of method 2 ($E_{\text{loss2total}}$) is half of method 1 ($E_{\text{loss1total}}$).

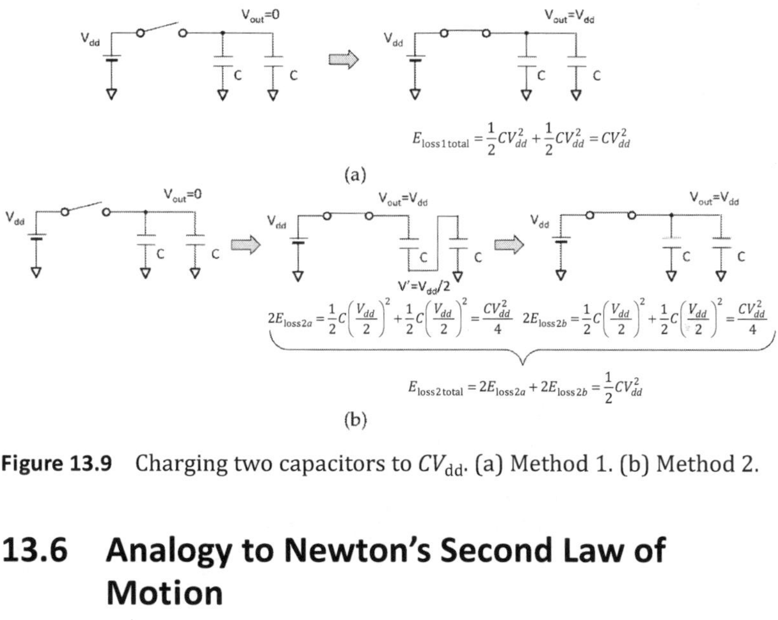

Figure 13.9 Charging two capacitors to CV_{dd}. (a) Method 1. (b) Method 2.

13.6 Analogy to Newton's Second Law of Motion

This section shows that theorem 2 has an analogy in Newton motion dynamics.

Suppose that there are two masses m_1 and m_2 whose velocities are v_1 and v_2 and whose momentums are p_1 and p_2, respectively.

$$p_1 = m_1 v_1,\ p_2 = m_2 v_2 \quad (13.40)$$

Their kinetic energies E_1 and E_2 are given by

$$E_1 = \frac{1}{2} m_1 v_1^2, \; E_2 = \frac{1}{2} m_2 v_2^2 \tag{13.41}$$

Suppose that they collide (Fig. 13.10) and they move together with velocity v_3. According to the law of conservation of momentum

$$p_1 + p_2 = p_3, \; p_3 = (m_1 + m_2)\, v_3 \tag{13.42}$$

Then we have the following:

$$v_3 = \frac{m_1 v_1 + m_2 v_2}{m_1 + m_2} \tag{13.43}$$

The total energy after collision is given by

$$E_3 = \frac{1}{2}(m_1 + m_2) v_3^2 \tag{13.44}$$

Table 13.1 shows the analogy between a switched-capacitor circuit and Newton's second law of motion.

Table 13.1 Analogy between a switched-capacitor circuit and Newton's second law of motion

Switched-Capacitor Circuit	Newton's 2nd law
Capacitor C_1, C_2	Mass m_1, m_2
Voltage V_1, V_2	Velocity v_1, v_2
Charge Q_1, Q_2	Momentum p_1, p_2
Charge conservation law	Conservation of momentum law

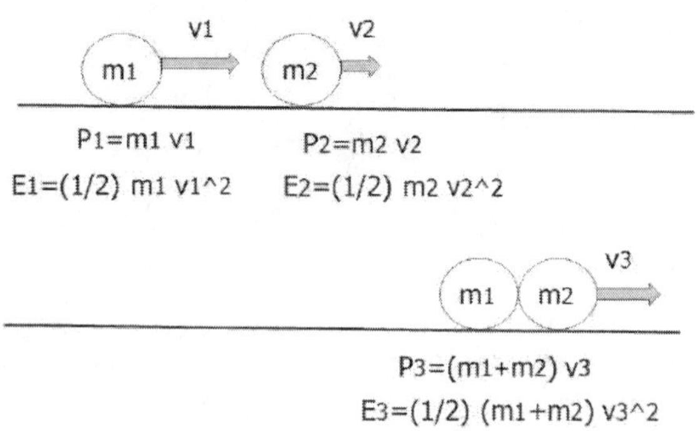

Figure 13.10 Collision of two masses.

Remark: There is some similarity between electrical circuit laws and Newton's second law, but the analogy is not perfect. The total capacitance can be made smaller than C_1 or C_2 by connecting capacitors C_1 and C_2 in series. However, the total mass of m_1 and m_2 cannot be smaller than the sum of m_1 and m_2 for any configuration of m_1 and m_2.

13.7 Digital CMOS Circuit Dynamic Power Dissipation

13.7.1 Dynamic Power Dissipation Formula

The dynamic power dissipation of a static digital CMOS circuit is given by

$$P_{\text{dyanamic}} = fC_{\text{load}}V^2{}_{\text{dd}} \tag{13.45}$$

Here f is the toggle frequency of the output node, C_{load} is the output load capacitance, and $V^2{}_{\text{dd}}$ is the supply voltage (Fig. 13.11).

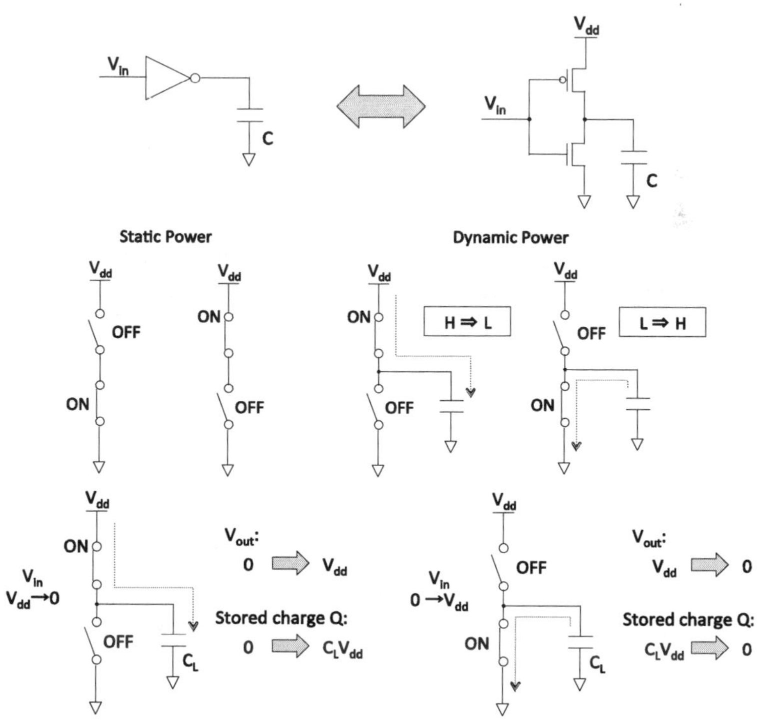

Figure 13.11 Dynamic power dissipation CMOS circuit.

This can be derived as follows:

1. Suppose the output voltage V_{out} is 0, the input falls from V_{dd} to 0, and the p-channel metal–oxide–semiconductor (PMOS) turns on. Then C_{load} is charged to V_{dd} and the energy loss is

$$E_{rise} = \frac{1}{2} C_{load} V_{dd}^2. \tag{13.46}$$

2. Next suppose V_{in} rises from 0 to V_{dd} and the n-channel metal–oxide–semiconductor (NMOS) turns on. Then V_{out} falls to 0 and C_{load} is discharged, and here the energy loss is

$$E_{fall} = \frac{1}{2} C_{load} V_{dd}^2 \tag{13.47}$$

3. We see from the above that when the output node toggles once, the total energy loss is

$$E_{dynamic} = E_{rise} + E_{fall} = C_{load} V^2{}_{dd} \tag{13.48}$$

4. Then the power loss (energy loss per unit time) is

$$P_{dynamic} = fE_{dynamic} = fC_{load} V^2{}_{dd} \tag{13.49}$$

13.7.2 Adiabatic Digital CMOS Circuit

The principle of the adiabatic digital CMOS circuit can be explained as follows: Suppose the output voltage V_{out} is 0, the input falls from V_{dd} to 0, and the PMOS turns on. Also the supply voltage rises from 0 to V_{dd} *very slowly* (Fig. 13.12). Then C_{loss} is charged to V_{dd} and the energy loss is close to 0. This is an extreme case of the efficient capacitor charge method, and such an adiabatic digital CMOS circuit may be used for low-speed ultra-low-power applications.

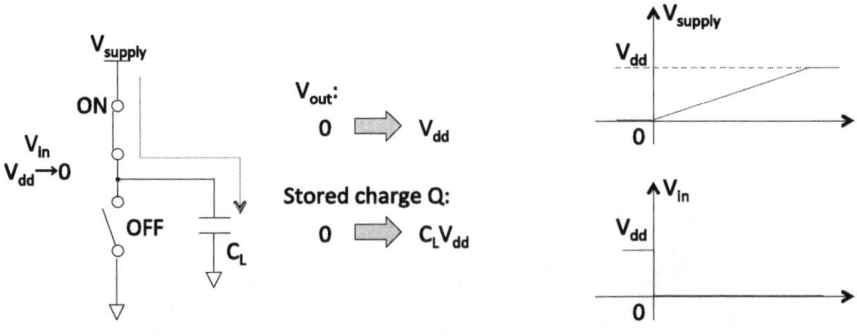

Figure 13.12 Principle of adiabatic digital CMOS circuit.

The principle of the ultra-low-power mechanism can be explained intuitively as follows: The power loss is proportional to i^2, and i^2 is roughly proportional to the slew rate of the supply voltage (here i is the current that flows in the NMOS or PMOS). The total energy loss is proportional to

$$\frac{i^2}{\text{Supply voltage slew rate}} \propto \text{Supply voltage slew rate} \tag{13.50}$$

Thus we see that as the slew rate decreases, the energy loss decreases.

13.8 Switched-Capacitor ADC

Analog-to-digital converters (ADCs) and digital-to-analog converters (DACs) are often implemented with switched-capacitor circuits, such as charge-distributed successive approximation register (SAR) ADCs (Fig. 13.13) [5]. Recently there are increasing demands for low-power implementations [6, 7], and the energy dissipated in this case can be calculated based on theorems 1 and 2. Several low-power techniques, mostly based on the efficient capacitor-charging method, have been proposed.

Figure 13.13 shows a charge-distributed SAR ADC architecture [5], where the charge is supplied from the reference voltage V_R for each step, and energy is dissipated. At first, all capacitors are connected to the input voltage V_{in} and charged (Fig. 13.13a). Next C_{M-1} is connected to the reference voltage V_R, and the other capacitors are connected to ground, and in this case some charge flows into the capacitors from V_R, which causes some energy dissipation. In the following steps, the capacitors are connected to V_R or ground, depending on the comparator output in the previous step, and again this causes some energy dissipation.

Figure 13.14 shows a charge-sharing SAR ADC, where all capacitors are charged by the reference voltage during the initial step, capacitors are not connected to the reference voltage in the following steps, and thus low power consumption can be achieved.

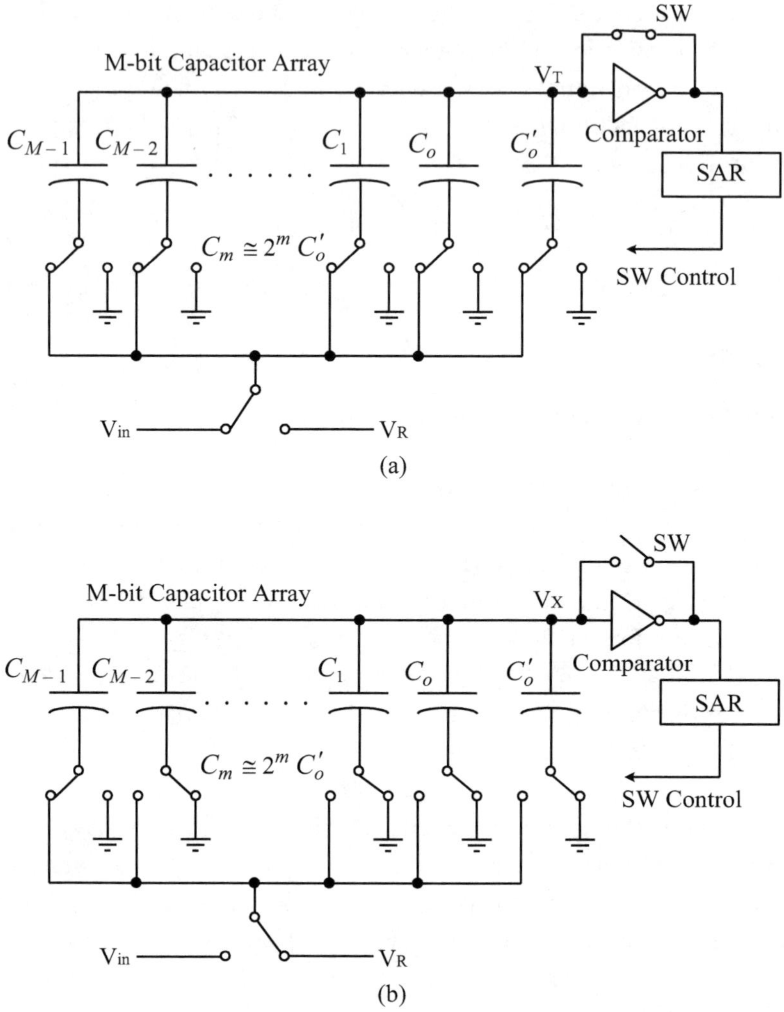

Figure 13.13 Charge-distributed SAR ADC.

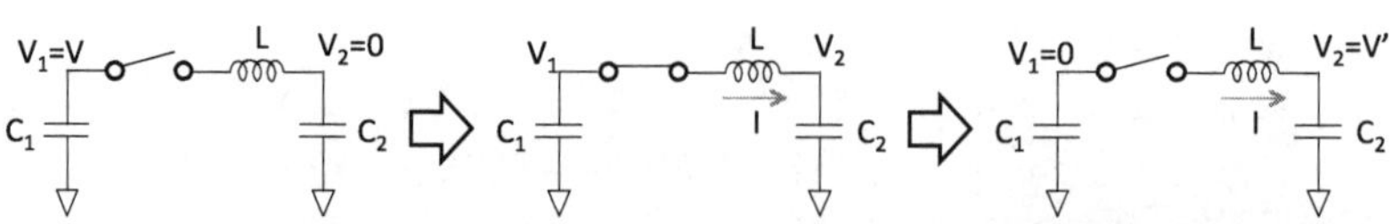

Figure 13.14 Charge-sharing SAR ADC.

13.9 Capacitor Charge Transfer through an Inductor

Let us consider the circuit in Fig. 13.15, which consists of capacitors C_1 and C_2 and a switch in series with an inductor L. At first the switch is off, and the voltages across C_1 and C_2 are, respectively, V_1 and 0. Suppose the on resistance of the switch is 0 and we turn on the switch at time 0. Then we turn off the switch at time

$$t_0 = \frac{\pi}{\sqrt{2LC}}$$

Let the voltages across C_1 and C_2 be V_1 (t) and C_2 (t) and the current through L be I_L (t) for $0 \le t \le t_0$. Then we have the following:

$$t_0 = \frac{\pi}{\sqrt{2LC}} \tag{13.51}$$

$$V_1(t) = V_1 - \frac{1}{C_1}\int_0^t I_L(\tau)d\tau. \tag{13.52}$$

$$V_2(t) = \frac{1}{C_2}\int_0^t I_L(\tau)d\tau. \tag{13.53}$$

Here, V_1, V_2 $(0) = 0$, $I_L(0) = 0$ and suppose C_1 that $=C_2 = C$

$$V_1 - 2\frac{1}{C_1}\int_0^t I_L(\tau)d\tau = L\frac{d}{dt}I_L(t). \tag{13.54}$$

$$-2I_L(\tau) = LC\frac{d^2}{dt^2}I_L(t). \tag{13.55}$$

Then we have

$$I_L(t) = \frac{1}{2}\omega C V_1 \sin(\omega t) \tag{13.56}$$

where $\omega = \sqrt{\frac{2}{LC}}$. We have $V(t_0) = 0$, $V_2(t_0) = V_1$

where $t_0 = \frac{\pi}{\omega}\pi\sqrt{\frac{LC}{2}}$.

Then we turn off the switch at time t_0, All of the charge in C_1 at $t_0 = 0$ is transferred to C_2 at $t = t_0$. Such a miracle is realized thanks to the inductor L.

Note that the total energy $E_{total}(t)$ stored in C_1, C_2, and L is given by

$$E_{\text{total}}(t)=\frac{1}{2}C_1V_1(t)^2+\frac{1}{2}C_2V_2(t)^2+\frac{1}{2}LI(t)^2$$

$$=\frac{1}{2}C_1V_1\left[\left(\frac{1}{2}(1+\cos(\omega t))\right)^2+\left(\frac{1}{2}(1-\cos(\omega t))\right)^2\right]+\frac{1}{2}L\left(\frac{1}{2}\omega CV_1\sin(\omega t)\right)^2$$

$$=\frac{1}{2}CV_1^{\ 2}$$

We see that total energy in C_1, C_2, and L is preserved for $0 \le t \le t_0$.

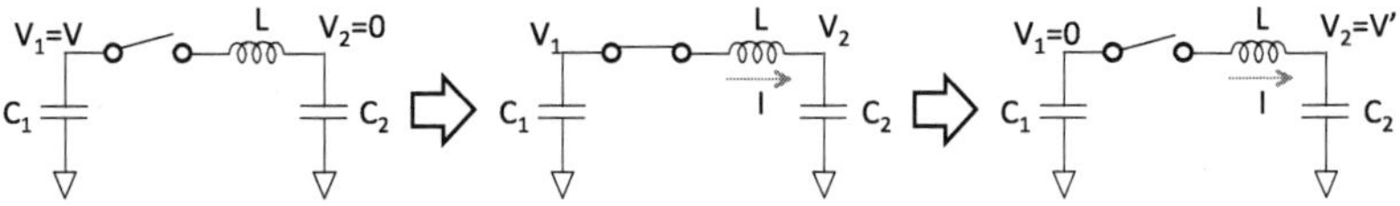

Figure 13.15 Charge transfer through an inductor.

13.10 Efficiency of Switched-Capacitor and Charge Pump Circuits

13.10.1 Dickson Charge Pump Circuit at Start-Up Time

Fig 13.16 shows a three-stage Dickson charge pump circuit, and we here derive a formula for the start-up-time node voltage transient of an ideal three-stage Dickson charge pump circuit in Fig.13.17 [8] and also show that one half of the energy E_{supply} supplied from V_{dd} and clock drivers for $t = 0$ to ∞ is stored in the capacitors Cs at $t = \infty$ and the other half is dissipated in the switches from $t = 0$ to ∞; this is an extension of theorem 1.

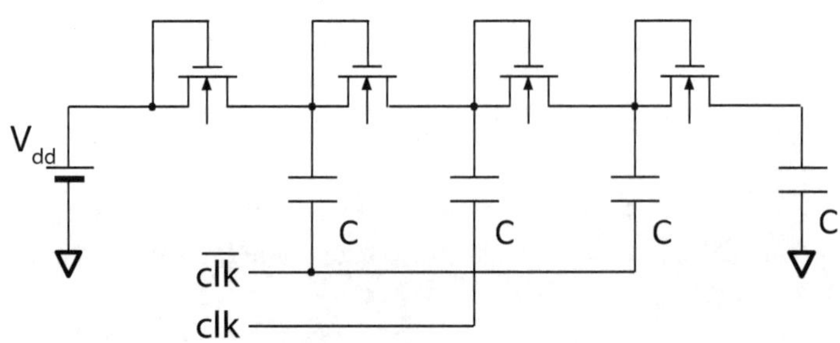

Figure 13.16 A three-stage Dickson charge pump.

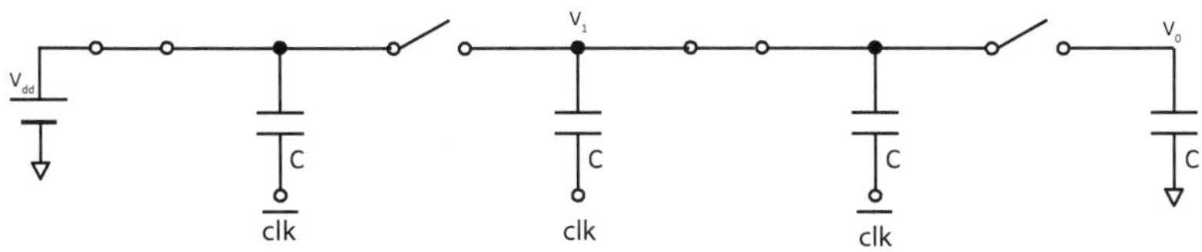

Figure 13.17 An ideal three-stage Dickson charge pump circuit.

Proposition 1: In the ideal three-stage charge pump circuit of Fig. 13.17, the output node voltage $V_0\ (n)$ and the internal node voltage $V_1\ (n)$ at the n-th clock cycle are given by

$$V_0(n) = \frac{1}{2}(\lambda_1^{n} + \lambda_2^{n})V_0(0) + \frac{1}{\sqrt{2}}(\lambda_1^{n} - \lambda_2^{n})V_1(0)$$
$$+\left[-\left(2+\frac{3}{\sqrt{2}}\right)\lambda_1^{n} - \left(-2+\frac{3}{\sqrt{2}}\right)\lambda_2^{n} + 4\right]V_{dd}, \tag{13.57}$$

$$V_1(n) = \frac{1}{\sqrt{2}}(\lambda_1^{n} - \lambda_2^{n})V_0(0) + \frac{1}{2}(\lambda_1^{n} + \lambda_2^{n})V_1(0)$$
$$+\left[-\left(2+\frac{3}{\sqrt{2}}\right)\lambda_1^{n} - \left(-2-\frac{3}{\sqrt{2}}\right)\lambda_2^{n} + 3\right]V_{dd}, \tag{13.58}$$

Here

$$\lambda_1 = \frac{2+\sqrt{2}}{4}, \lambda_2 = \frac{2-\sqrt{2}}{4}$$

Proof: Using the charge conservation law, we have the following state equation for Fig. 13.17:

$$v(n+1) = Av(n) + bV_{dd}$$

Here

$$v(n) := \begin{bmatrix} V_1(n) \\ V_2(n) \end{bmatrix}, \mathrm{A} := \begin{bmatrix} \frac{1}{4} & \frac{1}{2} \\ \frac{1}{2} & \frac{1}{2} \end{bmatrix}, \mathrm{b} := \begin{bmatrix} \frac{1}{2} \\ \frac{1}{2} \end{bmatrix}$$

Then

$$v(n) = A^n v(0) + \sum_{k=0}^{n-1} A^k b\, V_{dd}$$

and Eqs. 13.57 and 13.58 are obtained (QED).

Remarks:

(i) Since $|\lambda_1| < 1$ and $|\lambda_2| < 1$,

as $n \to \infty$, $\lambda''_1 \to 0$, $V_0(n) \to 4V_{dd}$, $V_1(n) \to 3V_{dd}$,

(ii) The proof of proposition 1 uses a so-called state-space approach. With this approach we can derive output node and internal node voltages at the n-th cycle in an N-stage charge pump circuit for any positive integer N.

Proposition 2: When the charges in all the capacitors are zero at $t = 0$ in Fig. 13.17, we have the following:

$$E_{supply} = E_C + E_R,\ E_C = E_R \tag{13.59}$$

Here E_{supply} is the energy supplied from V_{dd} and clock drivers for $t = 0$ to ∞, E_C is the energy stored in all the capacitors Cs in the steady state ($t = \infty$), and E_R is the power dissipated in the switches for $t = 0$ to ∞.

Proof: Using the charge conservation law for each cycle, the energy loss $E_{loss}(n)$ and the energy stored in all capacitors $E_C(n)$ at the n-th cycle are given by

$$E_{loss}(n) = \frac{C}{16}\left[\begin{array}{r} 10V_1(n)^2 + 5V_0(n)^2 + 74V_{dd}^2 - 8V_1(n)V_0(n) \\ -16V_0(n)V_{dd} - 28V_1(n)V_{dd} \end{array}\right]$$

$$E_c(n) = \frac{C}{2}[2V_1(n)^2 + 2V_{dd}^2 - 2V_1(n)V_{dd} + V_0(n)^2]$$

Then considering the initial condition $E_{loss()}$

$$E_R = \sum_{n=0}^{\infty} E_{loss}(n) + E_{loss()} = 15CV_{dd}^2$$

and Eq. 13.59 is obtained (QED).

Remark: Proposition 2 is a general case of theorem 1.

13.10.2 Steady-State Analysis of Dickson Charge Pump Circuit

Next we will consider the steady state for a charge pump circuit that includes circuit nonidealities (voltage drop across the switch, parasitic capacitance from each node to ground) and supplies the output load current.

13.10.2.1 Effects of voltage drop across switch

Let us consider the case that the switch is realized with a diode or diode-connected metal–oxide–semiconductor field-effect transistor (MOSFET) and that it has some voltage drop V_d when it is on (Fig. 13.18).

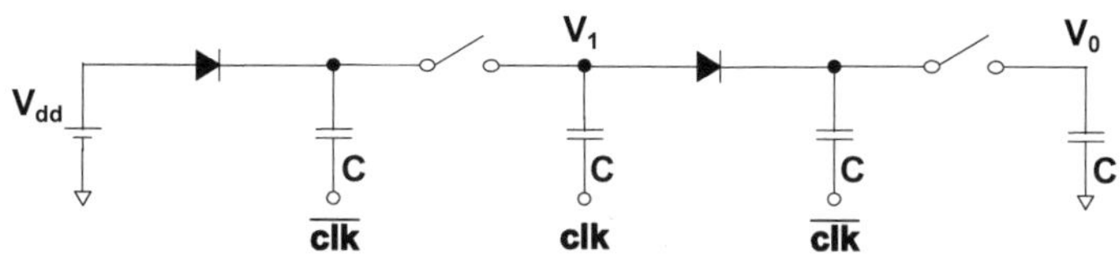

Figure 13.18 A non-ideal three-stage Dickson charge pump circuit where the voltage drop V_d across each switch is considered.

Proposition 3:

(i) For a three-stage charge pump circuit, the output voltage in the steady state is given by

$$V_0(\infty) = 4(V_{dd} - V_d) \tag{13.60}$$

(ii) The energy dissipation $P_{loss}(\infty)$ during one clock cycle T for the circuit in the steady state is given by

$$P_{loss}(\infty) = 0$$

Proof: We have the following state equation for Fig. 13.18:

$$v(n+1) = Av(n) + Bv_m$$

Here

$$B := \begin{bmatrix} \frac{1}{2} & 0 \\ \frac{1}{2} & -1 \end{bmatrix}, v_m = \begin{bmatrix} V_{dd} \\ V_d \end{bmatrix}.$$

$$v(\infty) = Av(\infty) + Bv_m$$

When $n \to \infty$, we have

$$v(\infty) = Av(\infty) + Bv_m$$

and Eq. 13.60 is obtained (QED).

13.10.2.2 Effects of parasitic capacitance

Let us consider the case that each node has parasitic capacitance C_p (Fig. 13.19).

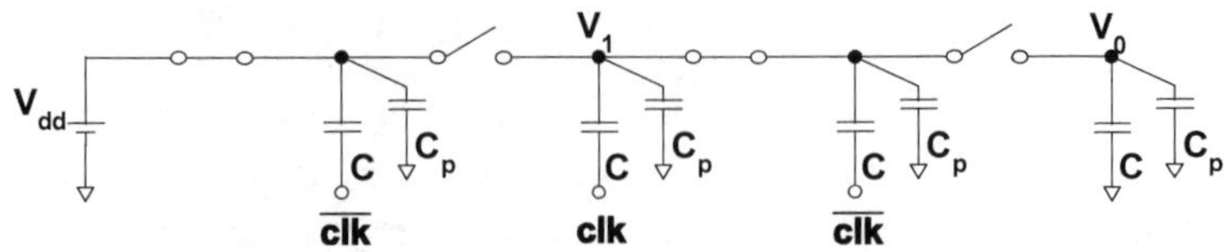

Figure 13.19 A non-ideal three-stage Dickson charge pump circuit where the parasitic capacitance C_p of each node is considered.

Proposition 4:

(i) For a three-stage charge pump circuit, the output voltage in the steady state is given by

$$V_0(\infty)=\left(4-\frac{3C_\mathrm{p}}{C+C_\mathrm{p}}\right)V_\mathrm{dd} \tag{13.61}$$

(ii) The energy dissipation $P_{\mathrm{loss}}(\infty)$during one clock cycle T for the circuit in the steady state is given by

$$P_{\mathrm{loss}}(\infty)=\frac{3CC_\mathrm{p}}{C+C_\mathrm{p}}V_\mathrm{dd}{}^2 \tag{13.62}$$

Proof: We have the following state equation for Fig. 13.19:

$$v(n+1) = Av(n) + cV_\mathrm{dd}$$

$$c := \begin{bmatrix} \dfrac{C}{2(C+C_\mathrm{p})} \\ \dfrac{C+C_\mathrm{p}}{4(C+C_\mathrm{p})} \end{bmatrix}.$$

When $n \to \infty$, we have

$$v(\infty) = Av(\infty) + cV_\mathrm{dd}$$

and Eq. 13.61 is obtained (QED).

Remark: Theorem 2 helps us understand intuitively why parasitic capacitance causes the energy loss given by Eq. 13.62.

13.10.2.3 Effects of output current

Let us consider the case that the output node provides load current I_{out} (Fig. 13.20).

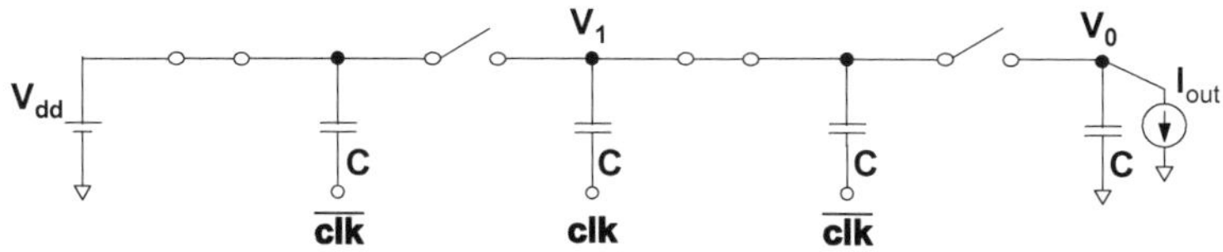

Figure 13.20 A non-ideal three-stage Dickson charge pump circuit where the output load current I_{out} is considered.

Proposition 5:

(i) For a three-stage charge pump circuit, the output voltage in the steady state is given by

$$V_0(\infty)=\left(4-\frac{7TI_{\text{out}}}{C}\right)V_{\text{dd}} \tag{13.63}$$

(ii) The energy dissipation $P_{\text{loss}}(\infty)$during one clock cycle T for the circuit in the steady state is given by

$$P_{\text{loss}}(\infty) = 8TI_{\text{out}}V_{\text{dd}} \tag{13.64}$$

Proof: We have the following state equation for Fig. 13.20.

$$\begin{bmatrix} V_1(n+1) \\ V_0(n+1) \end{bmatrix} = \begin{bmatrix} \frac{1}{4} & \frac{1}{2} \\ \frac{1}{2} & \frac{1}{2} \end{bmatrix} \begin{bmatrix} V_1(n) \\ V_0(n) \end{bmatrix} + \begin{bmatrix} -\frac{1}{2} & -\frac{3}{2} \\ \frac{1}{2} & -\frac{1}{4} \end{bmatrix} \begin{bmatrix} V_{\text{dd}} \\ \frac{TI_{\text{out}}}{C} \end{bmatrix}$$

When $n \to \infty$, we have

$$\begin{bmatrix} V_1(\infty) \\ V_0(\infty) \end{bmatrix} = \begin{bmatrix} \frac{1}{4} & \frac{1}{2} \\ \frac{1}{2} & \frac{1}{2} \end{bmatrix} \begin{bmatrix} V_1(n) \\ V_0(n) \end{bmatrix} + \begin{bmatrix} -\frac{1}{2} & -\frac{3}{2} \\ \frac{1}{2} & -\frac{1}{4} \end{bmatrix} \begin{bmatrix} V_{\text{dd}} \\ \frac{TI_{\text{out}}}{C} \end{bmatrix}$$

and Eq. 13.63 is obtained (QED).

Remark: Theorems 1 and 2 help us understand intuitively why parasitic capacitance causes the energy loss given by Eq. 13.64.

Proposition 6:

Efficiency η for a three-stage charge pump circuit in the steady state is given by

$$\eta=\frac{\text{Output power from } V_0}{\text{Supplied power from } V_{\text{dd}} \text{ and clock drivers}}=1-\frac{3TI_{\text{out}}}{2CV_{\text{dd}}} \tag{13.65}$$

13.10.2.4 Combined effects of switch voltage drop, parasitic capacitance and output current

Let us consider the case that the charge pump circuit has a switch voltage drop V_d , parasitic capacitances C_p, and output load current I_{out} (Fig. 13.21).

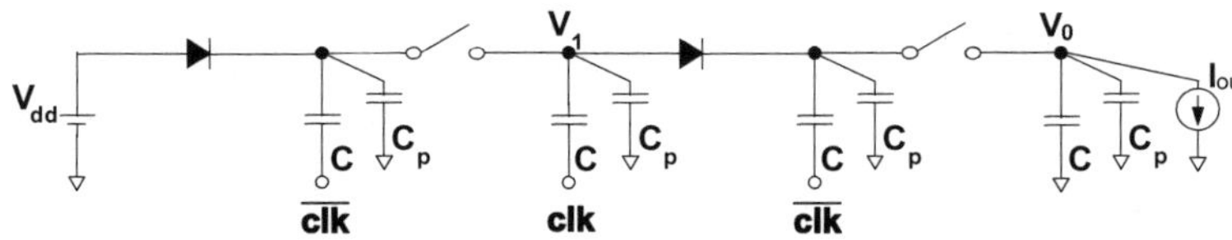

Figure 13.21 A non-ideal three-stage Dickson charge pump circuit where the voltage drop V_d across each switch capacitance C_p at each node and the output load current I_{out} are considered.

Proposition 7: For a three-stage charge pump circuit, the output voltage in the steady state is given by

$$V_0(\infty)=\left(1+3\frac{C}{C+C_P}\right)V_{dd}-4V_d-\frac{7TI_{out}}{C} \tag{13.66}$$

Proof: For Fig. 13.21 we have the following state equations:

$$V_1'(n)=\frac{1}{2}V_1(n)+\frac{1}{2}V_{dd}-V_d$$

$$V_0'(n)=\frac{1}{2}V_1(n)+\frac{1}{2}V_0(n)-V_d+\frac{CV_{dd}-TI_{out}}{2(C+C_p)}$$

$$V_1(n+1)=\frac{1}{2}V_1'(n)+\frac{1}{2}V_0'(n)+V_d$$

$$V_0(n+1)=\frac{1}{2}V_0'(n)-\frac{TI_{out}}{C+C_p}$$

Thus Eq. 13.66 is obtained (QED).

Proposition 8: Efficiency η for an N-stage charge pump circuit in the steady state is given by

$$\eta=1-\frac{NCC_pV_{dd}^2+2(N+1)(C+C_p)V_dTI_{out}+4NT^2I_{out}^2}{NCC_pV_{dd}^2+2(N+1)CV_{dd}TI_{out}+2C_pV_{dd}TI_{out}} \tag{13.67}$$

Proof: Consider the steady state for Fig. 13.21. Then the energy E_{load} supplied from the output node during one clock cycle T is given by

$$E_{\text{load}} = \left[2V_{\text{dd}} - 8V_{\text{d}} + 6\frac{C}{C+C_{\text{p}}}V_{\text{dd}} - 12\frac{TI_{\text{out}}}{C+C_{\text{p}}}\right]TI_{\text{out}}$$

Also the energy $E_{\text{s}}(\infty)$ supplied from V_{dd} and clock drivers during one clock cycle T is given by

$$E_{\text{s}}(\infty) = 3CV_{\text{dd}}^2 - \frac{3C^2}{C+C_{\text{p}}} + \frac{6CV_{\text{dd}}TI_{\text{out}}}{C+C_{\text{p}}} + 2V_{\text{dd}}TI_{\text{out}}$$

Then

$$\eta = \frac{E_{\text{load}}}{E_{\text{s}}(\infty)}$$

And Eqs. 13.65 and 13.67 are obtained (QED).
Note that Eq. 13.67 reduces to the following:

$$\eta = \frac{1 - N\alpha + 2(N+1)(1+\alpha)\beta x + 4Nx^2}{N\alpha + 2\gamma x} \tag{13.68}$$

where x, α, β, and γ are dimensionless and defined by

$$x := \frac{TI_{\text{out}}}{CV_{\text{dd}}}, \alpha := \frac{C_{\text{p}}}{C}, \beta := \frac{V_{\text{d}}}{V_{\text{dd}}}, \gamma := N + 1 + \alpha$$

Proposition 9:
Efficiency η given by Eq. 13.68 has a peak value when x has the following value:

$$x = \frac{-N\alpha + \sqrt{N^2\alpha^2 + [(1-\beta)\gamma - N\alpha\beta]\gamma\alpha}}{2\gamma} \tag{13.69}$$

Proof: By calculating $\partial\eta/\partial x = 0$ and $x > 0$, we obtain Eq. 13.69.
We conclude this section by noting that proposition 9 can theoretically explain our measurement result that the efficiency of our charge pump circuit [3] has a peak for a certain value of the output current I_{out} and also for a certain value of the clock period T when the other conditions are fixed.

13.11 Conclusions

This chapter has provided tutorials to help readers understand the efficiency of switched-capacitor power supply circuits and has shown the following:

- There is a theoretical limit to the efficiency of switched-capacitor circuits.
- The efficiency is strongly related to capacitor size, switching frequency, and load current.
- How and why parasitic capacitances decrease the efficiency was explained.
- An analogy between the switched-capacitor circuit and the collision of two masses was provided.
- There is a dual problem of switched-inductor circuits.
- The dynamic power dissipation of CMOS digital circuits can be explained by the switched-capacitor circuit.
- The low power consumption of adiabatic digital circuits can be explained the same way.
- The low-power design of switched-capacitor ADCs was explained.
- Dickson charge pump circuits and switched-capacitor power supply circuits were discussed.

References

1. J. F. Dickson, On-chip high-voltage generation in NMOS integrated circuits using an improved voltage multiplier technique, *IEEE J. Solid-State Circuits*, **11**, 374–378, 1976.
2. H. San, H. Kobayashi, T. Myono, T. Iijima, and N. Kuroiwa, Highly-efficient low-voltage-operation charge pump circuits using bootstrapped gate transfer switches, *IEEJ*, **120-C**(10), 1339–1345, 2000.
3. T. Myono, S. Kawai, A. Uemoto, T. Iizima, and H. Kobayashi, Highly efficient charge-pump circuits with large output current load for mobile equipment applications, *IEICE Trans. Electron.*, **E84-C**(10), 1602–1611, 2001.
4. T. Myono, Y. Onaya, K. Kashiwase, H. Kobayashi, T. Nishi, K. K obayashi, T. Suzuki, and K. Henmi, Reducing startup-time inrush current in charge pump circuits, *IEICE Trans. Fundam.*, **E87-A**(4), 787–791, 2004.

5. H. Sato and H. Kobayashi, Dynamic power dissipation of track/hold circuit, *IEICE Trans. Fundam.*, **83**(8), 1728–1731, 2000.

6. T. Ogawa, H. Kobayashi, S. Uemori, Y. Tan, S. Ito, N. Takai, T. J. Yamaguchi, and K. Niitsu, Design for testability that reduces linearity testing time of SAR ADCs, *IEICE Trans. Electron.*, **E 94-C**(6), 1061–1064, 2011.

7. V. Giannini, P. Nuzzo, V. Chironi, A. Baschirotto, G. V. Plas, and J. Craninckx, An 820μW 9b 40MS/s noise-tolerant dynamic SAR ADC in 90nm digital CMOS, *Tech. Dig. Int. Solid-State Circuits Conf.*, February 2008.

8. K. Kashiwase, H. Kobayashi, N. Kuroiwa, N. Hayasaka, S. Inaba, T. Myono, T. Suzuki, T. Iijima, and S. Kawai, Dynamics of parallel-type and serial-type charge pump circuits for high voltage generation, *IEEJ Int. Analog VLSI Workshop*, 14–19, Singapore, September 2002.

9. T. Myono, T. Suzuki, A. Uemoto, S. Kawai, T. Iizima, N. Kuroiwa, and H. Kobayashi, High-efficiency charge-pump circuits which use a 0.5Vdd-step pumping method, *IEICE Trans. Fundam.*, **E86-A**(2), 371–380, 2003.

10. J. T. Wu and K. L. Chang, MOS charge pumps for low-voltage operation, *IEEE J. Solid-State Circuits*, **33**(4), 592–597, 1998.

11. H. Lin, K.-H. Chang, and S.-C. Wong, Novel high positive pumping circuits for low power supply, *Proc. ISCAS*, 1238–1241, 1999.

12. K.-W. Min and S.-H. Kim, A high-voltage generator using charge pump circuit for low voltage flash memories, *IEICE Trans. Electron.*, **E82-C**(5), 781–784, 1999.

13. T. Tanzawa and S. Atsumi, Optimization of word-line booster circuit for low-voltage flash memories, *IEEE J. Solid-State Circuits*, **34**(8), 1091–1098, 1999.

14. J. S. Witters, G. Groesenken, and H. E. Maes, Analysis and modeling of on-chip high-voltage generator circuit for use in EEPROM circuits, *IEEE J. Solid-State Circuits*, **24**(5), 1372–1380, 1989.

15. A. Umezawa, S. Atsumi, M. Kuriyama, H. Banba, K. Imamiya, K. Naruke, S. Yamada, E. Obi, M. Oshikiri, T. Suzuki, and S. Tanaka, A 5-V-only operation 0.6-μm flash EEPROM with row decoder scheme in triple-well structure, *IEEE J. Solid-State Circuits*, **27**(11), 1540–1546, 1992.

16. C. Lauterbach, W. Webe, and D. Romer, Charge sharing concept and new clocking scheme for power efficiency and electromagnetic emission improvement of boosted charge pumps, *IEEE J. Solid-State Circuits*, **35**(5), 719–723, 2000.

17. M. A. Mohamed Zin, H. Kobayashi, K. Kobayashi, J. Ichimura, H. San, Y. Onaya, Y. Kimura, Y. Yuminaka, Y. Sasaki, K. Tanaka, and F. Abe, A high-speed CMOS track/hold circuit, *6th IEEE Int. Conf. Electron., Circuits Syst.*, 1709–1712, Paphos, Cyprus, September 1999.

18. Kh. Hadidi, M. Sasaki, T. Watanabe, D. Muramats, and T. Matsumoto, An open-loop full CMOS 103MHz -61dB THD S/H circuit, *Proc. IEEE Custom Integr. Circuits Conf.*, May 1998.

19. IEEE Standard for Digitizing Waveform Recorders, IEEE Standard 1057 (1994).

20. N. H. E. Weste and K. Eshraghian, *Principles of CMOS VLSI Design: A Systems Perspective, Second Edition*, Addison Wesley, 1993.

21. J. Craninckx and G. Van der Plas, A 65fJ/conversion-step 0-to-50Ms/s 0-to-0.7mW 9b charge-sharing SAR ADC in 90nm digital CMOS, *Tech. Dig. Int. Solid-State Circuits Conf.*, February 2007.

Index